AF182943

# New millennium fibers

## Related titles from Woodhead's textile technology list:

*Wearable electronics and photonics* (1 85573 605 5)
Building electronics into clothing is a major new concept, which opens up a whole array of multi-functional, wearable electro-textiles for sensing/monitoring body functions, delivering communication facilities, data transfer, individual environment control, and many other applications. A team of distinguished international experts consider the technical materials and processes that will facilitate all these new applications.

*New fibers 2nd edn* (1 85573 334 X)
Since the first edition of *New fibers* was published in 1990, new research and development has produced fibers with high tenacity and modulus resulting in the super-fibers now used as industrial materials. New chapters deal with 'Fibers for the next millennium', and examine the resurgence of synthetic cellulosics since 1990, in particular the various solvent-spun fibers of the Lyocell and Tencel families.

*Smart fibres, fabrics and clothing* (1 85573 546 6)
This important book provides a guide to the fundamentals and latest developments in smart technology for textiles and clothing. The contributors represent a distinguished international panel of experts and the book covers many aspects of cutting edge research and development. It examines the background to smart technology, and goes on to cover a wide range of the material and fibre science aspects of the technology.

*High-performance fibres* (1 85573 539 3)
This important handbook looks at how high-performance fibres are designed and manufactured and covers their capabilities and applications. The high-modulus, high-tenacity (HM-HT) fibres fall naturally into three groups – polymer fibres such as aramids and polyethylene fibres; carbon fibres such as Kevlar; and inorganic fibres based on glass and ceramic fibres. The book shows how high-performance fibres are being increasingly used for a wide range of applications including geotextiles and geomembranes, and for construction and civil engineering projects, as well as in specialist fibres within composite materials where their ability to fulfil demanding roles makes them an effective choice for the engineer and materials scientist.

Details of these books and a complete list of Woodhead's textile technology titles can be obtained by:

- visiting our web site at www.woodheadpublishing.com
- contacting Customer Services (e-mail: sales@woodhead-publishing.com; fax: +44 (0) 1223 893694; tel.: +44 (0) 1223 891358 ext.30; address: Woodhead Publishing Limited, Abington Hall, Abington, Cambridge CB1 6AH, England)

# New millennium fibers

Tatsuya Hongu, Glyn O. Phillips and
Machiko Takigami

The Textile Institute

CRC Press
Boca Raton Boston New York Washington, DC

WOODHEAD PUBLISHING LIMITED
Cambridge England

Published by Woodhead Publishing Limited in association with The Textile Institute
Abington Hall, Abington
Cambridge CB1 6AH
England
www.woodheadpublishing.com

Published in North America by CRC Press LLC
2000 Corporate Blvd, NW
Boca Raton FL 33431
USA

First published 2005, Woodhead Publishing Limited and CRC Press LLC
© 2005, Woodhead Publishing Limited
The authors have asserted their moral rights.

This book contains information obtained from authentic and highly regarded sources. Reprinted material is quoted with permission, and sources are indicated. Reasonable efforts have been made to publish reliable data and information, but the authors and the publishers cannot assume responsibility for the validity of all materials. Neither the authors nor the publishers, nor anyone else associated with this publication, shall be liable for any loss, damage or liability directly or indirectly caused or alleged to be caused by this book.

Neither this book nor any part may be reproduced or transmitted in any form or by any means, electronic or mechanical, including photocopying, microfilming and recording, or by any information storage or retrieval system, without permission in writing from the publishers.

The consent of Woodhead Publishing Limited and CRC Press LLC does not extend to copying for general distribution, for promotion, for creating new works, or for resale. Specific permission must be obtained in writing from Woodhead Publishing Limited or CRC Press LLC for such copying.

Trademark notice: Product or corporate names may be trademarks or registered trademarks, and are used only for identification and explanation, without intent to infringe.
British Library Cataloguing in Publication Data
A catalogue record for this book is available from the British Library.

Library of Congress Cataloging in Publication Data
A catalog record for this book is available from the Library of Congress.

Woodhead Publishing Limited ISBN 10: 1-85573-601-2
Woodhead Publishing Limited ISBN 13: 978-1-85573-601-6
CRC Press ISBN 0-8493-2598-6
CRC Press order number: WP2598

The publishers' policy is to use permanent paper from mills that operate a sustainable forestry policy, and which has been manufactured from pulp which is processed using acid-free and elementary chlorine-free practices. Furthermore, the publishers ensure that the text paper and cover board used have met acceptable environmental accreditation standards.

Project managed by Macfarlane Production Services, Markyate, Hertfordshire (macfart@ad.com)
Typeset in India by Replika Press Pvt Ltd
Printed by T J International Limited, Padstow, Cornwall, England

# Contents


| | | |
|---|---|---|
| *Preface* | | *ix* |
| *Author contact details* | | *xii* |
| 1 | Searching the roots of fibers | 1 |
| 1.1 | The importance of fiber in human life | 1 |
| 1.2 | What is high-tech fiber? | 4 |
| 1.3 | Natural versus synthetic fiber | 10 |
| 1.4 | Artificial fiber by biomimetics | 12 |
| 1.5 | Definition of fibrous materials | 15 |
| 1.6 | Fiber: characteristics and shapes | 16 |
| 1.7 | Fibers as hierarchical structures | 18 |
| 1.8 | What should we investigate in the field of fiber and textiles? | 19 |
| 1.9 | Bibliography | 28 |
| 2 | The new frontier fibers? | 32 |
| 2.1 | Enlargement of the frontier in a fiber competition age | 32 |
| 2.2 | 'Selection', 'concentration' and 'originality' development on a world-wide scale | 35 |
| 2.3 | New fibers for the next generation have arrived | 43 |
| 2.4 | The distinction between high-tech fiber, frontier fiber, and new frontier fiber | 47 |
| 2.5 | Key words for the near future | 47 |
| 2.6 | How to develop new application fields | 49 |
| 2.7 | New frontier field now growing | 51 |
| 2.8 | Bibliography | 63 |
| 3 | Superfibers | 65 |
| 3.1 | Description of superfibers | 65 |
| 3.2 | Development of superfiber in Europe, the United States and Japan | 68 |

3.3 Superfiber as a reinforcing material 69
3.4 Frontiers of superfiber applications 76
3.5 Nanofiber (carbon nanotube) 91
3.6 High polyketone fiber 92
3.7 Bibliography 95

4 Carbon fiber expands towards the twenty-first century 99

4.1 PAN-based and pitch-based carbon fiber lead the world 99
4.2 A step of development of carbon fiber 103
4.3 The future of PAN-based carbon fiber 127
4.4 Bibliography 128

5 High function fiber 130

5.1 Prospects for high function fiber development 130
5.2 Sportswear using the high function fiber 144
5.3 Comfort function fiber 151
5.4 Biomimetic and intelligent fibers 158
5.5 The new areas 166
5.6 Bibliography 168

6 Frontier of health and comfort fibers 173

6.1 Fibers for health 173
6.2 Development of medical care materials to learn from 'smart fiber' 179
6.3 Development trend of comfortable fiber for health 184
6.4 Trend to seek for cleanliness and comfortableness 186
6.5 Fiber to guard environment and health 204
6.6 Technical concentration to achieve comfort 212
6.7 Bibliography 216

7 Polymer fibers for health and nutrition 218

7.1 The concept and effects of dietary fiber 218
7.2 Hydrocolloid fibers 221
7.3 The main hydrocolloids 227
7.4 Dietary fiber – in health and disease 243
7.5 The appropriate molecular features to achieve 244
7.6 Bibliography 245

8 Fibers in medical healthcare 247

8.1 Nonwoven 247
8.2 Alginate fibers 248

8.3 Superabsorbent fibers 250
8.4 Wound healing and polysaccharide fibers 251
8.5 Hyaluronan – a new medical fiber 255
8.6 Other fibrous scaffolds for tissue engineering 261
8.7 Collagen: medical applications 261
8.8 Medical textiles 263
8.9 Bibliography 268

9 Developments in nanofibers for the new millennium 269

9.1 Background 269
9.2 Nanotechnology, materials and nanofiber 271
9.3 Creation of new industries 283
9.4 Researches and global developments of nanofiber 286
9.5 Further reading 287
9.6 References 287

*Index* *289*

# Preface

Looking forward into the twenty-first century, it is becoming clear that the narrow concept of a fiber as something capable of being used for clothing and related areas is becoming outmoded. Fibers in biological systems are driving the new fiber science and technology, which is flourishing in Asia and particularly in China. No longer can there be reliance on products based on petroleum, which is already proving unreliable in maintaining guaranteed supplies. Alternatives must be found, preferably from renewable sources. Energy saving, securing the environment along with personal health and safety will call for a greater emphasis on quality rather than quantity. It is the climate in which the nanofibers have appeared. Japan has been a pioneer in combining fiber design with biological function (biomimetics) and has pioneered this bottom-up type of development, but now more top-down type investigations are also called for and are slowly gaining momentum. One practical result of this approach has been the market launch of 'Morphotex®' with the layered structure of nanofiber as in the MORPHO butterfly.

It is these newer areas, which are likely to appear during the twenty-first century, which are given emphasis in this book. It follows and is meant to supplement our previous two books, both entitled ***New Fibers***. The first edition was published in 1990, then by Ellis Horwood Ltd. It was extremely well received and we were encouraged by Woodhead Publishing Limited (who acquired the assets of Ellis Horwood) to produce a revised edition in 1997. We have named this new volume *New Millennium Fibers* to preserve the same approach and to follow on from our other two volumes. We hope that it will be received just as enthusiastically. Once again the emphasis is on developments in fiber technology and applications, mainly in Japan. It opens up to Western countries the type of thinking which directs the fiber science and technology industry in Japan, and thus to a global perspective.

The Japanese fiber textile industry is characteristically technology-oriented. Its innovative approach led to Shingosen ultra-fine fiber, and the new target is now high-tech fibers. These are the fibrous materials produced by advanced technology, and include high performance fiber with ultra-strength, high function fiber with various functions for health care, comfort etc., and high touch fiber with superb hand feel.

Fiber textiles science and technology is moving forward positively into the twenty-first century, and the prospect is that the new fiber textile materials will be developed in association with other industrial fields. Fiber textile technology, for example, has closely followed and been greatly influenced by the information technology (IT) and biotechnology revolutions. Tension members for supporting optical fibers, integrated circuit boards for mobile phones, and wearable computers are examples where fiber textile technology has interfaced with the IT revolution. Bacteria are known to produce cellulose and polyester, and soon there could be more industrial bio-plants to produce fibers. Many chemical fibers and textiles in the twentieth century were developed by mimicking the structure of natural fibers, using the approach called 'biomimetics'. These fibers and textiles could well themselves possess bio-functions in the twenty-first century. This book projects the subject into this exciting future. Chapter 1 describes fibers and textiles in general, and identifies the scope of high-tech fibers. Chapter 2 reviews the present status and prospects of fiber textile technology, into what we have termed 'new frontier fiber'. Chapter 3 reviews the high performance fiber from its origin to its future application. Carbon fiber is used as an example of a typical high performance fiber in Chapter 4.

Thereafter, attention is focused on health care and the environment. We wish to convince the reader that fiber textile technology can be pivotal in helping to enrich our lives in the twenty-first century. Fibers in the twentieth century were perceived mainly in terms of clothing, ropes or nets by the consuming public. These were the visible areas which people could recognize. In the twenty-first century fibers will enter into novel and unexpected applications. We are approaching the age of the wearable computer and organic electroluminescent wearable displays. New potential is open to fibers by building on traditional fiber engineering which produced fiber composite materials for the amusement and car industries, civil engineering and construction, and the aerospace industry; separation membranes using hollow fibers for artificial organs, plastic optical fibers for information technology, biodegradable fiber for ecological conservation; and fibers with biological functions. A greater integration will ensure their increasing contribution to new aspects of the environment and human life.

The book moves away from a narrow interpretation of fiber. Surely the scope of fibers will be enlarged in the twenty-first century away from the visible fiber cloths of the twentieth century to unseen fiber composite materials, and from fibers for practical conventional use to molecular fiber, and nanofiber which can themselves control operations. It is an exciting future. The ideas have already been advanced in the Japanese language in the books by Dr Hongu: *High-Tech Fibers* (1999) and *New Frontier Fibers* (2000). We have received great support in initially transporting these ideas into English by Professor Dr Kanji Kajiwara, Otsuma Women's University, and valuable information about recent trends in fiber technology were conveyed by Professor Dr Akihiko Tanioka and Dr Masatoshi Tokita, Assistant Professor of Tokyo Institute of Technology. We thank them and also Nikkan Kogyo Sinbun-sha for permission to use information from the books mentioned, and member companies of the publicity committee of the Japan Chemical Fiber Association and other fiber-related companies for valuable photographs and data.

Tatsuya Hongu
Glyn O. Phillips
Machiko Takigami

# Author contact details

Tatsuya Hongu
Institute of Techno Strategy
7-39-3-chome Nukui
Nerima-ku, Tokyo 176-0021
Japan

Glyn O. Phillips
Phillips Hydrocolloids Research Limited
2 Plymouth Drive
Radyr
Cardiff
CF15 8BL
UK

Machiko Takigami
2-23 Higashihisakata 1-chome
Kiryu, Gunma 376-0053
Japan

# 1
# Searching the roots of fibers

## 1.1 The importance of fiber in human life

An instinctive reaction is that organic fiber is nothing like as strong as metal. However, nylon, a superfiber, appeared which was stronger than metal in terms of tensile strength per unit cross-section. Organic fibers are light, but since some organic fibers possess a high tenacity/high modulus and much-improved heat resistance, these fibers have expanded into new industrial uses. When these fibers are incorporated into a resin to produce a composite material, the resulting advanced composite material (ACM) is light, strong and deform-resilient. This material surpasses metal, to some extent, in its mechanical properties. These ACMs are applied widely in the aero and space industries to replace aluminum alloys. ACMs are also being used increasingly in the civil engineering and construction industries. For example, in Kansai International Airport, which was built on land reclaimed from the sea, the ACMs have been used as geotextiles following the Kobe earthquake. Although the term 'fiber' tends to have old-fashioned connotations, the fact is that these new fibers have reached new performance levels and found new functions, which match social needs. Now fiber is applied widely in so many fields, which could not have been imagined even a decade ago. This book will describe this new range of fibers.

Tech-textile denotes the textiles applied in the high technology area, including the aero/space industry (as primary and secondary structural materials), the transportation industry (as tyres), the marine industry (as ropes and fishing nets), the civil engineering and construction industry (as reinforcing materials) and the sports industry (as tennis rackets, golf shafts and ski plates). The term 'Tech-textile', implying integrating technology into textiles, was first introduced at the EXPO for industrial textile materials at Frankfurt, and later at the EXPO in Japan. Figure 1.1 shows the development prospects of the fiber industry for the next generation. Figures 1.2 and 1.3 demonstrate the importance of fiber as materials and the expansion of advanced fiber technology into diverse industrial areas. New fibers have appeared in

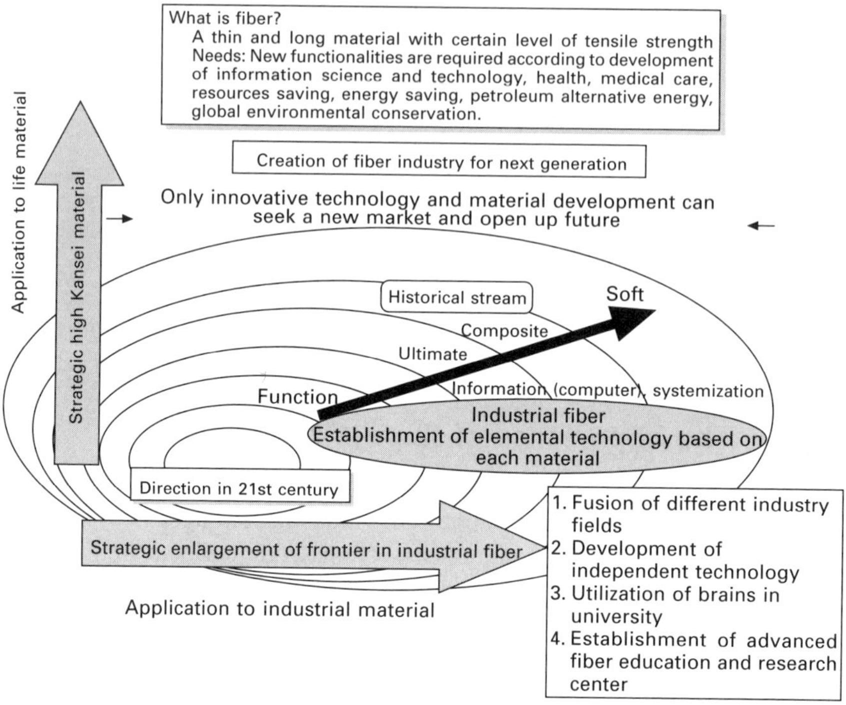


*1.1* Development prospects of the fiber industry for the next generation.

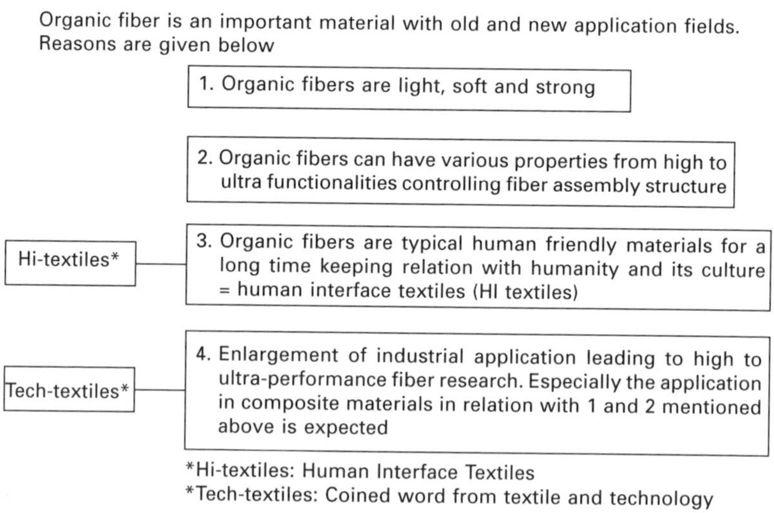


*1.2* Importance of fibers as materials.

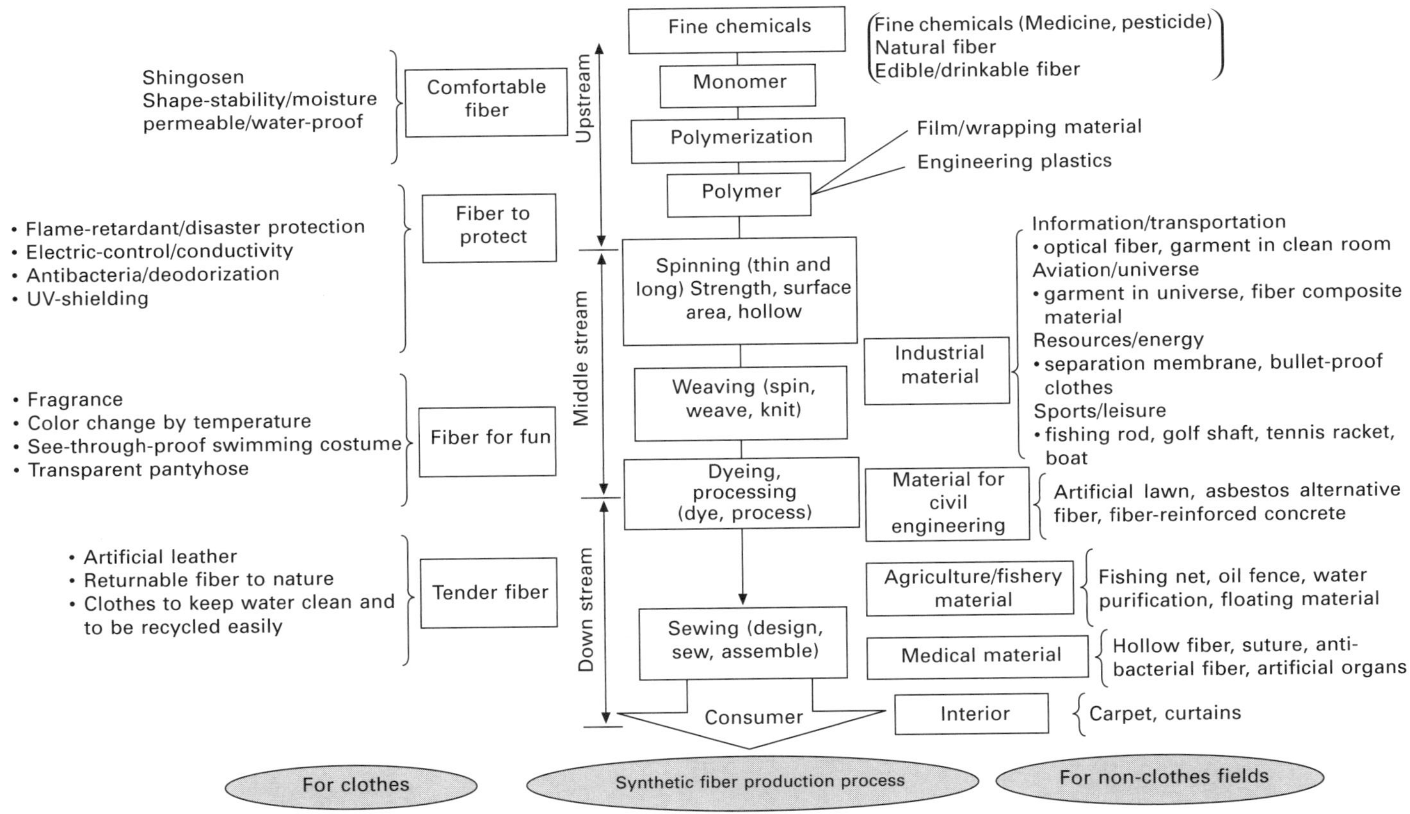


*1.3* Expansion of synthetic fiber technology in diverse industrial areas.

order to meet the changing social needs over the past thirty years, including superfiber, advanced composite materials, optical fiber, shingosen (ultra fine fiber), various functional fibers, comfort/healthcare fiber, etc.

## 1.2 What is high-tech fiber?

High-tech fiber is an expressive way to indicate that advanced science and technology have been used to produce this fiber. High performance fibers, such as the superfiber, high function fiber, which functions as a sensor and actuator, and high touch fiber which possesses a new hand feel, as exemplified by shingosen, are examples of high-tech fiber. The concept of new fibers is summarized in Fig. 1.4, and their applications are found in diverse areas as shown in Fig. 1.5 and summarized in Fig. 1.6. Figure 1.7 illustrates examples of high-tech fiber.

High performance fiber, which has improved physical properties compared with conventional fibers, needs to be distinguished from superfibers. In general,

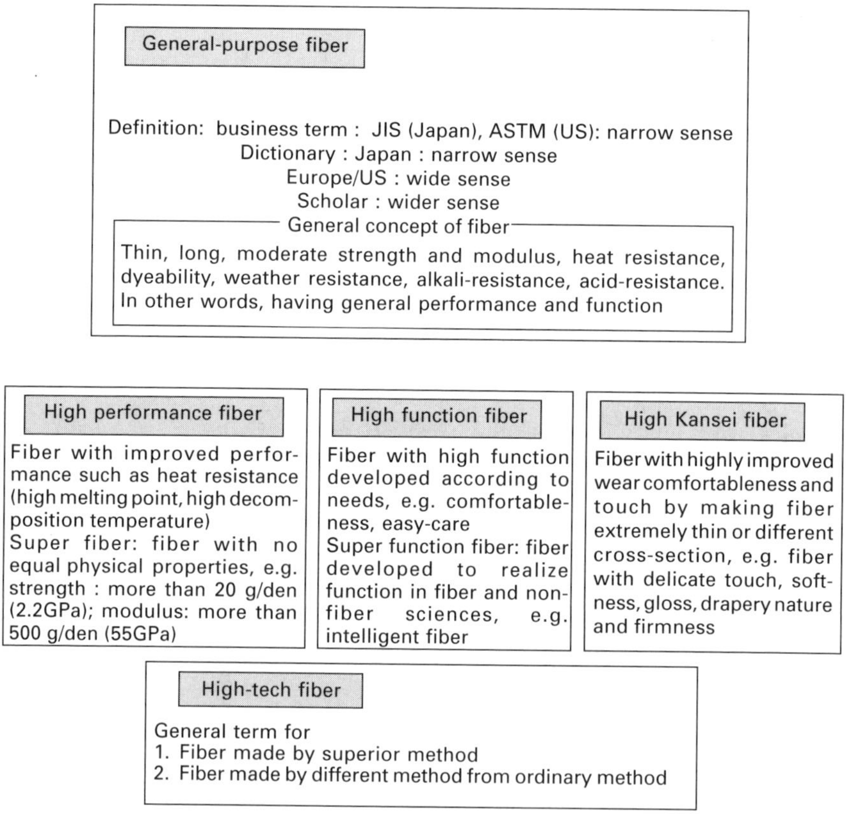


*1.4* Concept of new fibers.

*1.5* Utilization of fiber in various fields.

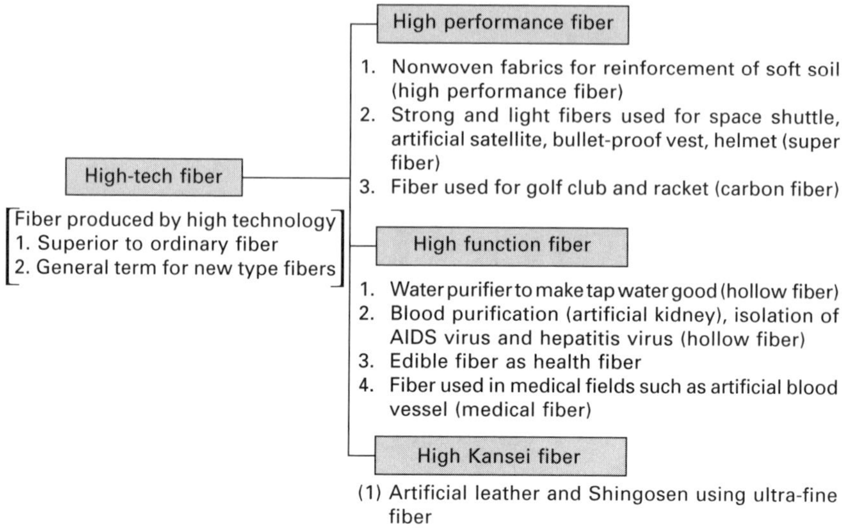


*1.6* Applications of high-tech fibers.

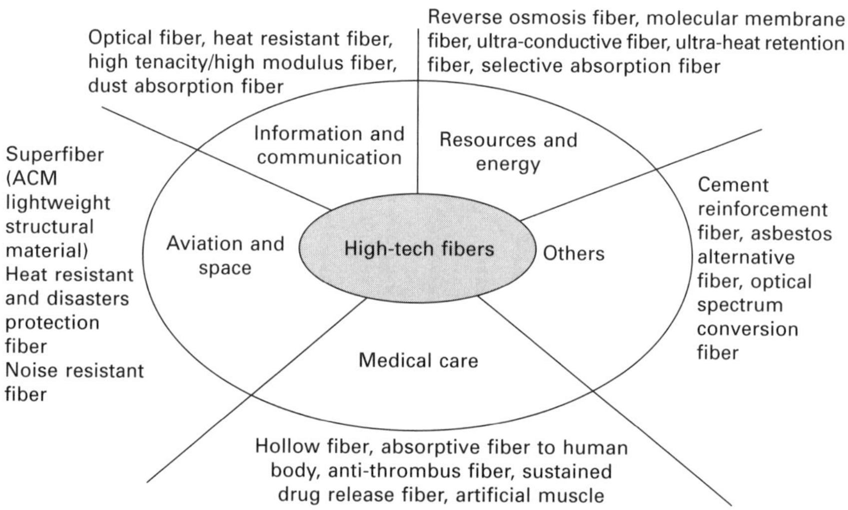


*1.7* Applied examples of high-tech fiber (from short to middle term).

fibers for commodity purposes need to possess appropriate physical properties including such mechanical properties as tensile strength and Young's modulus, heat resistance, dyeability, weather resistance, alkaline resistance and acid resistance. A superfiber, which needs to be superior to high performance fiber, must have a tensile strength larger than 20 g/denier (or 2.2 GPa) and a Young's modulus higher than 500 g/denier (or 55 GPa). Typical examples of superfiber are listed in Table 1.1. GPa (giga Pascal) is the international unit

*Table 1.1* Typical examples of superfiber

| Fiber | Trade name | Strength (g/den) | Modulus (g/den) |
|---|---|---|---|
| Measure of superfiber | | More thn 20 Satisfy both at the same time | More than 500 |
| *para*-aramid*1 | Kevlar 49 (Du Pont) | 22 | 850 |
| | Twaron*2 (Teijin Twaron) | 22 | 850 |
| | Technora (Teijin) | 28 | 560 |
| All aromatic polyester | Bectran (Kuraray) | 29 | 670 |
| Polyethylene fiber | Dyneema (Toyobo) SK60 | 30–40 | 1000–1400 |
| | High tenacity product | 40–45 | 1200–1600 |
| PAN-based carbon fiber*2 (liquid crystal) | TORAYCA (Toray) Besfight (Toho Tenax) Pyrofil (Mitsubishi Rayon) | 20–45 | 1400–3500 |
| Pitch-based carbon fiber (liquid crystal pitch) | Glanoc (NGF) Dialead (Mitsubishi Chemical) | 13–19 | 700–4500 |
| PBO fiber | ZYLON® (Toyobo) | 42 | 2000 |

(*Note*)
*1There are *para*-aramid (high tenacity and high modulus) and *meta*-aramid (heat resistance). *para*- and *meta*-types belong to super and high function fibers, respectively.
*2There are isotropic (for general use) and liquid crystalline pitch (high performance). Liquid crystalline pitch belongs to superfiber.

to express the strength and modulus of a material, and 1 GPa corresponds to the load of approximately 100 kg per 1 $mm^2$, which can suspend two persons. Currently much effort is focused on producing superfibers from polyvinyl alcohol, poly acrylonitrile, polyacetal, nylon and polyester. The tensile strength of various super fibers is compared in Fig. 1.8.

Generally superfibers are defined as the fiber whose strength is more than *ca.* 2 GPa, and elastic constant more than *ca.* 50 GPa. The strength and the elastic constant of the general-purpose fibers are usually represented in units of cN/dtex (centi-Newton/deci-tex). For the superfibers, a unit of GPa is often used. The value in cN/dtex represents the load per unit line density. On the other hand, the value in GPa represents the load per unit sectional area and is larger than that in cN/dtex. For the Para-aramid superfiber a strength of 20 cN/dtex and elastic constant 500 cN/dtex (density = *ca.* 1.44 $g/cm^3$) corresponds to 2.9 GPa and 72 GPa, respectively.

Conventional nylon or polyester fibers for apparels have tensile strength 4.5~6.5 g/denier, and that for industrial use 6.5~10.0 g/denier, both of which

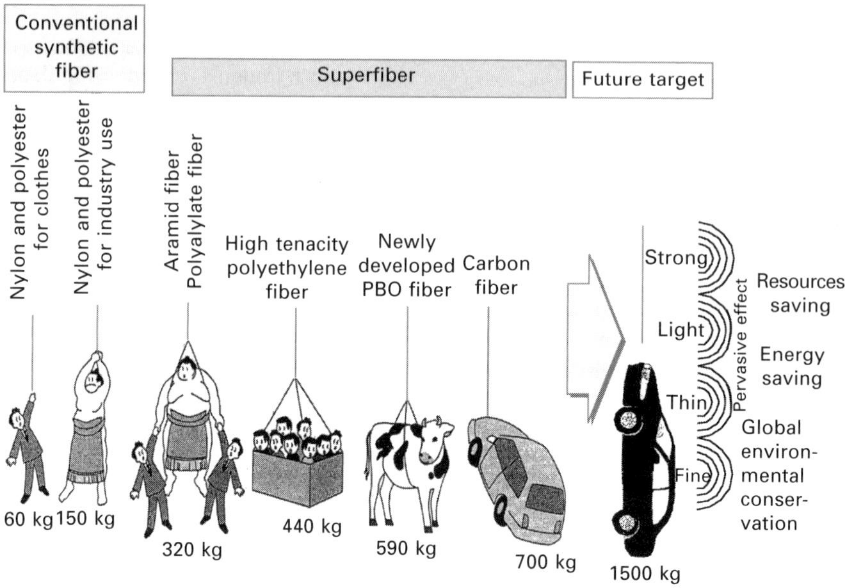


*1.8* Strength of superfibers (How many kilograms can a fiber with 1 $mm^2$ cross-section support?).

are well below the superfiber. Superfiber is used for ropes, monofilament fishing string, in the composite for helmets, tennis rackets and golf shafts, and is also used as a structural advanced composite material for airplanes and space shuttles. Since the superfibers are generally used on the inside of structures, they are rarely evident to the public eye.

Although all fibers have a role within products, functional fiber should additionally possess some distinctive chemical functions. Fiber is usually defined as a thin and long structure. Three features are generally used to characterize a fiber:

(1) The material which makes up the fiber, which could be organic or inorganic, a biomaterial, a nanofiber, such as DNA or a synthetic polymer.
(2) The shape of cross-section can be circular or non-circular.
(3) The fiber microstructure: homogeneous or non-homogeneous, a hollow fiber or a composite fiber.

As the three elements of melody, rhythm and harmony can amalgamate to produce music and move the human heart, so also can the three elements of materials, shape and microstructure in fiber stimulate human emotion through the five senses. For example, shingosen has a characteristic feel which people had never previously experienced and fall in love with. We may be able to produce something even more advanced than shingosen, since no synthetic fiber has yet been produced with a similar feel to animal hair, such as alpaca, angora, cashmere, vicuna, mohair, and camel. The characteristics of these animal hairs are summarized in Table 1.2. Figure 1.9 shows how three elements of fiber are amalgamated to reveal new functions.

Chemical fiber is produced by extruding a polymer through a nozzle, so that the fiber cross-section is more homogeneous. Natural fibers, on the other hand, possess non-homogeneous cross-sections, indicating that we need

*Table 1.2* Animal hair and characteristics

| Variety | Characteristics (touch) | The habitat |
|---|---|---|
| Alpaca | Glossy and smooth | Peru |
| Angora | Glossy and soft<br>Impossible to spin because of no crimp<br>Spun with wool | Hair from Angora rabbit |
| Cashmere | Glossy, thin and soft<br>Glossy long hair like silk | Native goat in Indian Kashmir and Tibet |
| Vicuna | Thinnest and softest among animal hair<br>Expensive | Hair of vicuna living in Cordillera de los Andes |
| Mohair | Glossy as silk<br>Strong and elastic | Hair of Angora goat in Turkey |
| Camel | Soft as cashmere<br>Dark brown<br>Impossible to decolorize | Hair of camel in Asia |

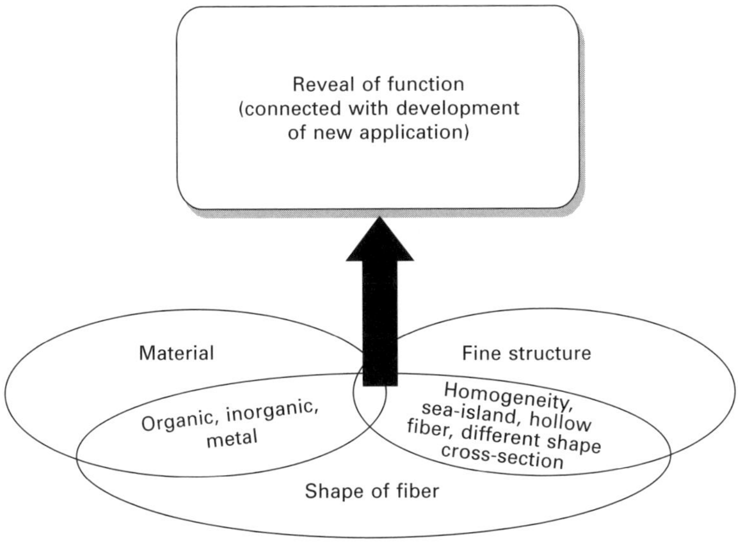


*1.9* Three elements of fiber to reveal new functions.

to learn much more from nature. The requirements for fibers are diverse and beyond conventional concepts. Thus, it is likely that more complex and sophisticated fiber will appear in the future.

## 1.3 Natural versus synthetic fiber

The main difference between natural and synthetic fiber is in structure. Synthetic fiber is produced by extruding a polymer through a nozzle and subsequent drawing. The resulting fiber has a simple structure. The fiber structure of most synthetic fibers is characterized by the *shish kebab* structure, as in the example of polyethylene.

Natural fibers such as cotton, wool, and silk have a non-even non-homogeneous surface. Those fibers also possess a multi-phase structure, which results in specific functions. The simple structure of a synthetic fiber could be suitable for high performance, but not for high function applications.

The silkworm eats mulberry leaves, which are converted by enzymes into two proteins (fibroin and sericin) in its body. Recent research has described how it drags out silk thread to make a cocoon, but the mechanism for how it produces the two proteins is still unknown. Moreover, the chemical structure of silk is complicated and the cross-section of its filament is not circular. Silk possesses a specific luster, warm touch, deep color, and moisture-absorption characteristics, which no synthetic fiber possesses.

The chemical structure of wool, a protein produced by sheep, has a complicated and cunning bilateral structure. Wool is cool in summer and warm in winter. The complicated wool structure gives wool this property, and also good resilience, high bulkiness, and water repellency. No artificial fiber can compare with wool with respect to those properties, which makes wool so suitable for clothing.

A similar argument can be applied to cotton, made up of cellulose photosynthesized from carbon dioxide in air and water. It is a homopolysaccharide with a relatively simple chemical structure. However, its morphological structure is ingenious such that no synthetic fiber can compare with cotton with respect to its moisture absorbency, dyeability and moisture maintainability.

Lentinan, a similar simple but branched fibrous polysaccharide, can be extracted, separated, and purified from a type of fungi, a polypore. It has anti-tumor activity because it increases the level of body immunity. This polysaccharide (Lentinan®) is now commercially available from Taitro Pharmaceutical (Manufactured by Ajinomoto) as an anti-tumor agent.

Why is synthetic fiber different from natural fiber? Why cannot synthetic fibers emulate natural fibers? One of the reasons is that their mode and speed of formation are quite different. Silk thread is produced from the silkworm mouth at the rate of 1 m/min. Wool or cotton has a much slower growing rate of around $10^{-6}$ m/min. The spinning speed of a synthetic fiber is steadily increasing with the development of fiber manufacturing technology, and has now reached the speed of the jet plane. Silkworms, sheep or cotton produce natural fiber in order to protect their bodies, so need to maintain enough function to cope with the environment where those living creatures live. Synthetic fiber, on the other hand, beats natural fiber with respect to its physical performance (tensile strength, heat resistance, durability at extremely low temperatures, etc.) and production efficiency (spinning speed, etc.).

Table 1.3 lists the differences between natural (cotton) and synthetic fibers. As seen from Table 1.3, natural fiber and synthetic fiber can be classified as high-function fiber and high performance fiber, respectively. Synthetic fibers were developed initially as copies of silk, wool, or cotton. At present, there

*Table 1.3* Difference between natural and synthetic fibers

| | Natural fiber (cotton)* | Synthetic fiber |
|---|---|---|
| Raw materials | Water, light and carbon dioxide | Petroleum (monomer) |
| Production method | Photosynthesis | Polymerization and spinning |
| Rate | Less than $8 \times 10^{-7}$ m/min | More than $1 \times 10^{3}$ m/min |
| Characteristics: Structure | Inhomogeneous, precise, complicated (multiphase structure) | Homogeneous and simple, *shish kebab* structure |
| Property | Show excellent properties in daily environment | Show excellent structure even in specific environment |
| Usage | Mainly used for clothes | For clothes and industrial materials |
| Prospect for next generation | 1. High functionality<br>2. Reformation of production technology by biotechnology (fiber production after insects and spiders)<br>3. New natural fiber (spinning mechanism of spiders and silkworms) | 1. Precise control of molecular orientation produces higher dimensional structure materials and composite materials<br>2. Contribute to human technology and fibers with ultimate functions and environmental response<br>3. Contribute to clarification of tissue structure and function |

*Plants derived natural fibers with higher dimensional fibrous structures are synthesized from water, carbon dioxide and light by biochemical processes in plants. The excellent water absorbency/releasing properties are not easily copied by synthetic fibers, and will have different application fields from those of synthetic fibers even if superior development of synthetic fibers. Wool and silk are synthesized by biological processes in sheep and silkworms using plants as raw materials. The structures are more complicated and precise than those of plants derived from natural fibers.

is no way to produce high function natural fiber with a speed comparable with synthetic fiber.

Synthetic fibers are no longer copies of natural fibers such as silk, wool, and cotton. To create synthetic fiber possessing new functions that even natural fibers do not have, research needs to develop a new method of simultaneous polymerization and spinning since wool is produced in such a way. Synthetic fiber indeed surpasses natural fiber, to some extent, as exemplified by ultra-fine fiber, high tenacity/high modulus fiber, water-absorbent fiber, and heat-resistant fiber. A new paradigm shift such as the application of biotechnology is needed to develop super high function fiber.

## 1.4 Artificial fiber by biomimetics

Most synthesized materials including synthetic fibers have been developed by science and sometimes by chance in the past, whereas natural materials are produced as a consequence of biological processes. These approaches have now been integrated.

### 1.4.1 Plant fiber synthesized from carbon dioxide

Plants produce carbohydrates by photosynthesis. Air contains only about 0.3% carbon dioxide. Yet plants utilize this small amount of carbon dioxide with water to produce cellulose by photosynthesis. The structure of the resulting fiber cross-section is non-homogeneous, and is composed of complex multi-layers, whereas that of the artificial fiber is homogeneous. Cellulose could be termed 'carbon dioxide fiber', and gives a hint as to how to produce an environmentally friendly fiber without using fossil energy if we could learn from nature.

### 1.4.2 Lessons from the silkworm

Rayon appeared about a century ago as the first chemical fiber mimicking silk. Rayon filament is made from wood pulp that is dissolved and wet-spun. Rayon is thus chemically composed of the same component (cellulose) as wood pulp. Then nylon appeared about a half a century later. Nylon was aimed to mimic silk chemically, and has similar amide groups. Fifty years after the invention of nylon (around 1988), a synthetic fiber reached a new stage of development when the combined yarn processing technology (the blends of filaments of different shrinkage characteristics) was developed to produce high bulky polyester fiber fabric with a characteristic feel different from natural silk. However, not all of silk's features were reconstructed. For example, the characteristic luster, moisture-absorbent characteristic and bright dyeability of silk have not yet been achieved.

Dr J. Magoshi at the National Institute of Agrobiological Science (NIAS) in Japan has elucidated the mechanism of *in vivo* synthesis of silk in silkworm. It was not previously known how the silkworm makes silk from mulberry leaves. Mulberry leaves are digested to amino acids, which are concentrated in the silk gland where a two-layered silk protein is produced at room temperature. Gel-like fibroin solution is formed by calcium ions in the silk gland. Gel is transformed to sol by carbon dioxide in air, and becomes liquid crystalline in a narrow tube. Sol is transferred slowly to the spinning tube and is spun to silk filament.

The silk gland corresponds to the polymerization/spinning tank in terms of the synthetic fiber production. Silk filament is produced enzymatically at room temperature. This high technology required for the precise control of such a molecular assembly has not yet been achieved by human beings.

The process is now being reconstructed on an industrial scale. A bio-spinning factory is expected to develop to a commercial scale on the basis of the elucidated bio-mechanism of spinning soon as a substitute for oil-based fibers. The silkworm spins fibroin, not by extrusion, but by drawing. The silkworm fixes the end of fibroin on to the ground, and swings its head in the manner of a number '8' to draw fibroin. In conventional industrial spinning of synthetic fiber, the nozzle is fixed and extruded filaments are drawn, whereas the silkworm moves the nozzle (mouth) to draw out filament. Silk filament is crimpled, and its assembly is bulky. In effect, silk has good properties such as heat insulation, moisture absorption and a good feel. The silk filaments assemble and synchronize to achieve high functionality. Now the filaments can be designed to synchronize and promote the same specific functions artificially.

### 1.4.3 Learning simultaneous polymerization and spinning from nature

The process of human hair or wool growth is not well understood. However, human hair or wool grows simultaneously as it is polymerized from amino acids. Since human hair or wool is spun immediately when polymerized, no entanglement occurs during fiber formation. With synthetic fibers, the polymer melt is stored and then spun through a nozzle. We should learn how to spin a new type of synthetic fiber using a similar process to hair production in nature.

The regeneration of human hair is now being investigated. Modern biotechnology has made it possible to manipulate the cells responsible for hair to grow *in vivo*. If the hair growing mechanism can be duplicated, then wool can be produced artificially by biotechnology in the future.

Spider silk is another interesting material. For a synthetic fiber the tenacity is inversely proportional to the elongation at break. In order to improve the tenacity, molecules should be oriented in the direction of the fiber axis. When molecules are more oriented in a fiber, the tenacity increases but the elongation at break decreases. Spider silk in warp has a good tenacity close to Kevlar, and the elongation at break is as high as 35%. Spider silk in weft is coated with adhesive liquid to catch insects, and elongates surprisingly effectively when wet. The spider has the means to remove this liquid in order to walk to its prey without adhering. Now investigations are focused on explaining the structure of spider silk and its relation to its physical properties.

### 1.4.4 From homogeneous intelligent materials to non-homogeneous intelligent materials

As has been indicated 'biomimetics' (the art of learning from the bio-system) could be the key to developing new materials. Applying information from biomimetics has in fact led to the development of new chemical fibers. Biomimetics is expected to lead to the next generation of materials. Non-homogeneous materials can be developed with this technology, whereas only homogeneous material such as chemical fibers were the main target of the twentieth century. For example, new functions may emerge from mimicking the insect shell wing composed of liquid-crystal protein reinforced with chitin, which cuts out infra-red radiation in a hot desert. Bamboo is a natural fiber-reinforced composite material composed of alternating parts of stalk and joint. Its cross-section reveals the distribution of fibrous materials, where the outside is dense and hard while the inside is coarse and soft. A bamboo has a non-homogeneous structure (with density gradient) from the same material, and is thus resilient to very windy and heavy snow conditions. Professor T. Kikutani (Tokyo Institute of Technology) has succeeded in producing a composite with the same density gradient by mimicking the cross-section of bamboo.

There are many examples of materials having density gradients around us in nature. A cap of a turbo shell is an example of a composite reinforced with micro-particles. In this example, the composite is made of a protein matrix and calcium carbide micro-particles. The density of the cap decreases gradually from the surface to the inside. The cap should grow as a turbo (*Turbo cornutus*) grows, and protects it from enemies. The disk-shaped cap has an amorphous layer structure, and grows in its radial direction as the turbo grows. No artificial system yet follows this type of processing, but we may expect to develop new processing methods for plastic materials in this way. The control of the non-homogeneous structure seems a key technology to developing the intelligent fiber.

One of the most demanded characteristics is the ultimate strength of materials as exemplified by high tenacity/high modulus fiber. In order to explore the ideal potential of the polymer material, we should increase the molecular weight of the polymer to almost infinity and reduce the molecular defects. The new spinning and processing technology to achieve this should also be innovative enough to cope with the control of molecular orientation with predetermined precision. In nature, we find proteins of high molecular weight over 2,000,000, but the molecular weight of synthesized polyamide is at most 200,000. There is therefore much to gain by learning the mechanism where by nature synthesizes extremely high molecular weight polymer and spins high-oriented fiber with precision.

## 1.5 Definition of fibrous materials

The field of fibers and textiles covers a very broad range of science and technology. Since people generally regard this as a limited subject, it may be useful here to reconsider what fiber really is.

### 1.5.1 Narrow and broad definitions of fiber

Fiber can be defined in more than one way. A narrow definition of fiber can be found in JIS (Japanese Industrial Standards) L0204-1979, which specifies a fiber to be the structural units constituting yarn, fabric, etc., which are flexible, thin and long enough with respect to its thickness. The shape of fibrous materials is specified by the aspect ratio defined by L/D with L and D being the length and cross-sectional diameter, respectively, and the aspect ratio is over 1,000 in general for fiber. However, JIS does not specify this aspect ratio for fiber explicitly. The fiber characteristics are revealed when the aspect ratio exceeds 100 as exemplified by monofilaments of cotton linter or beaten wood pulp. ASTM (American Society for Testing and Materials) defines textiles to be, 'a generic term for any one of the various types of matter that form the basic elements of a textile and that is characterized by having a length at least 100 times its diameter' (D123-91a).

Although fiber and textile are not well distinguished in Japan, textile is defined explicitly in ASTM as a general term for the fabric or product composed of fiber or fiber assembly. The definition of fiber/textile in JIS or ASTM is an important matter for the textile industries in Japan or the US, and includes no concept of molecules. However, a more general concept of fiber will be found, for example, in the Oxford Advanced Learner's Dictionary:

- the part of food that one's body cannot digest but which helps the body to function well (for example, cellulose and pectin that stimulate peristalsis in the intestine)
- a material made from a mass of thin threads
- any of the thin threads from which many animals and plant tissues are formed.

### 1.5.2 The concept of fiber according to its thickness

Fiber can be classified according to its cross-sectional diameter. A practical fiber has a cross-sectional diameter over 1 μm, and fiber with a smaller diameter would be classified as a molecular fiber. A practical fiber includes an anchoring rope (a cross-sectional diameter over 100 mm) various cables/ropes (a cross-sectional diameter varying from 10 mm to 50 mm), a sewing thread (a cross-sectional diameter around 1 mm), synthetic and natural fiber (a cross-sectional diameter from 10 μm to 50 μm), and a microfiber (a cross-sectional diameter less than 0.1 μm). Dietary fiber and molecular fibers, such as DNA, can be regarded as a nanofiber of a cross-sectional diameter of the order of nm. When the fiber is defined as a material with a large aspect ratio, fiber and textile are hierarchically classified according to their diameter as shown in Fig. 1.10.

## 1.6 Fiber: characteristics and shapes

### 1.6.1 Three characteristics and three shapes

Professor Emeritus S. Ohya (Kyoto Institute of Technology) has his own concept of fiber. He considers the aspect ratio and freedom, energy and

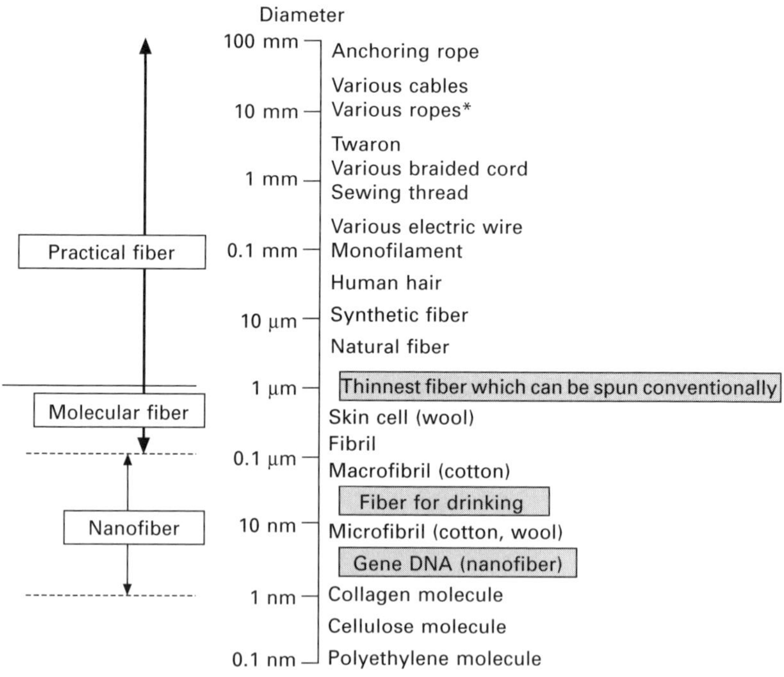


*: Distinction of fiber and assembly of fibers is necessary (Possible to make rope thicker)

*1.10* Classification of fiber according to diameter.

information as three characteristics, which are related to the shapes of fiber, specified as parallelism, branching, and fibrous wave. Fiber is defined as a slender solid with a large aspect ratio. He considers that the concept of fiber should now be extended to include energy, if energy is transmitted over a long distance through a long thin medium such as optical fiber or electric wire. In this concept, linearly propagating information such as music is a type of fiber. Music is thus treated as a fibrous wave. He also argues that fiber and textiles are tools to input/output information. This concept of fiber and textile is summarized in Fig. 1.11

## 1.6.2 Peace of mind and natural remedy

Fiber science and technology have been concerned mostly with the physical performance or function. Now, textiles for healthcare and comfortable clothing comfort have been developed. Here comfort is not only an important factor with regard to wear, but also people consider how they can appeal to other people. Thus, human sensibility is becoming a big issue to be investigated in developing new fibers and textiles.

Professor Emeritus T. Musha (Tokyo Institute of Technology) found a characteristic rhythm in the sound of birds and the wind, a river murmuring, a heart beating, and in brain waves when people were relaxed. These rhythmic sounds appear irregular at first sight, but really have a characteristics fluctuation inversely proportional to the frequency, specified as the 1/f fluctuation. This

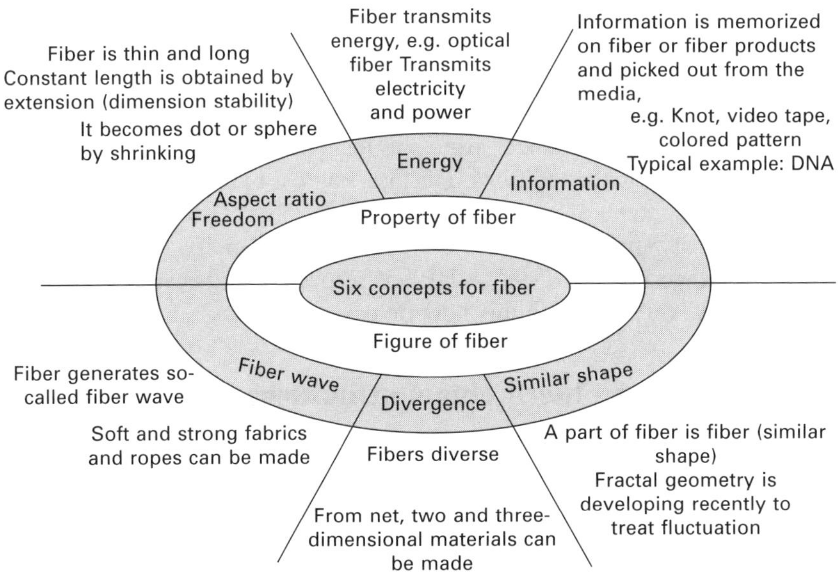


*1.11* Concept of fiber and textile.

fluctuation brings peace of mind. The natural rhythm was analyzed by computer and applied to an electric fan to produce a natural breeze.

Many people suffer from the stress caused by a constantly changing society. They show symptoms of mental disorder and are confused by the rhythm of everyday life. The natural rhythm applied to textile products can provide a compensating function, such as with deodorant or anti-bacteria clothes, but the function is to refresh the mind. Another type of textile with a similar effect is also available, where the function is produced by certain minerals blended in fiber and properly processed, and which generate a negative ion or anion as in the forest. Wavy magic® (Kurabo), Shizen-no-Yuragi® (Nisshinbo) and Biosound® (Toyobo) are examples of the application of the theory of 1/f fluctuation. Holic® (Shikibo) and Stayers® (Fujibo) are the commercial products which can refresh the human mind using the forest rhythm effect.

The theories of the 1/f fluctuation, the fractal, and the bio-sound seem to be applied in different ways to the knitted or woven fabrics by each manufacturer, but all these textiles have a certain effect to refresh the mind. Starting from comfort and health, the target for developing textiles has advanced to mental fulfillment much needed in the modern day and age.

### 1.6.3 Product value generated by human senses

The research target has now been expanded from the objective world to the subjective world. Production efficiency has been the most important factor in operating a factory. The change of lifestyle and values is forcing a shift from this product-led policy to a market-led one, since a comfortable and private life becomes the primary concern. Although in the initial stages, investigation has started into the quantitative evaluation of the in-cloth climate by monitoring the heat and moisture transfer from the inside to the outside of the cloth under various ambient conditions. The fast progress in this new field resulted in the establishment of two research units in Nara Women's University (Laboratory of Apparel Science) and Shinshu University (Department of Kansei Engineering), in 1993 and 1995, respectively. These two research units are now very active in this new field.

## 1.7 Fibers as hierarchical structures

Fiber, as noted, can extend from a molecular fiber to an anchoring rope (a cross-sectional diameter over 100 mm). When fiber is classified according to its diameter, fiber and textile products are included in a geometric similarity.

The fibrous system can be considered as a continuous phase. This concept implies that a series of hierarchical structures constitute a whole structure as observed often in ecosystems. A thread is the fibrous system constituted of fibers, which are the assemblies of fibrils composed of microfibrils. More examples will be found in natural fibers such as cotton, wool, and silk which have continuous hierarchical structures. This concept covers various systems from the universe and animals/plants to molecules, atoms and elementary particles as demonstrated in Fig. 1.12.

## 1.8 What should we investigate in the field of fiber and textiles?

### 1.8.1 Systematic fiber/textile science has a warp and weft

How we can activate fiber and textile science? Fiber/textile science can be regarded as a woven fabric composed of warp and weft. Figure 1.13 shows

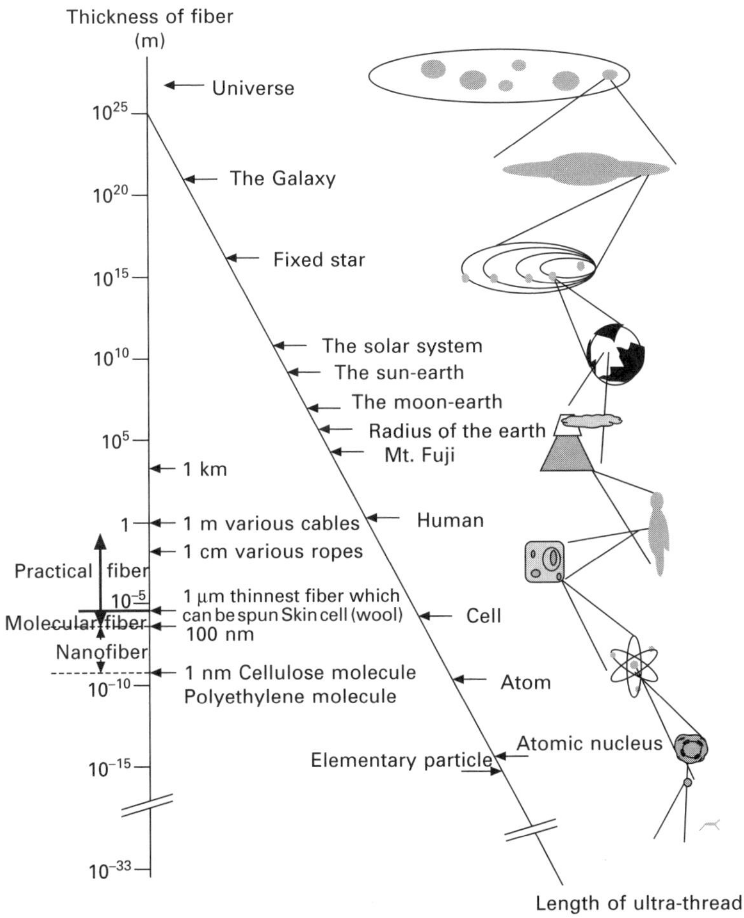


*1.12* Classification of fiber by thickness and hierarchical structures of materials.

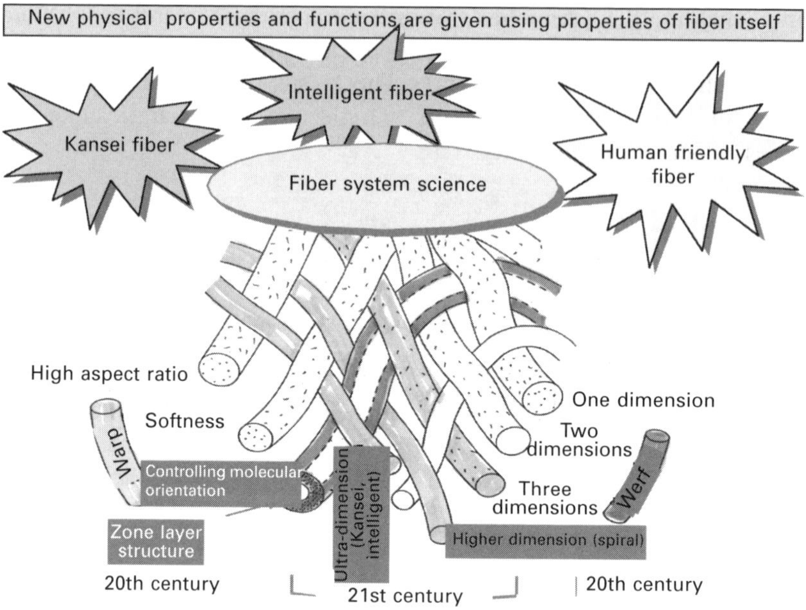


1.13 Dimensional fiber science.

the warp and weft within fiber/textile science. The warp represents the characteristics of fibrous materials, including high aspect ratio, flexibility and molecular orientation. A high aspect ratio is exemplified by a 4.16 g microfilament reaching from the earth to the moon (see Fig. 1.14), an optical

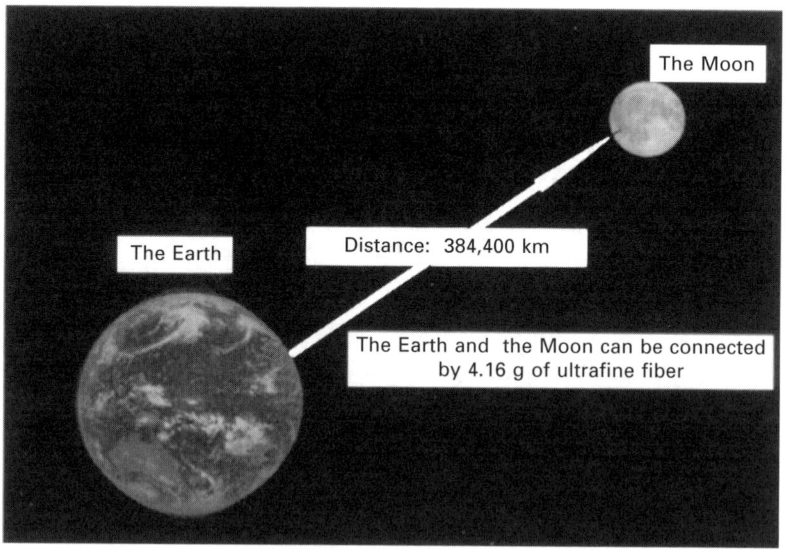


1.14 Image of ultrafine fiber thickness.

fiber or a nerve fiber. The hollow fiber applied for water purification, artificial kidney and artificial liver are other examples of a high aspect ratio. A genetic code can be stored in fibrous DNA (nanofiber) and another example is carbon-nanotube.

### 1.8.2 Flexibility

Since fibrous materials are long and thin, they are flexible. How can we apply this flexibility to the technology? Soft and flexible materials are not well adapted to the technology except for clothes. NASA (National Aeronautics and Space Administration) launched the Mars explorer Mars Pathfinder. This explorer had a unique landing device made of an airbag. The airbag was made of the superfiber Vectran® woven fabric (Kuraray), and had the shape of a beach ball. This soft and flexible ball was basically shapeless, making it possible for it to land on any surface.

Softness is one of the basic characteristics of human-friendly fiber/textile. Bulkiness and heat insulation of the fabric are also important characteristics induced by a synergistic effect of fiber and air. Thus the technology to utilize such materials requires a new 'shapeless' approach.

The physical properties of polymer materials depend on the molecular orientation of component polymers. The control of molecular orientation in the polymer materials is thus a key technology to develop high performance (e.g., superfiber) or high function (e.g., healthcare) fiber. The hybrid of natural and synthetic fiber could possess a high physical performance as well as environmentally-friendly characteristics, and the biodegradable fiber is developed in this context.

In conventional fibers, molecules are oriented relatively in one direction along the fiber axis. Professor C. Kajiyama (Kyushu University) proposed a fiber with the multi-layered structure at the NEDO (New Energy and Industrial Technology Development Organization) Meeting in 1999. When the polymer molecules are oriented in both lateral and radial directions with respect to the fiber axis, such a fiber can have a high tensile strength in both directions. The two-layer structure in the fiber axis can also improve the light-transmittance efficiency of optical fiber. The lateral orientation can be controlled during the spinning and drawing process. If the radial orientation can also be controlled, the potential of polymer materials is optimized in the form of fiber, and we can improve the mechanical, electric, magnetic, and optical properties of fibrous materials.

In this research on fibrous materials, one should keep in mind three factors as the characteristics of fibrous materials:

(1) a high aspect ratio to express the shape characteristic of long and thin fiber

(2) a softness and bulkiness to take into account a physical property of flexible and soft fiber as an industrial material
(3) a molecular design and molecular orientation to optimize the performance and function.

### 1.8.3 Structural control in the fiber cross-section

Fiber is a one-dimensional long, thin material. Now an ultra-microfiber can be produced, which reaches from the earth to the moon with a total weight of 4.16 g. This microfiber contains about 40,000 polymer molecules in its cross-section. When a fiber has a cross-section less than a certain value, a living body can no longer recognize it as a foreign object.

When spun through a nozzle, the fiber cross-section becomes circular because of surface tension. The non-circular cross-section became popular when shingosen first appeared. Various kinds of Microdenier (ultrafine) fibers are shown in Fig. 1.15.

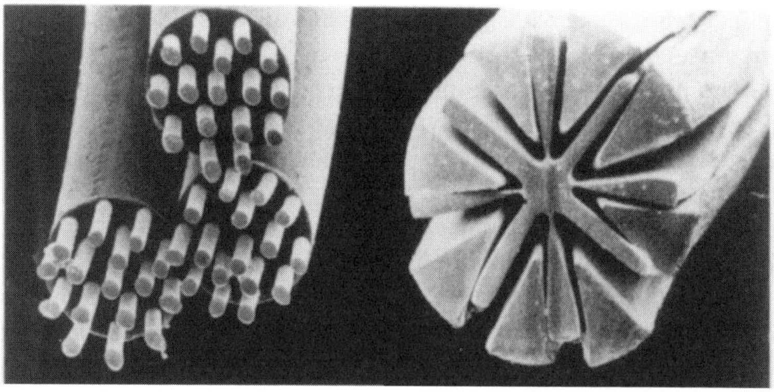

Sea-island type "Toraysee®" (Toray)
Separation type "Belima®" (Kanebo)

Multi-layer type "WRAMP®" (Kurarag)
Separation type "Micro Star®" (Teijin)

*1.15* Various kinds of microdenier (ultrafine) fibers.

*Non-transparent effect by a non-circular cross-section component incorporated into fibers*

A non-transparent effect was achieved by incorporating non-circular components in filaments. Toray developed a non-transparent white swimming costume (Bodyshell®) by combining conjugate spinning and non-circular cross-section technology. Why does a white swimming costume become transparent? When wet, fiber transmits light from inside and becomes transparent. Bodyshell® appeared on the market in summer 1994, and became a big seller. As shown in the cross-sectional view of Bodyshell® original filament in Fig. 1.16, the filament is composed of a star-shaped core containing white pigment (titanium oxide). Titanium oxide is a white powder, which reflects light (non-light transmitting), is stable against light, does not turn yellow, and can be processed into very fine particles. Since the core polymer containing white pigment has an eight-edge star shape, and screens light, incident light from any direction is randomly reflected, and in consequence the swimming costume becomes non-transparent.

A conventional swimming costume is made of two-layer knits, but Bodyshell® has a special three-layer structure as shown in Fig. 1.17. The extra layer prevents light transmittance through an opening of stretched knit fabric. This two-step device suppresses light transmittance to 40% in comparison with a conventional swimming costume. White pigment (titanium oxide) is also used in cosmetics because it has a UV-blocking effect.

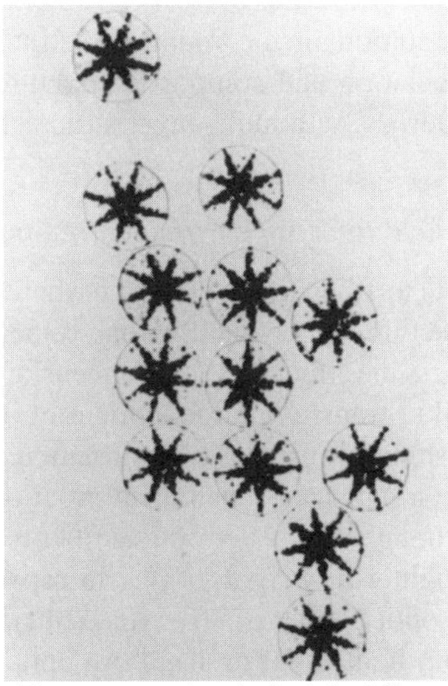

The cross-sectional of original fiber

*1.16* Bodyshell® (Toray).

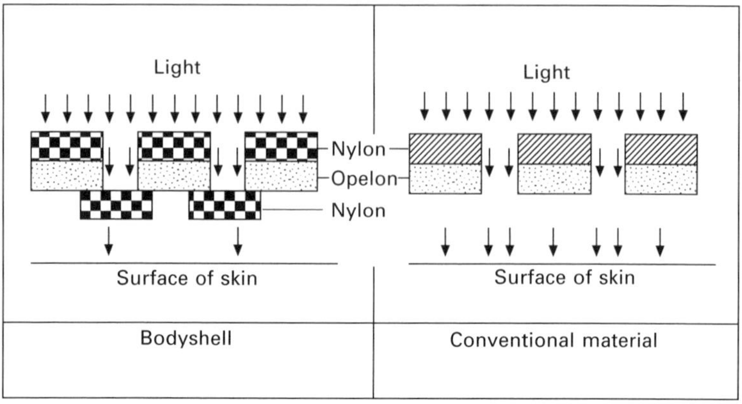


*1.17* Characteristics of Bodyshell® (Toray).

*Multi-layer structures in nature*

There are many examples of multi-layer structures in nature. Morphos in the upper Amazon in Brazil have metallic cobalt blue on their wings. This color is caused by light interference due to the multi-layer structure of scales on the wing. The principle of light interference was applied to Morphotex® (developed by Teijin), a polyester fiber with a multi-layer structure. Details about Morphotex® will be given in Chapter 9.

If we can introduce scales on to a fiber like wool, the friction coefficient becomes dependent on the direction of applied force. The mechanical and optical properties of fiber can be designed precisely by adjusting the molecular orientation in the radial direction. The precise control of the molecular orientation and composite structure enables the production of new fibrous materials with multi-layer structures in a radial direction.

*Optical fiber supporting today's information technology society*

Optical fiber transmits light where the refractive index varies in the radial direction. Fiber itself is one-dimensional, but light cannot be transmitted unless the structure is two-dimensionally controlled. Optical fiber is a powerful tool to transfer a large amount of information quickly, and plays a key role in supporting today's information technology society. A fine optical fiber like a hair can transmit information equivalent to 6000 telephone circuits. Although the cost of optical fiber is higher than copper wire, the optical fiber is lighter in weight, higher in capacity and lower in the transmittance loss. An optical fiber is a fine filament, 0.1 mm in diameter, and transmits 95% of input light as far as 1 km. An optical fiber has a two-layer structure of core and clad. A core part is composed of the material with a high refractive index, and a clad part with a low refractive index. Light input in the core part reflects at the circumference with the clad part and is transmitted without leaking outside, because of the refractive index difference. Three types of light transmittance are available and can be classified as:

- the step index (SI) type,
- the graded index (GI) type, and
- the single mode (SM) type (see Fig. 1.18).

Optical fiber is made of

- quartz,
- multi-component glass or
- plastics

which are applied to a suitable field depending on the transmittance scale and distance as shown in Fig. 1.19

Although quartz is expensive, its transmittance distance is long and is used for transmittance over medium to long distances. The light transmittance of plastic is not so good as quartz, but is easier for handling. Its cost is low, and it is used as a guideline over short distances in applications such as in the control equipment for a measuring instrument, and for office and factory automation. For example, the main optical highway from Asahikawa (Hokkaido) to Kagoshima (Kyushu) is laid with single mode quartz optical fiber, but superfiber is incorporated in order to protect the brittle quartz. Plastic optical fiber (POF) is easy to handle for branching and connecting, and a big market is expected for POF to replace the present in-house telephone circuit if its intrinsic performance is improved. A genuine multi-media society will be realized when the optical fiber network is widespread in all households. POF, for domestic use, will complement the glass-type (quartz or multi-component glass) optical fiber for long distance optical highways. 'Fiber to the home' has become the catchphrase of the POF network by fixing the peripheral technology complementary to the high-speed/large-scale transmittance technology using a single mode.

A SI-type POF (see Fig. 1.20) has a two-layer structure composed of high-purity poly(methylmethacrylate) resin with a high refractive index in the core part and fluorocarbon polymer with a low refractive index in the clad part. Since the clad part has a lower refractive index than the core part, the incident beam will reflect totally at the boundary and propagate to the other end.

Recently POF made of whole fluorocarbon polymer has been developed by Professor Y. Koike (Keio University) and appeared on the market (Lukina® from Asahi Glass). Whole fluorocarbon POF has an advantage over the quartz optical fiber in two respects. Fluorocarbon-type POF has a larger core size for light transmittance and is more flexible than the quartz optical fiber. The latest developments in a GI-type whole fluorocarbon POF are very remarkable, and have received much attention.

| Material | Core | Cladding | Way of transmission | Properties (db/km) |
|---|---|---|---|---|
| Quartz glass fiber | Quartz glass | Quartz glass | SI type, GI type, SM type {5-62.5 μm} | Advantage: good optical property, transparent, less loss<br>Disadvantage: expensive (10–15) |
| Multi components glass | Multi components glass | Multi components glass | | |
| Plastic fiber (POF) | Plastic | Plastic | (Fiber to the home) | |
| | PMMA | Special fluorine resin | SI type (100–1000 μm) | Advantage: soft and easy to process, large diameter<br>Disadvantage: (100–150) |
| | Fluorine resin | Fluorine resin | GI type (100–1000 μm) | Optimization of refractive index distribution cause equal arrival rate of information<br>Simple connector makes reduction of error (10–15) |

① Step Index (SI) type optical fiber

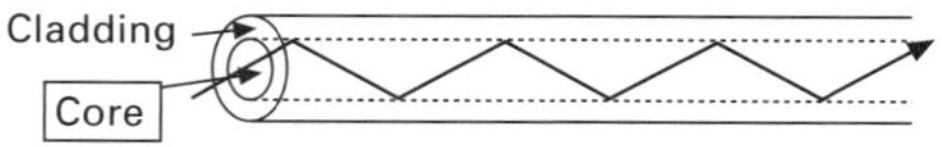


Incident light to core reflects at core-cladding border and moves forward (total reflection)

② Graded Index (GI) type optical fiber

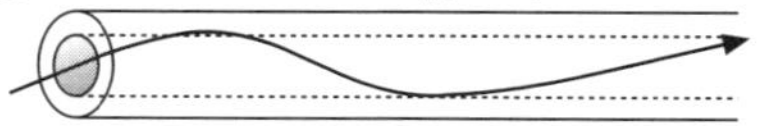

Refractive index of core is more than that of cladding. They are not uniform and gradually change from center to the surroundings. Light moves in a zigzag line. The moving rate is inversely proportional to refractive index of media

③ Single Mode (SM) type optical fiber

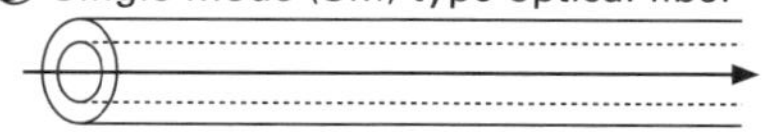

Light is transmitted as a simple beam along central axes (core)

*1.18* Kinds of optical fibers supporting information technology and way of transmission.

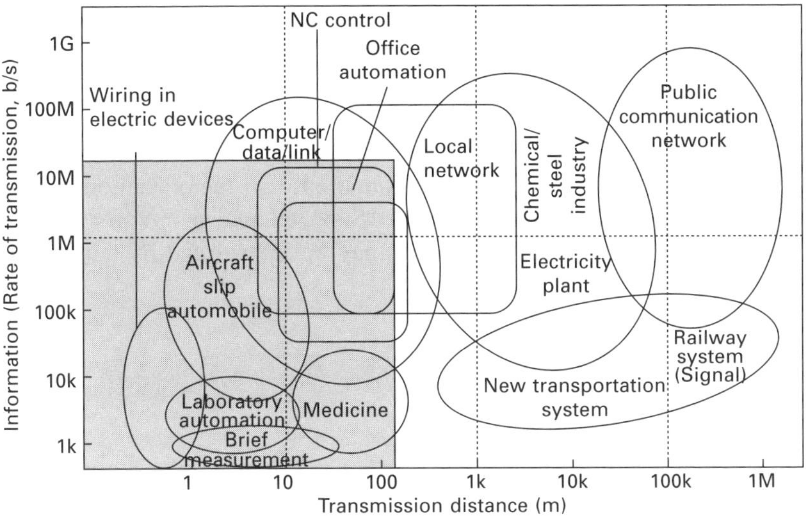


Note: Plastic optical fiber is available only in gray zone (bottom left) because of transmission loss.

*1.19* Application field of optical communication.

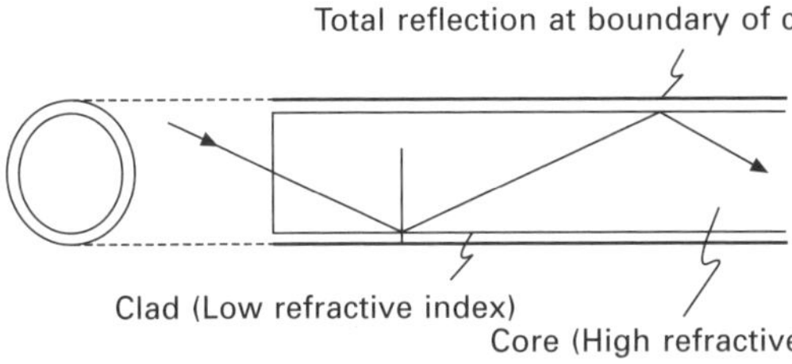


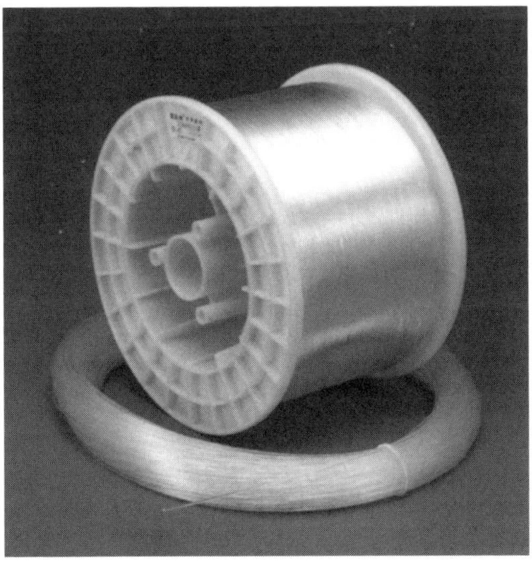

*1.20* Structure of plastic optical fiber and bulk fiber (Mitsubishi Rayon).

Although POF, in general, has a shorter transmittance distance than quartz optical fiber, it is light, flexible, easily processed, and durable. Thus POF can be widely applied in various fields from industrial use to all-purpose domestic use, including light guide (spot illumination, display), light sensor (optical measuring instruments, medical appliances), the short distance transmittance system (in-house automation, office automation, and factory automation), automobile/house electric appliance, etc. Soon telephone circuits will be replaced by POF and real-time pictures may be transmitted by telephone, thus improving the quality of our life.

### 1.8.4 New fiber science

*Three-dimensional textiles*

Arisawa Manuf. Co. and Shikishima Textile Co. produce three-dimensional fabrics for advanced composite materials applied in the aero-/space motor, machinery, and civil engineering industries.

*Human- and environmentally-friendly intelligent fiber*

Fibers can possess intelligence and can function as a sensor to detect external stimuli, as a processor to evaluate external stimuli, and as an actuator to respond/control actively according to the external stimuli. Figure 1.21 shows the concept of the human- and environmentally-friendly intelligent fiber.

Fibrous materials for intelligent fibers are being developed. Sportswear should have a good heat insulation with low conductivity when a body is still cold, but have a good sweat permeability to prevent the body from being steamed up when the body is hot and sweaty. Intelligent sportswear controls ventilation through its texture.

## 1.9 Bibliography

*1.1 The importance of fiber in human life*

Hongu T., (ed.), *New Fiber Science – Challenge for New Frontier,* Research Institute of Economy, Trade and Industry, Tokyo 1955.

Hongu T., (ed.), *Knowledge for Fibers,* Koshin-sha, Tokyo, 1996.

Hongu T. and Phillips G.O., (eds), *New Fibers,* 2nd edn., Woodhead Publishing Ltd, Cambridge, 1997.

Hongu T., *New Frontier for Fibers* in *Polyfile,* **34** (6), pp. 17–67, Taisei-sha, Tokyo, 1997.

*1.2 What is high-tech fiber?*

Hongu T. and Phillips G.O., in *New Fibers*, 2nd edn., T. Hongu and G.O. Phillips (eds), p. 168, Woodhead Publishing Ltd., Cambridge, 1997.

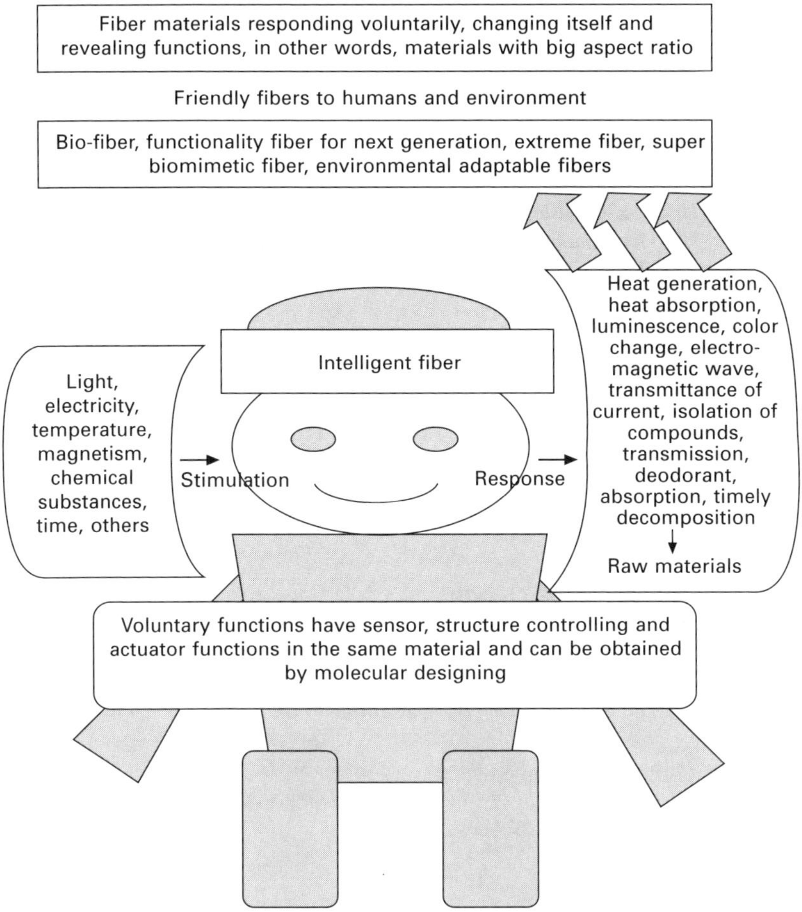


*1.21* Intelligent fiber friendly to humans and the environment.

Miyamoto T. and Hongu T., in *New Fiber Materials*, T. Miyamoto and T. Hongu (eds), p. 64, Nikkan Kogyo Shinbun-sha, Tokyo, 2002.
Hongu T., *High-Tech Fibers*, Nikkan Kogyo Shinbun-sha, Tokyo, 1999.
Okamoto H., *Sen-i Gakkaishi*, **44**, 81 (1988).

## *1.3 Natural versus synthetic fiber*

Miyamoto T. and Hongu T., in *New Fiber Materials*, T. Miyamoto and T. Hongu (eds), p. 19, Nikkan Kogyo Shinbun-sha, Tokyo, 2002.
Nagasawa N., *J. Text. Machine Soc. Jpn.*, **52** (8), 348 (1999).
Pennings A.J., *et al., Kolloid-Z.Z. Polym.*, **237**, 336 (1970).
Pennings A.J., *J. Polym Sci. Polym. Symp.*, **59**, 55 (1977).
Magoshi J., *et al., Sen'i Gakkaishi*, **53**, 202 (1997).
Hongu T., (ed.), *The New Fiber Science-challenge for New Frontier*, p. 234 Research Institute of Economy, Trade and Inductry, Tokyo, 1955.

### *1.4 Artificial fiber by biomimetics*

Miyamoto T. and Hongu T., in *New Fiber Materials*, T. Miyamoto and T. Hongu (eds), p. 128, Nikkan Kogyo Shinbun-sha, Tokyo, 2002.
Kumakura Y., *J. Text. Machine Soc. Jpn*, **52** (8), 321 (1999).
Tanimoto T., *J. Text. Machine Soc. Jpn*, **52** (8), 327 (1999).
Shoken G. and Fujie T., *J. Text. Machine Soc. Jpn*, **52** (8), 334 (1999).
Magoshi J., in *Handbook of Biomimetics*, Y. Osada (ed.), p. 1016, N.T.S., Tokyo, 2002.
Kikutani T., in *The 32th Summer Seminar Proceeding*, p. 82, The Society of Fiber Science and Technology, Japan, 2001.

### *1.5 Definition of fibrous materials*

JIS (Japanese Industrial Standards) L0204-1979.
ASTM (American Society for Testing and Materials) D123-91a.

### *1.6 Fiber: characteristics and shapes*

Ohya S., *Hokou*, **1**, 33 (1989).
Mandelbrout B.B., *Les Object Fractals: Forme et Dimension*, Flammmarion, Paris, 1975.
Mandelbrout B.B., *Fractals: From Chance and Dimension*, W.H. Freeman and Company, San Francisco, 1977.
Mandelbrout B.B., *The Fractal Geometry of Nature*, W.H. Freeman and Company, New York, 1982.
Musha T., *The Journal of Acoustical Society of Japan*, **50**, 6 (1994).
Musha T., *Yuragi no Sekai (World of Fluctuations)*, Kodansha, Tokyo, 1980.
Yanai U., *Sen'i Gakkaishi*, **51**, 252 (1995).

### Kansei (aesthetic) fiber

Shimizu Y., *J. Jpn. Soc. KANSEI Engineering*, **1**(1), 56 (1999) (in Japanese).
Nishimatsu T., Hayakawa H., Shimizu Y., Kamijoh M. and Toba E., *Kansei Engineering Int.*, **1**(1), 17 (1999).
Ohta K., Tanaka T. and Miyawaki F., *Kansei Engineering Int.*, **1**(1), 25 (1999).
Dai X., Mitsui S., Nomura K., Furukawa T., Takatera M. and Shmizu, Y. *Kansei Engineering Int.*, **1**(1), 41 (1999).
Tanaka M., Furukawa T., Shimizu Y., Kamijoh M., Hosoya S., Morisaki T. and Ohtake A., *Kansei Engineering Int.*, **1** (2), 1(2000).
Horiba Y., Kamijoh K., Hosoya S., Takatera M., Sadoyama T. and Shimizu Y., *Kansei Engineering Int.* **1**(2), 9 (2000).
Harada T., *Sen'i Gakkaishi*, **55**, 276 (1999).

### *1.7 Fibers as hierarchical structures*

Hongu T. and Kikutani T., *Polyfile*, **43**(6), 19 (1997).

## *1.8 What should we investigate in the field of fiber and textiles?*

### Human friendly fiber

Zimmerman N., Moore J.S. and Zimmerman S.C., *Chemistry & Industry*, 604 (1998).
Lawrence D.S., Jiang T. and Levett M., *Chem. Rev.*, **95**, 2229 (1995).
Sijbesma R.P., Beijer F.H., Brunsveld L., Folmer B.J.B., Ky Hirschberg J.H., Lange, R.F.M. Lowe J.K.L. and Meijer E.W., *Science*, **278**, 1601 (1997).
Kato T., Nakano M., Motei T., Uryu T. and Ujiie S., *Macromolecules*, **28**, 8875 (1995).
Lehn J.M., *Makromol. Chem. Macromol. Symp.*, **69**, 1 (1993).
Kato T., Kihara H., Kumar U., Uryu T. and Frechet J.M., *Angew. Chem., Int. Ed. Engl.*, **33**, 1644 (1994).
Sato A., Kato T. and Uryu T., *J. Polym. Sci. A. Polym. Chem.*, **34**, 503 (1996).

### Plastic optical fiber (POF) – SI type POF (Esuka)

Uozu Y., *Series of Basic Lectures on Fiber & Textile*, p. 52, The Society of Fiber Science and Technology, Japan, 2001.
Uozu Y., *The 32nd Summer Seminar Proceeding*, p. 145, The Society of Fiber Science and Technology, Japan, 2001.
Seidl D., Merget P., Schwarz J., Schneider J., Weniger R. and Zeep E., *Proc. of POF '98*, pp. 205–211, 1998.
Smyers S., *Proc. of POF '98*, pp. 69–76, 1998.
Mizofguchi T., *Proc. of POF '98*, pp. 77–80, 1998.

### Plastic optical fiber (POF) – GI type POF

Koike Y., *Polymer*, **32** (10), 1737 (1991).
Koike Y., Matusoka S. and Bair H.E., *Macromolecules*, **25** (18), 4807 (1992).
Koike Y., Ishigure T. and Nihei E., *J. Lightwave Tech.*, **13** (7), 1475 (1995).
Koike Y. and Ishigure T., *IEICE Trans. on Communications*, **E82-B** (8), 1287 (1999).

### Three-dimensional textiles

Suesada S., *Techtextile Symposium ASIA Outline of Speeches*, p. 204, Osaka International Trade Fair Commission, 2000.
Nishikawa A. and Shimizu Y., *IEICE*, **J72-B-II** (4), April 1989, Japan.
Shimizu Y. and Nagao H., *IEICE*, **J76-B-II** (10), Oct. 1993, Japan.
Suesada S. and Watanabe A., *SAMPE*, **44-2**, 2301 (1999).

### Morphos-structured fibrils

See references in Chapter 9.

# 2
# The new frontier fibers?

## 2.1 Enlargement of the frontier in a fiber competition age

General-purpose and high-tech fibers have been well developed in Japan. In Asia, these fibers and high-tech fibers are still under development. Food, clothing, housing, and energy are the national priorities in the United States. The status of fibers in this competitive age is shown in Fig. 2.1. Their application to industry is growing apace, fueled mainly by the high-tech fibers.

The high-tech fibers can be regarded as high function fiber, high aesthetic fiber having characteristic feel, and high performance fiber with specific properties to make them useful as industrial materials. Superfiber is one such high-tech fiber and is expected to grow in importance and application. The development of fiber usage in the twenty-first century is shown in Fig. 2.2.

### 2.1.1 Fiber as a major technical strategy in Japan

In June 1999, the Japanese government decided on 'the policy to stimulate employment and to strengthen industrial competition'. Fiber was selected as one of the main technical strategies, together with other technologies such as biotechnology, information technology, medical care, welfare, and the environment. The emphasis is not merely on fiber for textiles, and involves a fundamental review of the field. Fiber was announced as one of the national industrial technical strategies in March 2000 in Japan, twenty years later than in the United States. The key areas involved in this overall thrust are 'ageing society', 'health and medical care', 'stable supply of energy', 'global environmental', 'conservation', 'economical development', 'information technology (IT)', 'Kansei (aesthetic) information', and 'circulative economic society'. High-tech fiber can play an important role in all of them. In an ageing society, for example, high function fiber can contribute in the field of health and medical care, or as an optical fiber and a new material in the

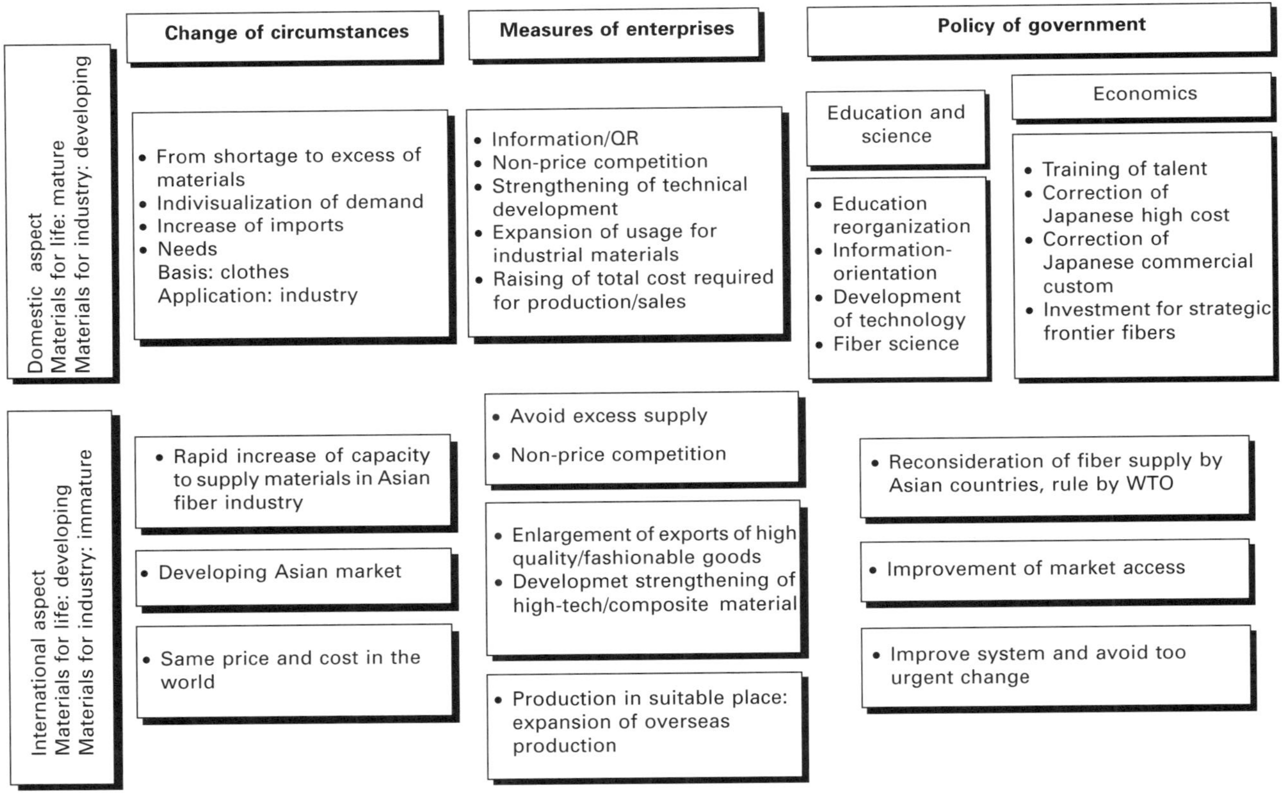


*2.1* Status of fibers in competitive age (Source: H. Ishige, Special Relay Symposium Proceedings, p. 23, The Society of Fiber Science and Technology, Japan, April, 1996).

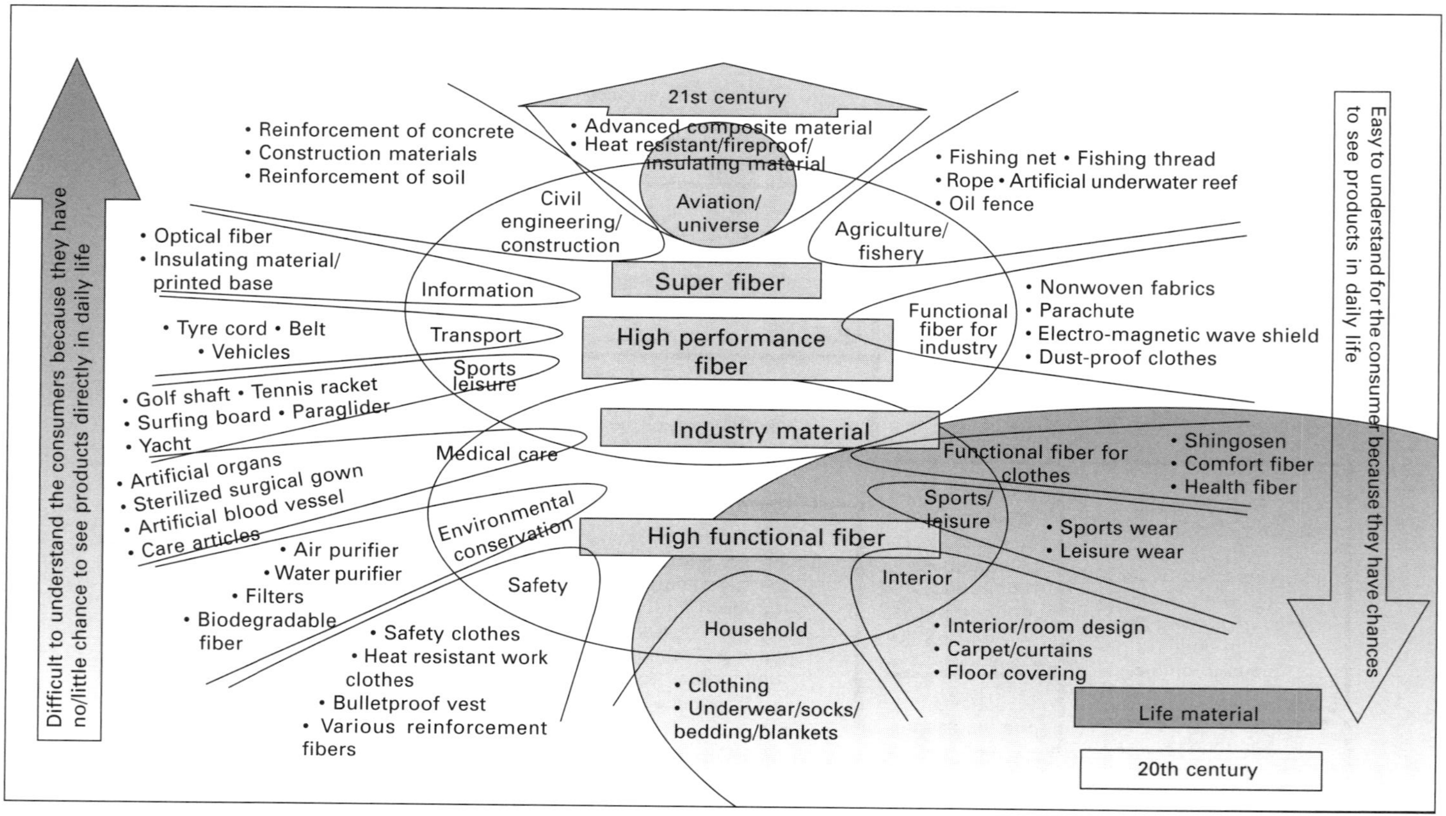


2.2 Development of fiber usage in the twenty-first century.

information technology revolution. Aramid fiber can decrease size and weight of a mobile phone, fiber can shield electromagnetic waves. Biofiber produced by microorganisms, biodegradable fiber, nanofiber and miracle fiber can exercise the pull needed by industry in the twenty-first century.

*Enlargement of usage for industry materials*

The enlargement and development of high-tech fiber as an industrial material depends on an understanding and the solving of market demand. Technology can be used to solve the demand and in the process requires new materials. The performance required for various application fields is shown in Table 2.1. New industry materials have many application fields in the market. The key words to launch fiber into the twenty-first century are: high performance, high function, high composite, and soft. Superfibers developed by many companies are good examples of pursuing these ultimate properties. The silk-like ultra-fine fibers are the result of pursuing biomimetics, copying the operating system of the silkworm. The strength of nylon and polyester fiber is only 3% of their theoretically ultimate value, but an increase of the strength to 5–6% of the theoretical value is required. It is seemingly a small improvement, but it calls for dedicated research over the next ten years. Cooperation among industry, academia and government will contribute to solving this problem and creating a new industry. Moreover, it will contribute to the fiber industry not only in Japan, but also in other parts of the world. These needs and development of the usage of superfiber are shown in Table 2.2.

## 2.2 'Selection', 'concentration' and 'originality' development on a world-wide scale

International competition in fibers became intense at the turn of the millennium. In the twentieth century, polyester filament was the most popular, but now selection, focusing and reorganization have already started.

### 2.2.1 Asian fiber manufacturers hotly pursue Europe/US manufacturers

In March 2000, the Japan Chemical Fibers Association reported their research on reorganization of synthetic fiber companies in Europe and USA. According to the report, three out of every five of the main manufacturers of synthetic fibers in the world changed their emphasis to other business in the past decade. The chemical giants of Europe and the United States moved their main product lines from synthetic fibers to life science and specialty chemicals, because of their stability and high profit margins. Accompanying this, and in

*Table 2.1* Performance required for various application fields

| | High strength | High modulus | High toughness | High tear strength | Shock resistance | Abrasion resistance | Fatigue resistance | Durability | Dimension stability | Dye stability | Lightweight | Transparency | Air permeability | Heat retention | Heat resistance | Insulation | Fire-proofing | Fire prevention | Fire resistance | Antiweatherability | Moisture absorption | Moisture permeability | Water absorption | Water-proofing | Water repellency | Electric controlling | Electric insulation | Bacteria-proofing | Fungi-proofing | Chemical-resistancce | High adhesive property | Ease of storage | Safety |
|---|---|---|---|---|---|---|---|---|---|---|---|---|---|---|---|---|---|---|---|---|---|---|---|---|---|---|---|---|---|---|---|---|---|
| Clothes | ○ | | | | | | | ○ | ○ | | ○ | | ○ | ○ | | ○ | ○ | | | | ○ | ○ | ○ | ○ | ○ | ○ | | | | | | | ○ |
| Bedding | ○ | | | | | | | ○ | | | ○ | | ○ | | | | | | | | ○ | | | | | | | ○ | ○ | | | ○ | |
| Interiors | | | | | | | | ○ | | | | | | ○ | | | | ○ | | | ○ | | ○ | | | ○ | | ○ | ○ | | | | ○ |
| Life materials | ○ | | | | | | | | | ○ | △ | △ | ○ | | | | | ○ | | | | ○ | | ○ | | | | | △ | | △ | | |
| Agriculture/ marine products | ○ | ○ | ○ | | | | | | | | | △ | | | ○ | | | | | ○ | | | | | | | | | | | | | |
| Industry | ○ | | ○ | | | | | ○ | | | | | | | | | | | | | | | | | | | △ | | | △ | | | |
| Traffic/transportation | ○ | | ○ | ○ | ○ | ○ | ○ | | ○ | | ○ | | | | | | | | | | | | | | | | | | | | | | |
| Civil engineering construction | ○ | ○ | ○ | ○ | ○ | ○ | ○ | | | | | | | | | | ○○ | | | ○○ | | | | ○○ | | | | | | ○○ | ○○ | | |
| Ocean development | ○ | | ○ | | ○ | | | ○ | | | ○ | | | | | | | | | | | | | | | | | | | | | | |
| Aviation/space | ○ | | ○ | | ○ | | | ○ | | | ○ | | | | ○ | ○ | ○ | | | | | | | | | | | | | | | | ○ |
| Energy development | | | | | | | | | | | | | | ○ | ○ | | ○ | | | | | | | | | | | | | | ○ | | ○ |
| Medical care | | | | | | | | | | | ○ | | | | | | | | | | | | | | | | | | | | | | |
| Information | ○ | △ | | | | | | ○ | ○ | | △ | | | | | | | | | | | | | | | | | | | | | | |
| Fire fighting | ○ | | ○ | | | ○ | | | | | ○ | | | | ○ | ○ | | ○ | ○ | ○ | | | | | | | | | | | | | |
| Defense/munitions | ○ | | ○ | | ○ | ○ | ○ | ○ | | | ○ | | ○ | | ○ | | ○ | ○ | | | △ | △ | | ○ | ○ | △ | △ | | | | | ○ | |

From K. Matsumoto, *Polymer Digest*, **36**(6), 4 (1984)
○ = Very important property.
△ = Less important property.

*Table 2.2* Needs and development of the usage of superfiber

| | |
|---|---|
| Needs | Lightening and high function are necessary for protection of global environment, resources saving, and energy saving<br>Enlargement of industry frontier |
| Application development | Pursuit of limit performance (strength, high modulus)<br>High performance with specific function (plus alpha function)<br>(Surface property, shrinkage property, durability)<br>New molding and processing technology, development of new matrix<br>Development of evaluation technology<br>Decrease of cost (balance between cost and performance)<br>More severe after collapse of bubble economy<br>Development plans for disposal and recycling |
| Expansion of application | Not only required functions but also plus-alpha functions are important. Acceptable price to consumers |

hot pursuit, Asian fiber manufacturers have developed to occupy 60% of the world production of synthetic fiber. Of course, for special fields the Western manufacturers continue to maintain their share of production in speciality fields, for example acrylic fiber (Bayer, German), polyester and acrylic fiber (Monte Fiber, Italy), polyester and nylon (Allied Signal, USA), and polyester and nylon (Welman, USA). Four companies in Europe, ICI in the UK, Hoechst in Germany, Rhone Poulenc in France, and Akzo in Netherlands are included in the top ten manufacturers of polyester production in the world in 1998. However, in 2003 Asian capital companies, such as Nan Ya in Taiwan/USA, Tuntex/Xiang Lu in Taiwan/Thailand/China, and Reliance in India, occupied the top spots. The market share by Asian companies will continue to increase due to the rise of Chinese companies, such as Shanoxing Yuandong and Yizheng as shown in Fig. 2.3.

Rayon was invented by Chardonnet (France) in the latter half of the nineteenth century, nylon by Carothers (USA) in 1932, polyester by Whinfield and Dickson (UK) in 1940, and Shingosen in 1988 by the Japanese, but the report indicates that the new streams of fiber in the twenty-first century will flow from Europe, USA, to Asia, including Japan. The new wave which will bloom in the twenty-first century is shown in Fig. 2.3. Fiber science is still studied actively in Europe and the USA and completely new fibers can be developed in the twenty-first century.

## 2.2.2 Studies in Japan now watched by the world

In Japan, fiber science is still being studied, not only at universities with Faculties of Fiber Science, but also at universities without these faculties. For example, plastic optical fiber (POF) from acrylic fiber and fluorine containing fibers are investigated by Professor Y. Koike (Keio University),

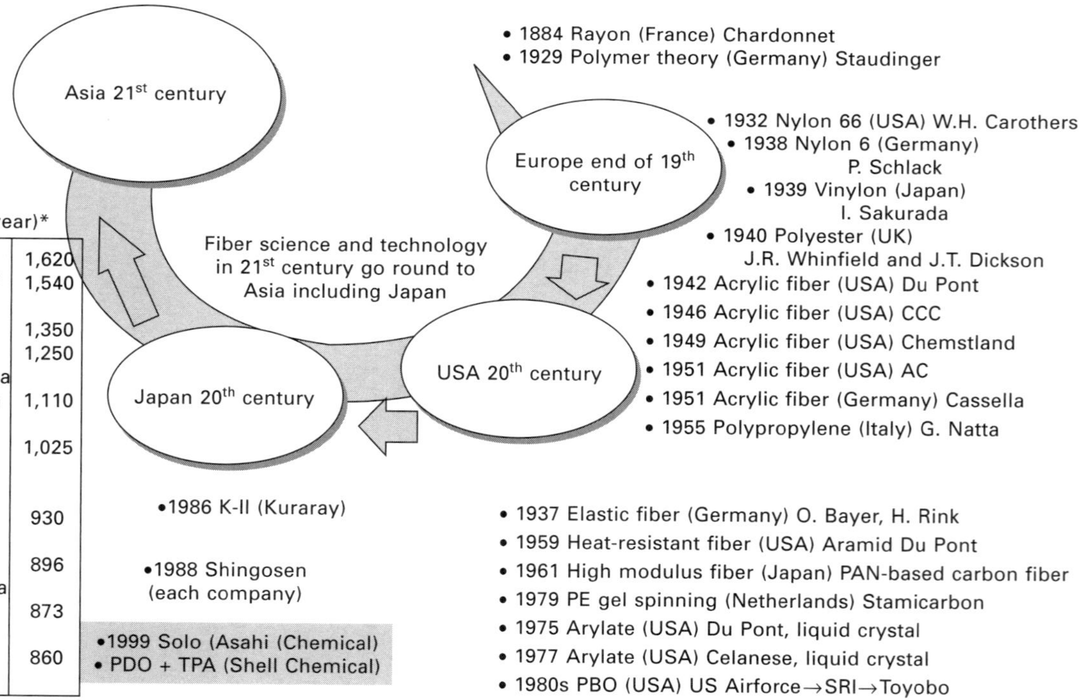


| | | |
|---|---|---|
| Shaoxing | China | 1,620 |
| Yizheng | China (Part of Sinopec) | 1,540 |
| Reliance | India | 1,350 |
| Nan Ya | Taiwan, USA, Vietnam, China | 1,250 |
| Far Eastern Textile | Taiwan, China | 1,110 |
| Tuntex/ Xiang Lu | Taiwan Thailand, China | 1,025 |
| Zhejiang Rongsheng | China | 930 |
| IndoRama • Ass. | Indonesia, Thailand, India | 896 |
| Huvis | Korea, Indonesia, | 873 |
| Sabfabgxiang Group | China China | 860 |

Polylactic acid
Biodegradable fiber
• 1998 TERRAMAC (Unitika)
• 2000 LACTRON
* From *Chemical Fiber International*, 53, 395 (2003)

*2.3* New wave expected to bloom in the twenty-first century.

synthesis of ultra-thin fibers having nano-structure are now investigated by Professor T. Aida (University of Tokyo), and strong fiber both in lateral and vertical directions is being developed by Professor C. Kajiyama (Kyushu University). Glass-based optical fiber is suitable for key communication networks requiring large capacity and high speed. However, POF is suitable for the connection of main lines in each home because it is cheap and easy to connect. In the twenty-first century, POF could be used in circuitry in the home instead of copper wire. Professor Aida has synthesized ultra-thin and strong fiber, which corresponds in properties to the spider's thread. Professor Kajiyama has made fiber strong in a lateral direction to produce new optical fiber and fiber with progressively increasing function.

### 2.2.3 Major US companies advance in fiber industry

Two major US companies have advanced in the fiber market. One is Shell Chemicals, part of the international petroleum company, and another is a joint enterprise of Cargill and Dow Chemicals, 'Cargill Dow Polymers' (CDP). Shell Chemicals has started production of the 3GT polyester, PTT. Although PTT fiber had already been produced by Asahi Kasei under the trade name of SOLO, Shell Chemicals gave a fashion show in Paris in March 2000 appealing to the world fiber manufacturers. CDP produce polylactic acid (PLA) from carbohydrate in corn and is a biodegradable polymer. Kanebo and Unitika import PLA from CDP and produce 'Lactron' and 'Terramac', respectively.

### 2.2.4 Development of biodegradable fiber

In the last century, synthesized polymer material products made from fossil resources resulted in mass production and mass consumption, which caused the two serious problems of exhaustion of resources and waste material. The problem of waste material takes place since the synthetic fibers drop out of material circulation (carbon circulation) since they are not biodegradable. The problem of resource exhaustion is caused by the consumption of raw materials, for synthetic fibers are dependent on fossil resources such as petroleum. It is expected that the extent of petroleum resources will tend to decrease after the peak of mass production at the beginning of this century, and it is said that it could be exhausted in the medium term. It is a problem for synthetic fibers whose raw material depends on the fossil resources.

In the relationship between humans and plants, natural fibers, such as silk, cotton, and wool, have their own excellent properties. The synthetic fibers do not have these excellent properties. Nylon, polyester and polyacrylonitrile have appeared as a substitutes for silk, cotton, and wool, respectively, since they are tougher than the natural fibers. However, they do not have functions as excellent as those of the natural fibers. So natural and synthetic fibers have to be used in the fields where they can show their best properties. In addition to functions and performance, the environment should be also considered when one develops materials and technologies. The most important thing that one should consider is a 'sustainable society'. In developments of ecological products and technologies in the synthetic fiber industry, there are, of course, 'problem areas' and 'defensive measures'.

'Problem areas' are UV-cut fibers for the prevention of the depletion of the ozone layer, several filters for water and air cleaning, oil barriers for prevention of sea pollution, water-swelling fibers for prevention of tropical forest decrease and desertification, etc.

'Defensive measures' are recycled fibers, counter penetration membranes for global warming prevention, biodegradable fibers for waste material decrease, environment beautification, etc. Much thought and effort will be required to introduce them into society.

### *Natural recycle system of biodegradable polymers*

Biodegradable polymers are expected to be adaptable to carbon recycling systems like the natural organic materials in the last century. Recyclable carbohydrates obtained from plant resources such as corn as a raw material are degraded by enzymes to glucose, which can be fermented by bacteria into lactic acid. The resulting lactic acid can be polymerized to poly(lactic acid). This polymer can be shaped into fibers – films which are biodegradable materials. Hydrolysis takes place in the polymer in compost or at the beginning of the degradation process in the natural environment. In the second half of the process, hydrolysis by enzymes secreted by microbes changes the polymer into water-soluble oligo-lactic acid or lactic acid monomer, which enters the microbe cell to change finally into carbon dioxide and water. The carbon dioxide can be used to synthesize carbohydrates in plants. Figure 2.4 shows the natural recycle system of the biodegradable polymer, Terramac® by Unitika.

### *Development of biodegradable fiber, Lactron™*

Table 2.3 lists companies who manufacture and develop goods using polylactic acid. Lactron™ is a poly(lactic acid) fiber produced by Kanebo Gohsen and was given the Technology Award of The Society of Fiber Science and Technology, Japan in 1999.

### *Processes for the development of Lactron™*

Kanebo Gohsen began technological studies for the production of biodegradable polymer fiber. They found that poly(lactic acid) is the most

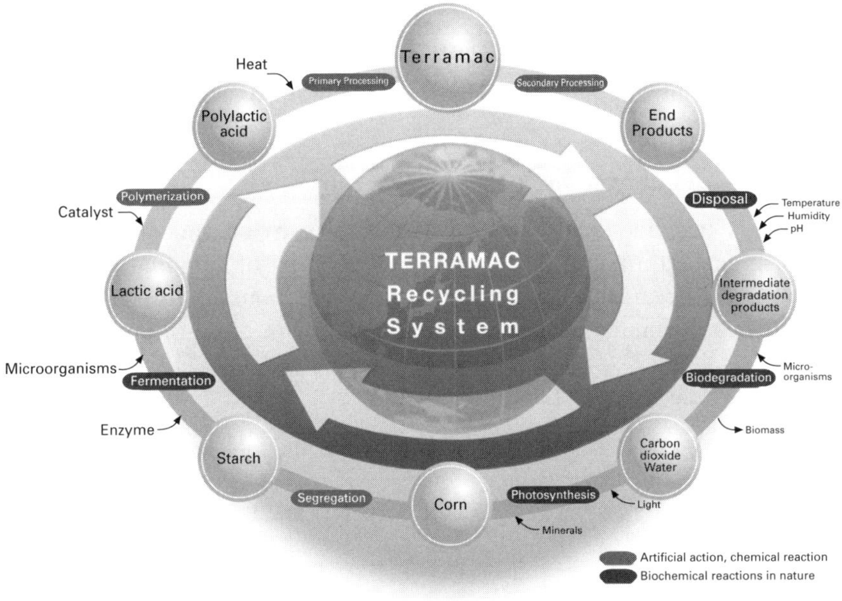


*2.4* Natural recycling system of biodegradable polymer (Unitika).

*Table 2.3* Manufacturers and developments of goods using polylactic acid

| Manufacturer | Trade name |
|---|---|
| Polylactic acid products | |
| Kanebo Gohsen | Lactron* |
| Unitika | Terramac |
| Mitsubishi Plastics | Ecoloju |
| Kuraray | Plastarch |
| Toray | Ingeo |
| Polylactic acid | |
| Cargill Dow (USA) | Nature Works |
| Mitsui Chemicals | Lacea |

* Production finished in 2004.

suitable material for manufacture with regard to transparency, strength, and costs.

In January 2000, Kanebo Synthetic Fibers announced an association with Cargill Dow Polymers (USA). Now they are developing markets associated with the building and operation of production plants for poly(lactic acid). They selected poly(lactic acid) because:

1. it is a non-petroleum-derived recyclable resource (the raw materials are plants such as corn and sugar beets)

2. it has correspondence with environmental problems (biodegradable means compostable, low combustion heat, no harmful gas), and
3. it has excellent properties and processability for materials (fibers) with melting point, strength, crystallinity suitable for practical use.

### *Characteristics of Lactron™*

Lactron is manufactured by melt spinning like nylon and polyesters. To obtain high performance fiber, a high technology of spinning as well as quality and strict specification of raw plastics are required. Its characteristic properties are summarized in Table 2.4, along with those of polyester and nylon.

The characteristics of Lactron™ include:

- fiber strength is as great as that of nylon and polyester, sufficient to produce fiber material.
- Young's modulus is between that of polyester and nylon, just a little more than nylon.
- soft touch
- good water diffusion, sweat-absorbable and rapid-dry
- interim-twisted processable
- low refractive index, mild gloss
- stainable by dispersed dye at 98°C, ambient pressure
- melting point is 175°C, higher than other biodegradable polymers.
- chemical structure is polyester, so absorbance is low.
- anti-bacterial, weak acid, retains humidity.

*Table 2.4* Characteristics of Lactron fiber (Kanebo)

| | Lactron | Polyester | Nylon |
|---|---|---|---|
| Physical properties | | | |
| Density | 1.27 | 1.38 | 1.14 |
| Melting point (°C) | 175 | 260 | 215 |
| Glass transition temperature (°C) | 57 | 70 | 40 |
| Absorption (%) | 0.5 | 0.4 | 4.5 |
| Combustion heat (cal/g) | 4500 | 5500 | 7400 |
| Fiber character | | | |
| Strength (cN/dtex) | 4.5–5.5 | 4.5–5.5 | 4.5–6.0 |
| Stretch (%) | 30 | 30 | 40 |
| Young's modulus ($kg/mm^2$) | 400–600 | 12000 | 300 |
| Stain | | | |
| Dye | dispersed | dispersed | acid |
| Staining temperature (°C) | 98 | 130 | 98 |

Thus Lactron™ has fiber properties and processability comparable to those of conventional polyester and nylon.

*Biodegradability of Lactron™*

Biodegradability of Lactron™ is estimated by various practical methods, assuming the various locations where the products are used.

The strength of Lactron™ decreases within a few years in soil. Decrease in weight follows a decrease in strength. In particular microbes and bacteria in active sludge decrease the strength immediately. The strength becomes almost zero within one year. According to results obtained by the standard compost method (ISO 148550), Lactron™ is completely degraded in about 70 days.

*Development of goods using Lactron™*

Lactron™ is processed in various shapes such as filament, staple, monofilament, spanbond, flat yarn spinning fiber, textile, knitting, non-weaved cloth and made into industrial materials and general wearing materials. Table 2.5 shows examples of the development of uses for Lactron™.

In keeping, with concept of biodegradation, Kanebo Gohsen released a mixture of Lactron™ and natural fibers such as cotton or wool in January 1998, and developed the goods widely. The main merit of 'corn fiber' is that it can solve the problem of dumping while keeping the advantages of conventional mixtures of synthetic and natural fibers.

## 2.3 New fibers for the next generation have arrived

The new fiber material 3GT polyester fiber (PTT fiber) has attracted attention in Japan, Europe and the United States since the spring of 1998. The raw

*Table 2.5* Examples of development of uses for Lactron (Kanebo)

| Classification | Use |
|---|---|
| Non-clothes | |
| Civil engineering and construction | Plant nets, non-weaved cloth, mat, plant soil, reinforcing material for weak soil |
| Agriculture<br>Fishery | Easy covering material, anti-weed net, bags, net for vines, bind tape, fishing nets, fishing line |
| Life | Packaging, sanitation, convenience goods, leisure goods, suture thread, absorbance material |
| Clothes | Inner wear, outer wear, wearing goods |

materials for 3GT fiber are somewhat different from those for polyester or PET fiber and are shown in Table 2.6. PET fiber is made of ethylene glycol and terephthalic acid and called 2GT. 3GT is called PTT fiber and made of 1,3-propane-diol (PDO) and terephthalic acid. Ethylene glycol in 2GT is exchanged for PDO in 3GT. 3GT itself is a well-known material and is not at all new. However, the yield of PDO in synthesis was low, so that the cost of production of PDO was high and could not compete with ethylene glycol. However, now a method to produce PDO cheaply has been developed by Shell Chemicals and Du Pont.

### 2.3.1 Petrochemical companies advance in fiber industry

Cos, an affiliated company of Texaco, produced the PTT fiber Cortera® in Italy and Spain, and launched the fiber at a fashion show in Moulin Rouge, Paris, which attracted attention from all over the world. PTT fiber is also produced by Asahi Kasei, Japan, and other manufacturers in Korea and Taiwan. Asahi Kasei decided to produce Solo® in 1998 using PTT provided by Shell, and started production in the latter half of 1999 at 1000 tons per year. Production will increase to 5000 tons per year. Asahi Kasei has applied for patents concerning Solo® production. Based on 127 patents, they will advance an intellectual property right strategy. Most of the patents are concerned with dyeing and production of fabrics. When the users of Solo develop new products, they can use these patents and decrease the time required for production.

*Table 2.6* Raw materials for 2–4 GT fibers

| Name of fiber | Abbreviation | Raw materials | Note |
|---|---|---|---|
| Polyethylene terephthalate fiber | 2GT (PET) | Ethylene glycol and terephthalic acid | Polyester fiber Produced in many countries |
| Polytrimethylene terephthalate fiber | 3GT (PTT) | 1,3-propane-diol and terephthalic acid | Shell Chemicals 'Cortera®' → Asahi Kasei 'Solo' Du Pont: Biotechnology to produce raw materials |
| Polybutylene terephthalate fiber | 4GT (PBT) | Butane-diol and terephthalic acid | Used as ester component in highly elastic fiber 'REXE' |

## 2.3.2 Development of PTT fiber 'Solo'

Asahi Kasei succeeded in synthesizing a new fiber composed of poly(trimethylene terephthalate). Solo™ was released in 1999. PTT (3G) is prepared by condensation of terephthalic acid or dimethyl terephthalate and 1,3-propane diol (PDO). PTT is a polyester homologous to PET, but it has different properties from that of PET. There were many difficulties to overcome in the manufacture and processing of Solo, these but were overcome and the technology has now been established. PTT is different in the crystal and amorphous structure, so the fiber properties of Solo are: new soft touch, reversible stretch, stainability at low temperature, form stability, and light and adhesive proof.

Asahi Kasei will develop Solo further for textiles and materials applications where Solo's characteristics are best displayed, as well as for raw fiber, processed fiber, combined fibers with natural fibers or cellulose fiber. The growth of Solo is so spectacular that it has been classified as one of four master fibers in use today in the goods field, exceed spandex in cost and performance.

### *Polymer structure and fiber performance of PTT*

Figure 2.5 shows the chemical structure of PTT. PTT has a crystal structure as shown in Fig. 2.6.

Crystal elasticity and elongation energy of the crystals of PTT are much smaller than those of PET. As shown in Fig. 2.6, PTT molecules bend in a Z shape. This is the part of the molecule that can be bent with a small amount of stress, resulting in low elasticity (or new soft touch) and reversible stretch. The Z-shaped structure of PTT produces mechanical properties such as large limit of elasticity and high recovery of stretch. The reversible stretch is significant in the processed fibers.

These properties have not appeared in conventional fibers. Most applications of PTT are in the areas of cloths and industrial materials. Stretch cloth is a new trend, so Solo is expected to be used widely here. In industrial materials, developments are making use of Solo's elastic properties. The characteristics of Solo are summarized in Table 2.7.

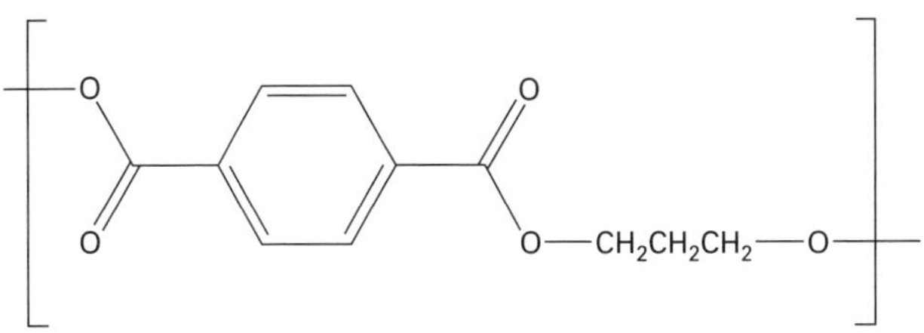


*2.5* Chemical structure of PTT (Asahi Kasei).

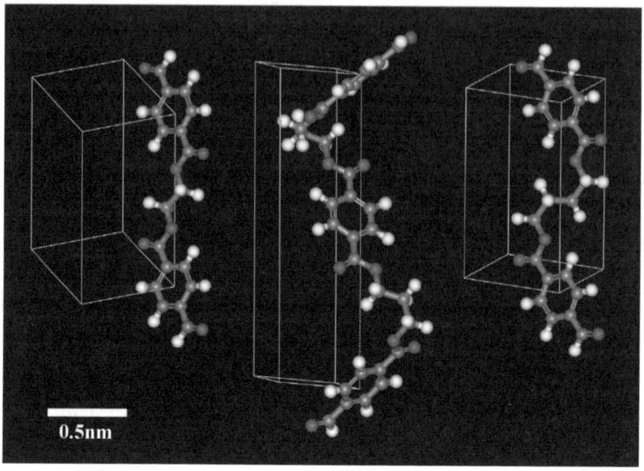


*2.6* Crystal structure of PET, PTT and PBT (Asahi Kasei).

*Table 2.7* Characteristics of Solo (Asahi Kasei)

| Characteristics | Unit | Solo | PET | PBT | Nylon 6 |
|---|---|---|---|---|---|
| Draw strength | cN/dtex | 3.5~4.0 | 3.7~4.4 | 3.5~4.0 | 3.5~5.3 |
| Stretch | % | 42~48 | 30~38 | 30~40 | 30~50 |
| Elasticity | cN/dtex | 22 | 90 | 22 | 20–30 |
| Recovery from 20% stretch | % | 85 | 29 | 40 | 62 |
| Density | g $cm^{-3}$ | 1.34 | 1.38 | 1.35 | 1.14 |
| Contraction in boiling water | % | 12~13 | 7 | 7 | 17 |
| Melting point | °C | 230 | 260 | 230 | 220 |
| Weather proof | | good | good | good | strength degradation sometimes color changes to yellow |
| Color stability | | good | good | good | sometimes not enough |

The peak temperature of the loss tangent (tan δ) which characterizes molecular motion in the amorphous part in the PTT fiber is reached at around 110°C, which is 30°C lower than that for PET fiber. It results in good dyeing properties at low temperature. If an appropriate dye is selected, PTT can be stained even at ambient temperature. Complex PTT with natural fiber or regeneration cellulose requires a low temperature dyeing ability. This and the stretch of Solo provide applications in fibers and textiles when complexed with wool, cotton, and Benberg®, etc.

Asahi Kasei and Teijin established a joint company, SOLOTEX Corporation and started production of SOLOTEX®. SOLOTEX® is used for clothing because of its stretch and soft properties and dyeing ability at low temperature, without losing the properties of the combined material. In the materials field, because of its high recovery, it can be used for cars, interiors, and carpets.

### 2.3.3 Fiber from corn

A biologist at Du Pont developed a new type of bacterium by combining the DNA of two kinds of bacteria. When the bacterium was fed with corn, it produced milky liquid known as 3G, which could react with terephthalic acid to produce 3GT resin. Then the resin is spun to give 3GT Fiber. Du Pont started commercial production of 3G on a scale of 3000 l/day from October 2000 to produce a windbreaker. They intend to construct a 3G plant with the capacity to produce 300 000 l/day. The product is soft, good in elastic recovery, can be dyed at room temperature, and the cost is cheap enough to compete with nylon and polyester. Future usage will be extended from the field of life materials to industrial uses. Each country now gives attention to this new fiber.

A joint venture company, Du Pont Tate & Lyle Bioproducts, LIC, plans to construct a manufacturing plant to produce 50 000 tonnes of PDO per annum from renewable resources such as corn by the middle of 2005. Accordingly, the production of PTT polymer will increase from 1200 to 60 000 tonnes per annum.

## 2.4 The distinction between high-tech fiber, frontier fiber, and new frontier fiber

*High-tech fiber* is made by high technology using advanced science and technology. High-tech fiber is a general term used for fibers made by highly advanced methods or methods which differ significantly from conventional ones. Therefore high-tech fiber is mainly a new fiber. However, *frontier fiber* is based on the development of a new application field, so it is not necessary that it should be a new fiber. To select the fiber and to devise the new application and open up new demand is the criterion which defines *new frontier fiber*, and this is regardless of whether the fiber is old or new. The development of the three fibers is shown in Fig. 2.7.

## 2.5 Key words for the near future

To develop a new application field, it is necessary to know future needs. High technology, health, comfort, environment, care, optical fiber, sense, special fiber and superfiber are now the key words. Social change and

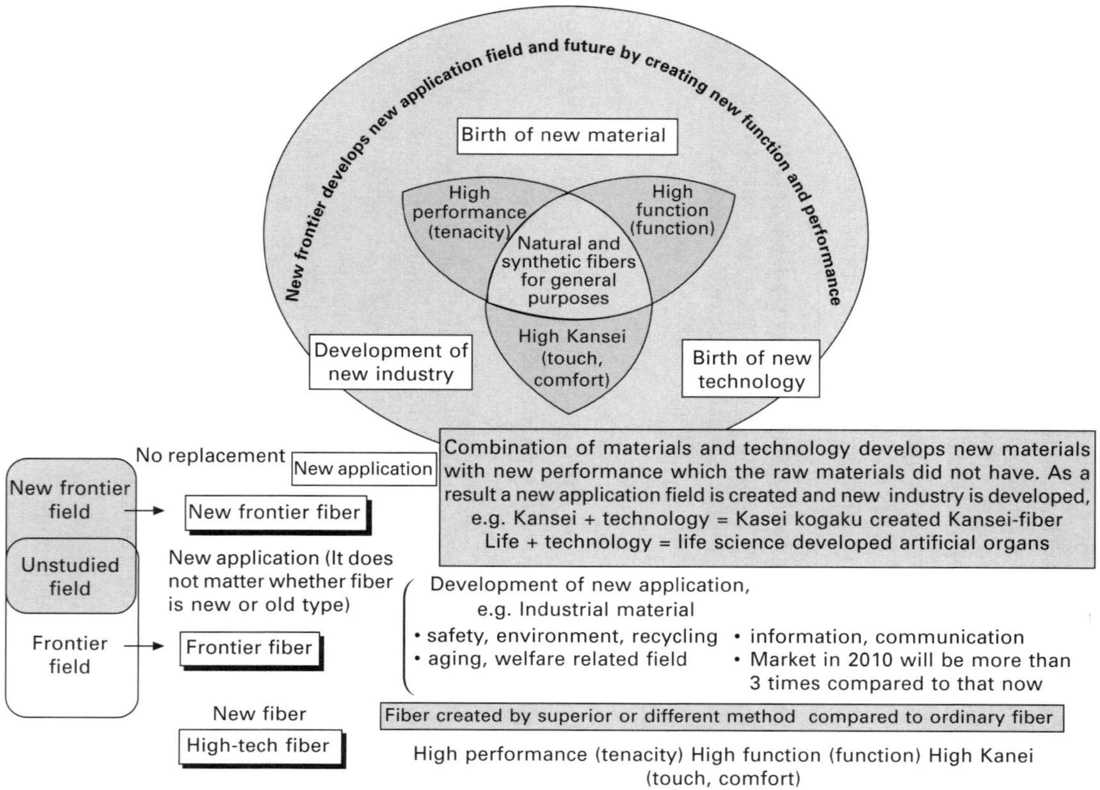


*2.7* Concept of general, high-tech, frontier and new frontier fibers.

development leads to the enlargement of the fiber market. Change in society creates a new market. New fiber development in response to the needs of the twenty-first century is shown in Fig. 2.8.

The direction of needs depends on creative technical development. History has taught us that only innovative technology and manufacture can break through into future new markets. The relation between the needs required to activate industry and innovative technology is shown in Fig. 2.9. Innovative technology and products related to development of new markets are included in the idea of 'new frontier'. Fiber as a mature industry cannot be revived as a new frontier without a change in ideas.

## 2.6 How to develop new application fields

It is essential for new frontier fibers that there is not only a developing fiber market but also new developments in application fields associated with existing industries which have no relation to fiber, and particularly in emerging industries. How fiber can be accepted in these new application fields as a material or as a system technology is important. In other words, to be ahead

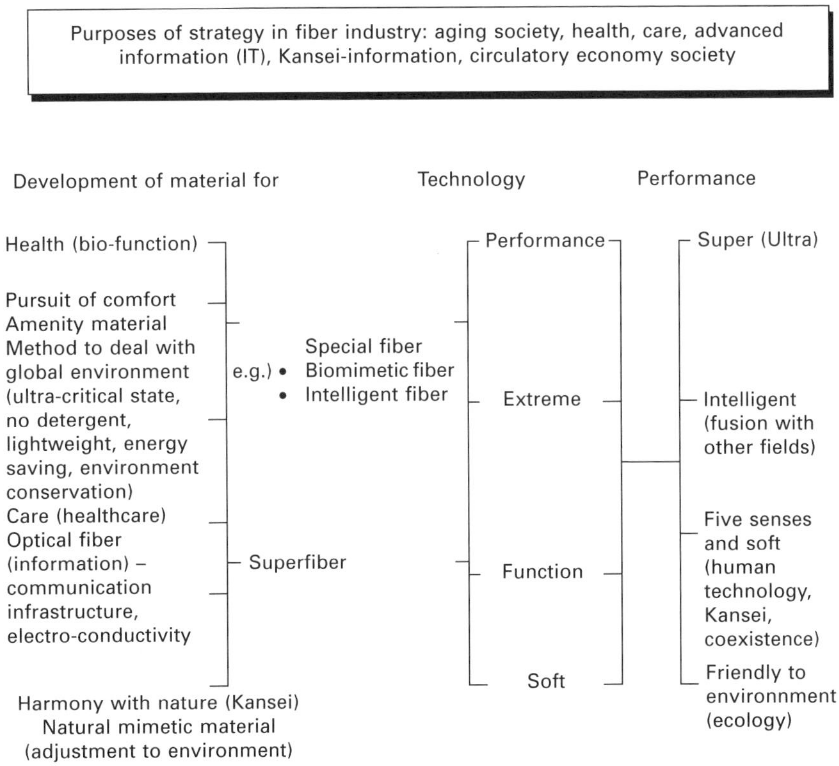


2.8 Development of material in response to the needs of the twenty-first century.

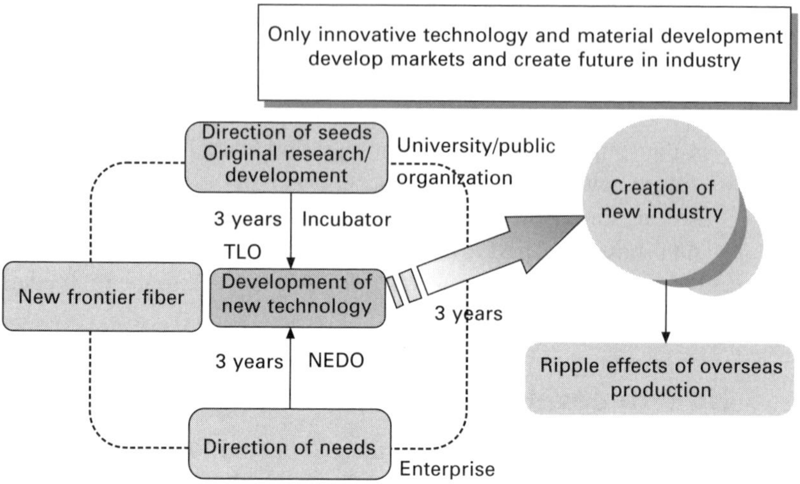


• Develop demands required for health, welfare, environment, energy
• Form new market corresponding to change in society (necessary to find needs)

Note: NEDO (New Energy Industrial Technology Development Organization) and TLO (Technology Licensing Organization) are organizations supported by Japanese government.

*2.9* Relation between needs, seeds and innovative technologies to activate industry.

of others in needs and seed ideas and to develop a new future for industry is the role of a new frontier fiber. An indication of how to develop a new field is shown in Fig. 2.10.

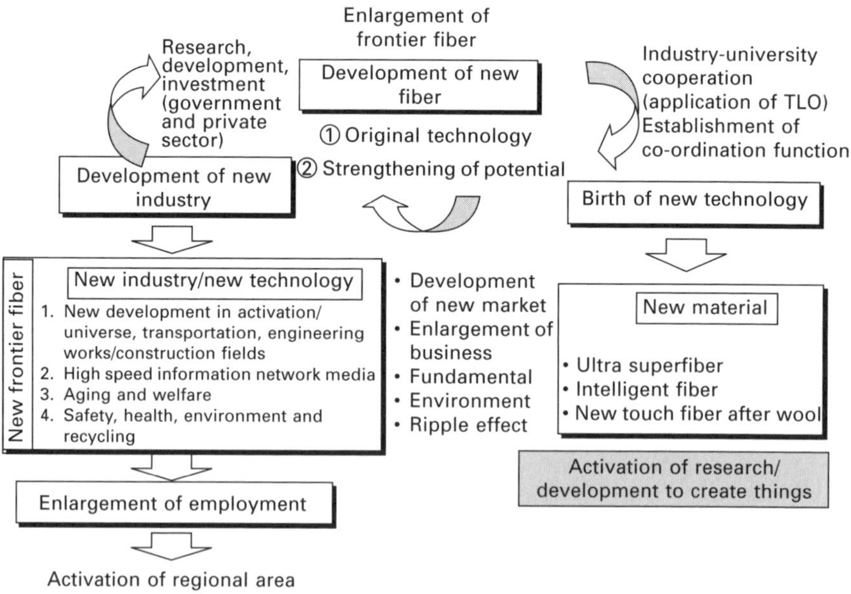


*2.10* How to develop a new field.

For example, suppose that a deodorant fiber and UV-cut fiber are developed. Each is a high-tech fiber and a frontier fiber when developed. Deodorant socks and apron with flavor are such examples of processing of fiber to add a function. However, this is too naïve a way of thinking. Similar products are already on the market. New frontier fibers require such functions, certainly, but they also need an added dimension to use fiber positively to meet social needs and restrictions.

On current thinking, fiber is a thread that is thin and long. Therefore, basic research and development mainly pursues developments related to the fiber's primary structural nature. New frontier fibers require a connection with different fields of industry, with development and enlargement in both vertical and horizontal directions.

## 2.7 New frontier field now growing

What types of new frontier are currently growing? Ocean development, space exploration, atomic energy, and clean energy fields are good examples (see Fig. 2.11).

How can fibers be applied in the information/communication field? There are optical fibers made of quartz and plastics. Fiber manufacturers make only plastic optical fiber and Japanese enterprises stand out above the others in plastic optical fiber production. However, plastic optical fiber occupies only minor markets compared to quartz optical fiber. Since plastic optical fiber is suitable for connections over short distances and costs only a quarter of quartz optical fiber, it will have superiority in local area networks used at home and in the car.

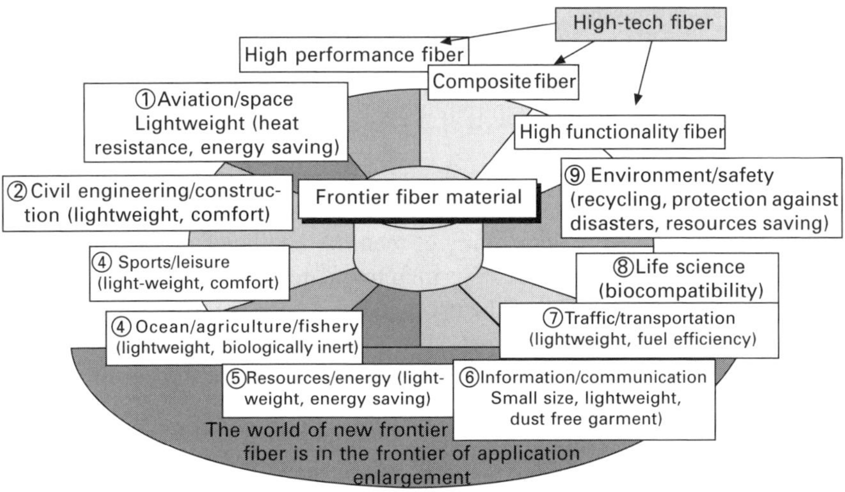


Note: European countries are very keen to develop tech-textile

*2.11* Enlargement of frontier fiber applications.

Fiber technology will also be used in the field of biotechnology and healthcare. How are artificial skin, tendon, and internal organs going to be developed? There is now an artificial organ which works outside the body like the artificial kidney. Embedded type organs need to be developed in the future.

## 2.7.1 Developments in the medical field

Major fiber manufacturers in Japan are already participating in the medical/care field. In the United States single use disposable materials are used in hospitals, so that waste products have increased many times compared with the past. In Japan linen goods are used, and used repeatedly after washing. Whether or not this system will change to a disposable one remains uncertain.

The development of artificial kidneys depends on the development of hollow fiber membrane. Polymer can be spun into hollow fiber membrane as shown in Fig. 2.12.

### *Artificial kidney (artificial dialysis membrane) and artificial liver*

Artificial dialysis membrane is the most advanced fiber in the medical industry. The total number of patients who suffer chronic kidney disease and undergo kidney dialysis in Japan was 167,000 in 1996 and some 170,000 in 1997. The number of patients who undergo kidney dialysis continuously for more than 20 years was 5812 in 1996. Artificial kidney dialysis is accepted as reliable. The purpose of the development of artificial kidney dialysis membrane is to mimic the ability of kidney to completely remove wastes like urea and albumin. The production process for the artificial kidney and dialyser are shown in Fig. 2.13. One of the side effects of long-term dialysis is a shoulder injury caused by β-2-microalbumin accumulating so that the joint cannot move. Big pores are effective in removing waste. However, other necessary components are also removed.

The problem can be solved by making a monomolecular layer fiber, which is controlled by the relation of surface structure to waste blood. There remain problems to be solved in controlling of material and holes on the surface of the hollow fiber. To improve dialysis membrane development, it is necessary to make fiber more identical to the organ itself. At present, heparin is used to prevent clotting of blood. If the patient who undergoes dialysis is a diabetic, the amount of heparin used must be decreased. Therefore, biocompatibility of the material needs to be achieved.

The metabolism of the liver is very complicated which poses problems for the artificial liver. This can be solved by using a double lumen structure with a hollow fiber within a hollow fiber. Blood is run outside and in contact with liver cells and blood, and after purification is run inside the fiber.

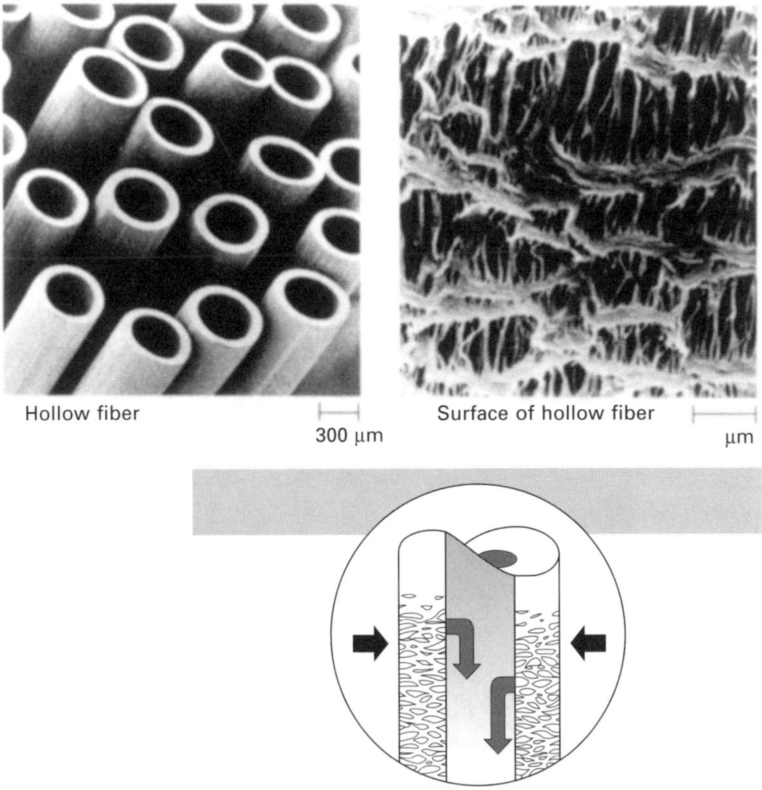


*2.12* Hollow fiber and its surface structure (Mitsubishi Rayon).

*Balloon catheter*

A catheter is a tube used for clinical tests and treatment. Catheter treatment was initiated in the United States and subsequently used in Japan. For example, adhesions within blood vessels can be removed by leading a balloon catheter into the vessel; alternatively a balloon catheter with a drug delivery system can be directed to an affected part. The balloon catheter is composed of a guide wire and catheter. The guide wire is used to guide the catheter to the affected part. The catheter manufactured from polyvinylchloride is led into the blood vessel and the balloon, constructed from latex and polyester micromesh, at the distal end is enlarged by a contrast medium introduced by a syringe to open the blocking adhesion and so enlarge the blood vessel.

In mitral valve disease, wire is led into the heart and the blocked part is opened. In pancreas extraction, several blood vessels leading to the pancreas need to be connected quickly. However, if the operation takes time, the blood vessels are connected with tubes coated with heparin to prevent clotting of blood.

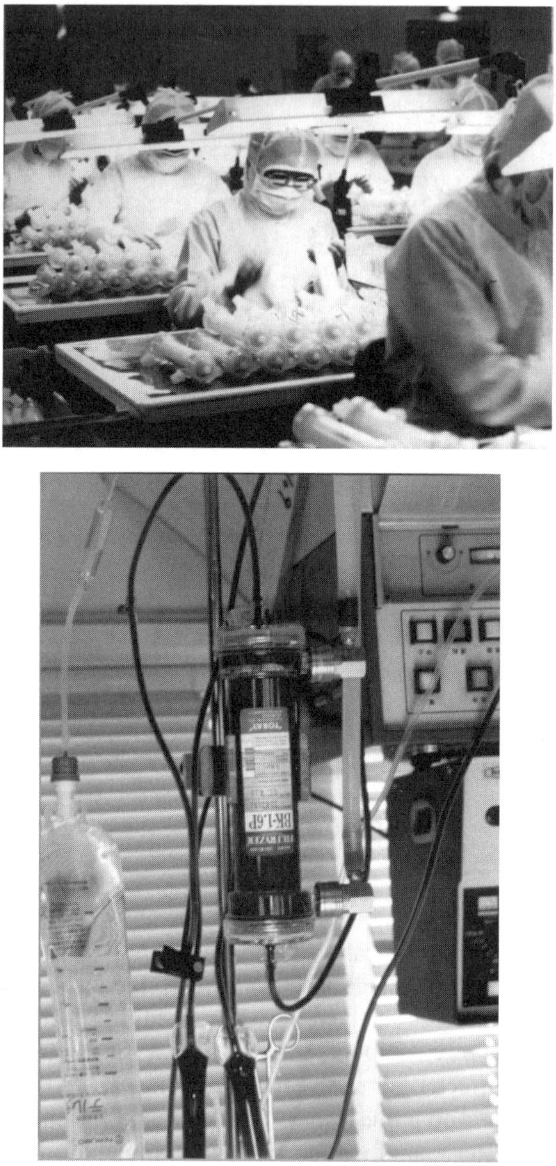


*2.13* Manufacturing process of artificial kidney and the product (dialyzer) (Toray).

The fine balloon catheter for medical use and balloon catheter developed for the mitral valve operation are shown in Fig 2.14, respectively. The treatment method for the left atrium of the heart using a balloon catheter is shown in Fig. 2.15.

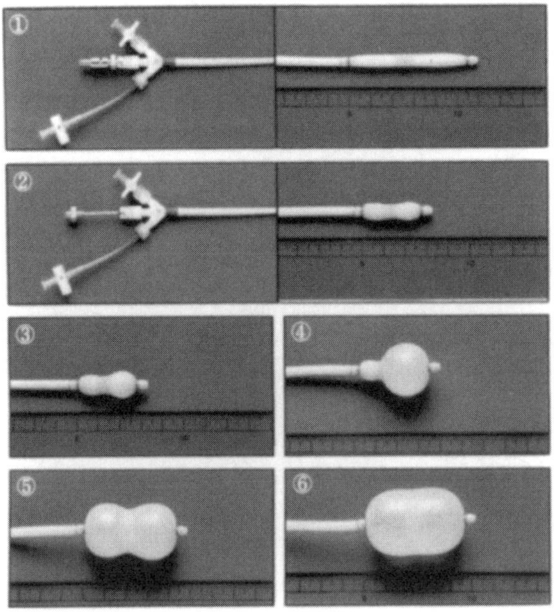

*2.14* Balloon catheter developed to treat percutaneous transvenous mitral commissurotomy. Numerical figures show the order of balloon enlargement (Toray).

## 2.7.2 Enlargement of use for industrial materials

Energy saving, resources saving, lightening, miniaturizing, substitution, and performance improvement are the key words in fiber applications for industrial materials. High performance fiber with superior mechanical properties, complex material made of composite material of high performance fibers with function and base material such as resin, rubber and cement, and composite are the ways to achieve these objectives. Advanced composite material is used in high-tech fields. New application fields can grow by developing advanced fiber composite materials and integrating with other the materials. The relationship between enlargement of fiber application and development is shown in Fig. 2.16.

### *Life and culture-related fields*

Teijin developed a lightweight, easy handling large air membrane, which was given the name *Aeroshelter*®. For outdoor events such as concerts, weddings and exhibitions, there is a need for a large air dome, which can house people comfortably, protect them from sunlight and rain, and which is easy to handle and set up. *Aeroshelter*® is ideal for use over one to seven days, and is easy to handle and store (15 m × 10 m in size). The weight of a conventional tent

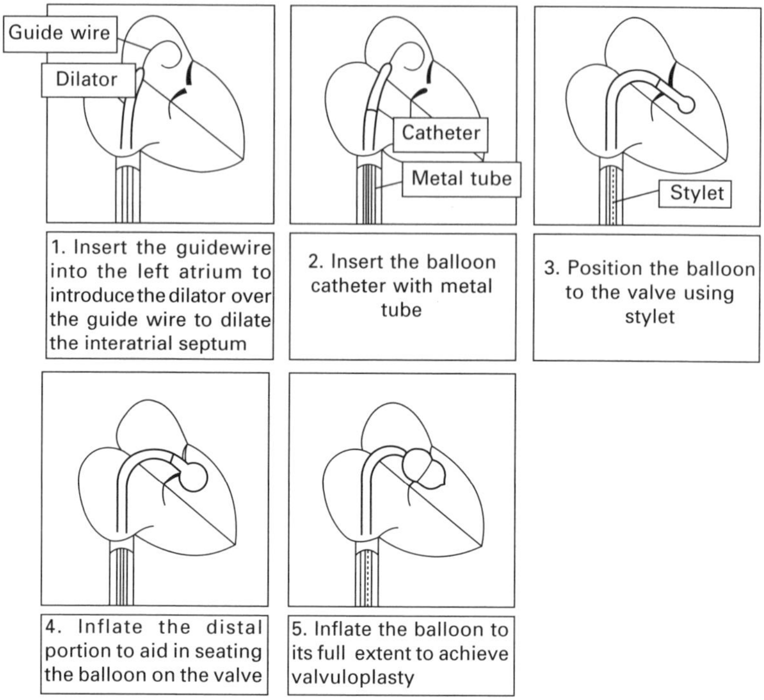


*2.15* Directions of balloon catheter to treat the left atrium of the heart (Toray).

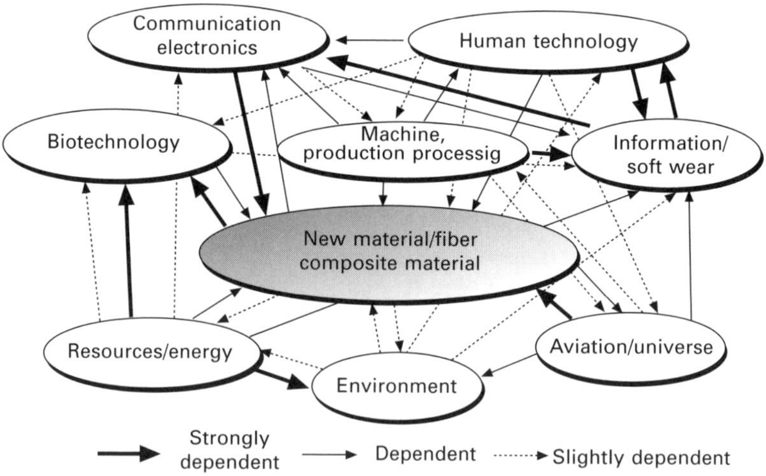


*2.16* Interdependence of fiber application enlargement and development.

dome is 650 g/m$^2$, but this can be decreased 40 g/m$^2$ by using Tetoron. The shape of this air dome is completely different from that of a conventional one. It needs only ten minutes to set up and twenty minutes to remove all the air. The total weight is 70 kg for a dome of 15 × 11 m in size, so that two people can transport the dome, which is shown Fig. 2.17. The normal dome can cope with a wind velocity of 5–6 m/sec, and stronger ones with velocities of 12 m/sec. An air beam is used as a structure, made from aramid fiber in a lateral direction and polyester fiber in the longitudinal direction. The inner wall is covered with urethane resin and air fills the inside of the tube. A 25 m × 30 m air dome was used to celebrate the 120th anniversary of Tokyo University and cost about 100 million yen (ca US$ 1 million), which includes the cost of the base construction. Alternatively, *Aeroshelter*® with the size of 15 m × 11 m can reduce the cost to 3 million yen. The Tokyo Dome is of the highest quality. It is made of glass fiber with fluorine finish, and the weight of the membrane is 800 g/m$^2$, and durable for semi-permanent use. The tension of the dome is 200 kg per 3 cm width. The United States has a long history of using air domes, but the Tokyo dome is the only one recognized as a building in Japan.

Mitsubishi Heavy Industry constructed a cornice capable of being towed to a shipyard in Nagasaki and made out of aramid fiber composite material. Because the cornice is light, one person alone can open and shut the cornice having an area of 3000 m$^2$.

*2.17* Large-sized air dome, easy to handle (Teijin).

*Domestic and civil engineering*

The Winter Olympics were held in Nagano in 1999 and a windbreak and snowbreak nets were used to enable the jumping to be possible in high winds. Generally high mountain jumping can be performed safely only when the wind velocity is less than 3 m/sec. However, when the jump location is surrounded by mountains, the wind blows consistently at not less than 5–6 m/sec. To allow jumping, therefore, the wind velocity must decrease by a half. Thus the Olympic Committee asked Teijin to develop a material which is resistant to wind velocities of 40 m/sec with a lifetime of more than ten years. Moreover, it must have enough light permeability to prevent a sense of gloom when setup. To do this they developed a new material, KINGLIGHT, which is now used as a windbreak or snowbreak net for roads and highways in Hokkaido. The net used in the Olympic Games is shown in Fig. 2.18. The strength of the material is 300 kg/10 cm, and the porosity is 60%. Producing the material in sheet form is not suitable because of a lack of strength to wind pressure and lack of transparency and light permeability.

Thus nets have now found a variety of practical uses. There is a regulation in Japan that the light from an oncoming car must be banned if the angle of the light is less than 11°. A rough net can shut out light from the opposite carriageway and it is used on highways in Japan.

Snowbreak nets are used in Kushiro city, Hokkaido, to prevent snowdrift on roads. Wind velocity is the slowest at heights of 4–5 times higher that of the net. For example, with nets with a height of 2 m, snow does not drift

*2.18* Wind/snow proof net set near jump shanze at Nagano Olympic Games (Teijin).

7–8 m from net and starts drifting only at 10 m. Even thin nets are effective in this respect.

Geo-membrane is used for civil engineering works. It is used at the bottom of the waste works to prevent leakage of wastewater. An example is ST LINUS developed by Teijin. The sheet is made of strong polyethylene terephthalate with high elongation. When water leaks, a sensor will detect the leakage and sets in action work to mend the point of damage. This is also used for bank protection as shown in Fig. 2.19. Common polyester fiber elongates only 30–40%, but Teijin developed a polyester which can elongate up to 140%.

*Information and communication fields*

Fiber materials are used for producing plastic optical fiber and as a subsidiary material for quartz optical fiber. Formerly, composite material of glass fiber and resin was used as a printed base for computers or mobile telephones, but their role has now been taken over by fibers. The use of chopped aramid fiber has decreased the size and price of mobile telephones, which is a market which has increased phenomenally in recent years, and merits the description 'new frontier'. A printed base using Technora for mobile telephones is shown in Fig. 2.20.

*2.19* Construction site using geo-membrane (Teijin).

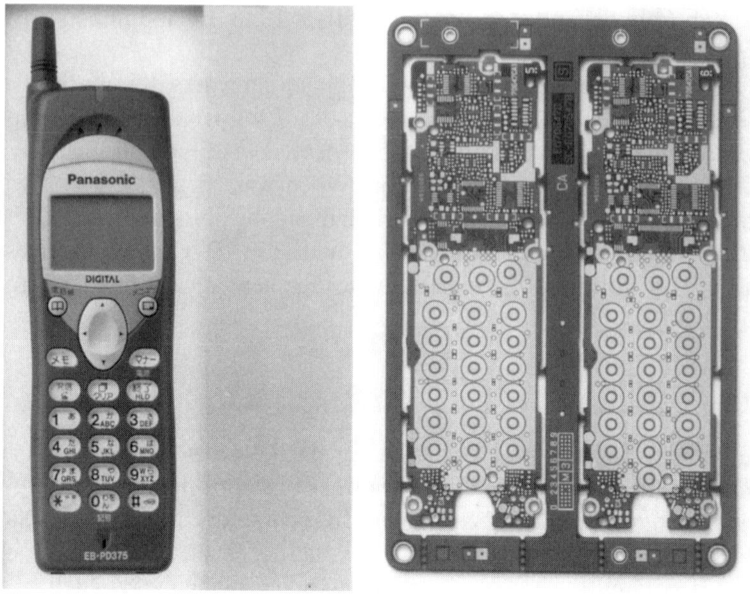


*2.20* Printed base for mobile phone using aramid fiber (Teijin).

Development of a stratosphere platform

The Ministry of Education, Culture, Sports, Science and Technology and Ministry of Public Management, Home Affairs, Posts and Telecommunications plan an airship to stay in the stratosphere as a substitute for an artificial satellite. A stratosphere platform was planned, as shown in Fig. 2.21. It will consist of a communication/broadcasting division, an airship division and an earth observation division. The airship division uses fiber as membrane and for this an ultra-strong, yet lightweight material using high-tech fiber is selected. An image of the sky net plan is shown in Fig. 2.22. An airship 200 m in length will stand still in the stratosphere at a height of 20,000 m, and be used to serve ultra high speed internet and digital broadcasting services and

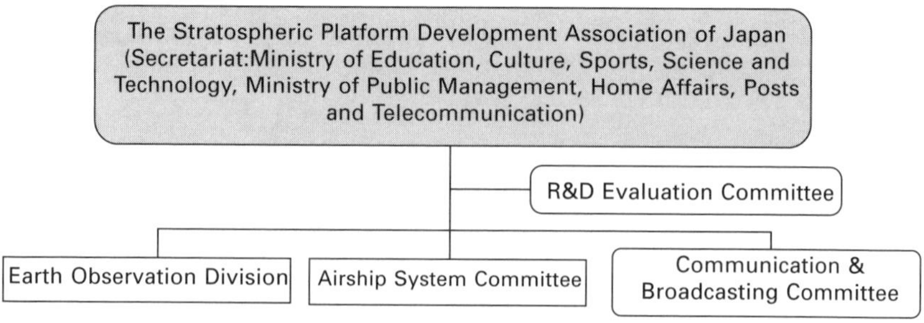


*2.21* The stratosphere platform development committee.

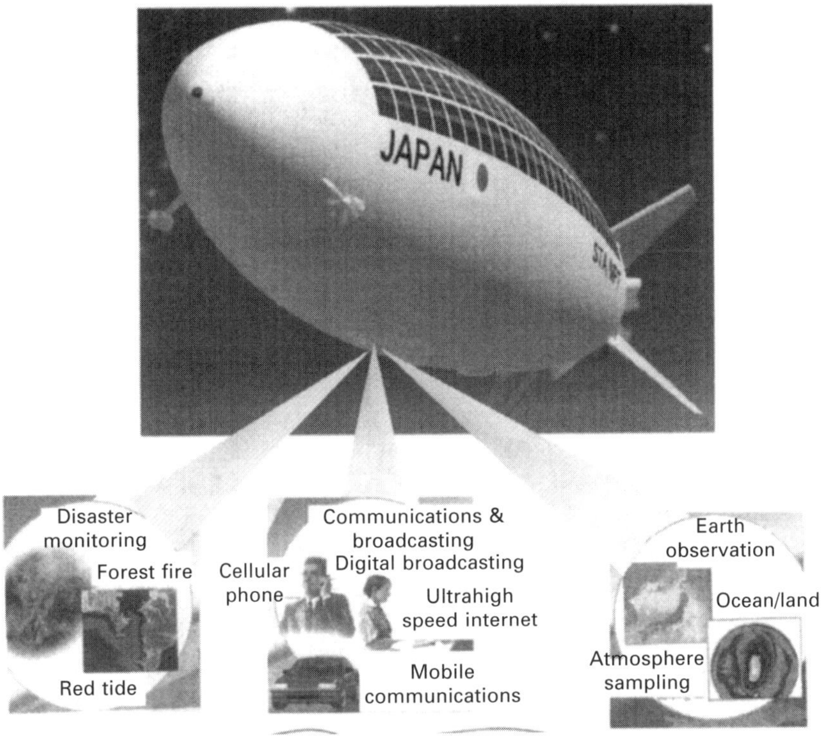


*2.22* Stratospheric platform.

high function mobile phones. At present satellites go around the earth at the height of 500–1000 km, so that even the radio wave, which can go round the earth seven and half times takes a second to arrive on the ground. Thus the stationary satellite at a height of 20,000 m will contribute to increases in the velocity of internet service.

The fiber used should be light, strong, and durable to temperature changes from 100°C to –100°C depending on sunlight. Aramid fiber is suitable for this purpose. The weight of fiber used for the airship will be 17.5 tons. Two hundreds airships are enough to cover all the mobile phones used in Japan. As it will cost 5 billion yen for an airship, the total cost will be 1 trillion yen. The plan at present suggests that an experimental airship could be launched in three years and fully operational airships within five years. At present, the development of the membrane materials is ongoing. The success of the plan will enlarge the new frontier fiber market.

### Fabrics with five-layered structure having a gas-barrier property

Teijin and Dimension Polyant (D/P) have developed a fabric suitable for unmanned balloons, such as the Ultra Long Duration Balloon of the National

Aeronautics and Space Administration (NASA). The base material for the balloon is Teijin's lightweight but industrial-strength polyester Tetoron Powerip, laminated with polyester and polyethylene films, creating a five-layered material with UV-resistant, high gas-barrier and low moisture absorption properties. The finest high strength fiber was used to lighten and increase tear strength of the double rip structure. The 140 m diameter balloons will orbit the earth six times in about 100 days from 52 km up in the stratosphere, and will carry about 3.5 tons of instrumentation. During the stay in the stratosphere, the balloon will collect information concerned with atmosphere, space, the sun, and the global environment. The balloon is shown in Fig 2.23.

A trial balloon was launched in October 1999 and first tested in December 2001. The cost required for the project will be 1/100 of achieving the same performance with an artificial satellite. The observation height is lower than that of a satellite so that clearer information can be obtained. Moreover, the balloon is unmanned, so that there is no risk to human life, which is one of its biggest advantages. NASA plans to extend the project for exploring Mars, Venus, and Jupiter.

*2.23* Balloon made of five-layered fabric with gas-barrier property (Teijin).

## 2.8 Bibliography

### *2.1 Enlargement of the frontier in a fiber competition age*

Matsumoto K., Monthly Report of Japan Spinners' Association, p. 51, Dec. 1993.
McIntyre J.E., *Sen'i Gakkaishi*, **51**, 88 (1995).
Höcker H., *Sen'i Gakkaishi*, **51**, 96 (1995).
Tateishi J., *Sen'i Gakkaishi*, **56**, 338 (2000).
Hongu T., *Sen'i Gakkaishi*, **57**, 7 (2001).
Kajiwara K., *Sen'i Gakkaishi*, **57**, 80 (2001).
Sakurai M., *Sen'i Gakkaishi*, **58**, 118 (2002).

### *2.2 'Selection', 'concentration' and 'originality' development on a world-wide scale*

Japan Chemical Fibers Association, Report of Investigation, No. 384, March, 2000.
Koike Y., *Sen'i Gakkaishi*, **56**, 277 (2000).
Koike Y. and Ishigure T., *IEICE Transactions on Communications*, **E82-B**, 8, pp. 1287–1295.
Aida T., *et al.*, *Science*, 285, 2113–2115 (1999). (Science Compass/ C & EN).
Takahara A. and Kajiya T., *et al.*, *Macromolecules*, **29**, 3040 (1996).
Kajiyama H., *Sen'i Gakkaishi*, **55**, 397 (2000).
Kajiyama H., *The 31st Summer Seminar Proceeding*, p. 41, The Society of Fiber Science and Technology, Japan.
Honda S., *The 33rd Annual Meeting of UCRS in Chubu Area, Japan*, p. 81, 2002.

### *2.3 New fibers for the next generation have arrived*

Hongu T., '*High-Tech Fibers*, p. 14, Nikkan-Kogyo Shinbun-sha, Tokyo, 1999.
Shima T., Matsumoto M., Kato J., Fujimoto K. and Koyanagi T., *Sen'i Gakkaishi*, **58**, 268 (2002).
E.J.C.M. Wely (Dupont), *Proceedings International Conference on Advanced Fiber Materials (Plenary Lecture)*, iii, 4 October 1999.
*Japan Chemical Fibers Monthly*, p. 96, December 1999.

### *2.4 The distinction between high-tech fiber, frontier fiber, and new frontier fiber*

Kajiwara K. and Hongu T., *New Frontier Fibers*, p. 17, Nikkann-Kyogyo Shinbun-sha, Tokyo, 2000.

### *2.5 Key words for the near future*

Kajiwara K. and Hongu T., *New Frontier Fibers*, p. 20, Nikkann-Kyogyo Shinbun-sha, Tokyo, 2000.

## *2.7 New frontier field now growing*


Kamo J., *Sen'i Gakkaishi*, **49**, 201 (1993).

Kamo J., *et al.*, *Artificial Organs*, **14**, 369 (1989).

Sakurai H., *The 32nd Summer Seminar Proceeding*, p. 95, The Society of Fiber Science and Technology, Japan, 2001.

Hiyoshi T., *Sen'i Gakkaishi*, **54**, 350 (1998).

Toray group Corporate Profile, English edition, 2002.

Saiga I., *The 30th Summer Seminar Proceeding*, p. 126, The Society of Fiber Science and Technology, Japan, 1999.

Fujinaga Y., *Sen'i Gakkaishi*, **56**, 287 (2000).

Ukikawa T., *Sen'i Gakkaishi*, **56**, 290 (2000).

Murayama S., Murata M. and Hiraoka K., *Sen'i-Gakkaishi*, **56**, 268 (2000).

Onda M., *The 33rd Summer Seminar Proceeding*, p. 9, The Society of Fiber Science and Technology, Japan, 2002.

# 3
# Superfibers

## 3.1 Description of superfibers

Superfiber has been produced by intensive research and development to achieve the ultimate strength of fibrous materials, and is mainly applied for industrial uses in the non-clothing field.

It is defined as fiber possessing a tenacity of over 20 g/d (2.2 GPa) and a modulus of over 500 g/d (55 GPa). (d denotes denier and denier corresponds to the weight in g of a fiber of 9000 m in length.) The mechanical performance of the second generation superfibers will soon be increased to a tenacity of more than 40 g/d and a modulus over 1,000 g/d. Zylon® (PBO fiber commercially available from Toyobo) is classified as a second generation superfiber. Presently available superfibers are summarized in Table 3.1.

Gel spinning and liquid crystalline spinning, developed for polyethylene and Kevlar, respectively, are two important technologies for producing superfibers. Kevlar molecules are rigid and rod-like because of hexagonal benzene rings connected by amino linkage, whereas polyethylene chains are flexible and bend almost freely.

In Japan, all superfibers are commercially produced by gel spinning and liquid crystalline spinning, and are used in various industrial fields, including the railway, civil engineering, automobile, sports goods, office automation, machinery/machine parts, and the space/aeronautical industries.

### 3.1.1 Conventional fibers and superfibers

The maximum theoretical mechanical strength of fibrous materials (the tenacity and modulus) can be estimated from the chemical structure of the parent polymer molecules assembling in an ideal extended form. However, the mechanical strength of conventional fibers is much less than the theoretically expected value as shown in Table 3.2. For example, the ideal polyester fiber should have a tenacity and a modulus 232 g/d, and 1023 g/d, respectively, whereas a conventional polyester filament has a tenacity 9.0 g/d and a modulus

*Table 3.1* Presently available superfibers

| Type | Polymer | Trade name | Manufacturer | Strength (g/d) | Modulus (g/d) | Melting point/ decomposition temperature (°C) |
|---|---|---|---|---|---|---|
| Rigid polymer | *para*-type aramid[*1] | Kevlar | Du Pont | 22–27 | 430–1110 | 560 |
| | | Twaron | Teijin Twaron | 22 | 850 | ↑ |
| | | Trevar | Hoechst (USA) | 23 | 657 | |
| | | Technora | Teijin | 28 | 560 | 500 |
| | Polyallylate | Vectran | Kuraray | 20–25 | 680–840 | 330 |
| | Heterocycle containing polymer | Zylon (PBO) | Toyobo | 42 | 2000 | 650 |
| Bending polymer | Polyethylene | Dyneema<br>Spectra | Toyobo DSM<br>Honeywell (USA) | 30–45 | 900–2000 | 145–155 |
| PAN-based CF | Carbon fiber | Torayca<br>Besfight<br>Pyrofil | Toray<br>Toho Tenax<br>Mitsubishi Rayon | 20–45 | 1400–3500 | |
| Pitch-based CF | Carbon fiber[*2] | Granoc<br>Dialead | NGF (Japan)<br>Mitsubishi Chemical | 13–19 | 700–4500 | |

Note: *1 *meta*–type (heat resistant) and *2 isotropic pitch (conventional use) are not included in superfiber

*Table 3.2* Mechanical strength and modulus of fibers

| | Theoretical value (g/d) | | Products on market (g/d) | | High strength product (g/d) | |
|---|---|---|---|---|---|---|
| Fiber | Tenacity | Modulus | Tenacity | Modulus | Tenacity | Modulus |
| Polyethylene | 372 | 2775 | 9 | 100 | 71 | 2400 |
| Nylon | 316 | 1406 | 9 | 50 | 17 | 50 |
| Polyester | 232 | 1023 | 9 | 160 | 10 | 203 |
| Polyacrylonitrile | 196 | 833 | 5 | 85 | 25 | |
| Polyvinyl alcohol | 236 | 2251 | 9 | 250 | 19 | 546 |
| Cellulose | 133 | 1010 | 5 | 160 | | |
| Kevlar | 235 | 1500 | 23 | 1000 | | |

*Source:* Kunigi, Ota, Yabuki, *High-strength high tenacity fibers,* p58, Kyoritsu Shuppan, Tokya (1988).

160 g/d, respectively. Other conventional fibers such as polyethylene, nylon, polyvinyl alcohol and polypropylene are in the same position with respect to the gap between the theoretical and practical mechanical strength.

## 3.1.2 Ultimate strength of conventional synthetic fiber

*High performance fiber*

Active investigation is being carried out to develop fibers possessing the tenacity of about half that of the superfiber from polyacetal, polyvinyl alcohol, polyacrylonitrile, polyethylene terephthalate and nylon. Superfiber was developed by a new spinning technology such as liquid crystalline spinning (Kevlar®) or gel spinning (Dyneema®). Inspired by these new methods, a similar attempt has been made to improve the mechanical properties of conventional fibers to their ultimate limits for the past ten years but with minimum success. Conventional fibers for industrial use have a tenacity of about 10 g/d and a further improvement in the mechanical properties has not been successful, despite the large expense on this project. The tenacity and modulus of fiber improvement is shown in Fig. 3.1, where the polyester filament for tire cord is taken as a reference and set to 1.

*Extensive applications of high performance fiber*

If the mechanical performance of the conventional fibers such as polyester and nylon can be improved by a factor of three, conventional fibers can be applied not only for clothing but also for industrial uses. At present only 4.1% and 15.6% of the theoretical strength and modulus, respectively, are

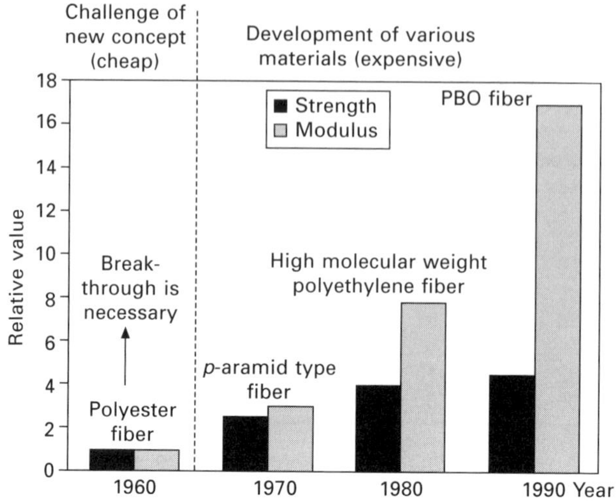


*3.1* Improvement of tenacity and modulus of fibers.

realized in conventional polyester fiber. Here we should understand the correlation between mechanical characteristics and high-order fiber structure.

As high-tenacity polyethylene fiber, the *shish-kebab* structure may be a key for nylon or polyester. If the *shish-kebab* structure is well controlled during spinning, the tenacity will increase considerably. The flexibility and/or large surface area of fiber are suitable for membrane and filter materials.

## 3.2 Development of superfiber in Europe, the United States and Japan

There are three main directions in the development of superfibers. In Europe, the development of superfiber started from the purely academic research on the *shish-kebab* type crystals of high molecular weight polyethylene as emerged from dilute solution. The results led Professor Pennings (Groningen University, The Netherlands) and his co-workers to develop the gel spinning of flexible polymer of high molecular weight that forms a gel-like solution.

The development of heat-resistant polymer has been one of the main research targets in the United States since the 1950s because of the Cold War. In the process of developing heat-resistant polymer, polyaramid fiber and PBO fiber emerged. Carbon fiber was developed initially from rayon as a starting material in the 1950s in the United States, while PAN-based carbon fiber was developed in the early 1960s in Japan by Dr A. Shindo (former Government Industrial Research Institute, Osaka, now changed to National Institute of Advanced Industrial Science and Technology (AIST)). Pitch-based carbon fiber was also developed in Japan by Emeritus Professor S. Otani (Gunma University). Global production volumes of carbon fibers combining PAN-based and pitch-based are estimated at around 17,000 tons/year, of which around 50% is produced in Japan. Both PAN-based and pitch-based carbon fibers have expanded in their industrial applications. An example of carbon fiber composites usage is in the weight-saving of trucks where carbon fiber composite plays a part in achieving weight reduction of the floor and wing-roof (Figure 3.2). Among these composites, consumption of PAN-based carbon fibers is remarkable in their use in civil engineering and construction and pitch-based carbon fibers are used for the anode of secondary cell of lithium batteries. Those carbon fibers exceed the performance of rayon-based carbon fibers, and account for most of market. Carbon fiber, aramid fiber and polyethylene fiber constitute three main currents in the development of superfiber, developed respectively in Japan from 1959 to 1960, in the United States in 1964 and in Europe from 1966 to 1969.

*3.2* A truck using carbon fiber composites (Toray)

The first satellite (Sputnik 1) was launched from the USSR in October 1957. The post-Sputnik space development in the United States promoted joint research among government institutes, universities, and industries. There were many important innovations in the field of fiber and textiles. An early result was polyaramid fiber produced by liquid crystalline spinning, developed by Dr Kuolek of Du Pont. Later came PBO fiber, first developed in the United States and now produced commercially in Japan. In these 30 years, vast quantities of heat-resistant polymers were synthesized, and the main concern has been how to achieve the theoretically expected values of the tenacity and modulus of polymer molecules when polymer is made into fiber.

## 3.3 Superfiber as a reinforcing material

The application of superfiber for advanced materials is summarized in Fig. 3.3. Superfiber is applied more for industrial use than for clothing. Thus we

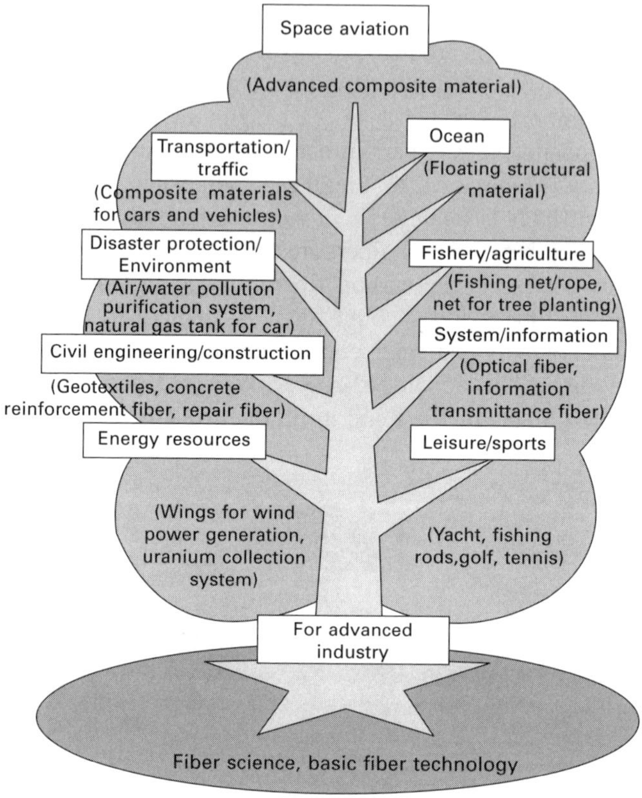


*3.3* Application of superfibers for advanced materials.

do not perceive superfibers directly, since they are used mostly in composite materials to reinforce, for example, rubber, resin, and concrete. Superfiber is incorporated in those materials in various shapes and forms to improve their mechanical performance. Figure 3.4 shows the high performance composite materials as metal replacements. Superfiber is used in the form of short staple, cord, meshed fabric or 3D woven fabric for reinforcement. The global need for synthetic fiber in the non-clothing field is expected to expand as shown in Figs 3.5 and 3.6. Table 3.3 summarizes the expected application areas of superfiber.

## 3.3.1 Application for transportation (bicycle and car)

Superfiber-reinforced rubber is used as the spring belt to replace the bicycle chain, and for the tire of mountain bikes. Some lightweight racing bicycles have composite frames.

In order to improve the petrol consumption efficiency, cars have become lighter and lighter, as superfiber-reinforced rubber is used for tires, belts, and structure. Steel cord was common for tires, but as highway networks extended, and faster driving conditions prevailed, heat friction led to many of the steel wire-lined tires bursting. Thus the reinforcing materials for tires are now composed mostly of nylon and polyester fiber.

Tires are good examples of mass-produced continuous fiber composites. They are made of a rubber polymer matrix, reinforced by continuous fiber, which could be nylon cord, polyester cord, or superfiber. This composite is called RMC (rubber matrix composite). The tire is made up of sections: tread, carcass, belt, and bead. Figure 3.7 shows the position of fiber, materials in the carcass section. Continuous fiber, such as aramid fiber, is used for the structural material of the carcass, which is the basic foundation of the tire, and is made of the laminated aligned cord fabrics, whereas cross-ply tires use cross-laminated fabrics and the radial tires radial-laminated fabrics. As

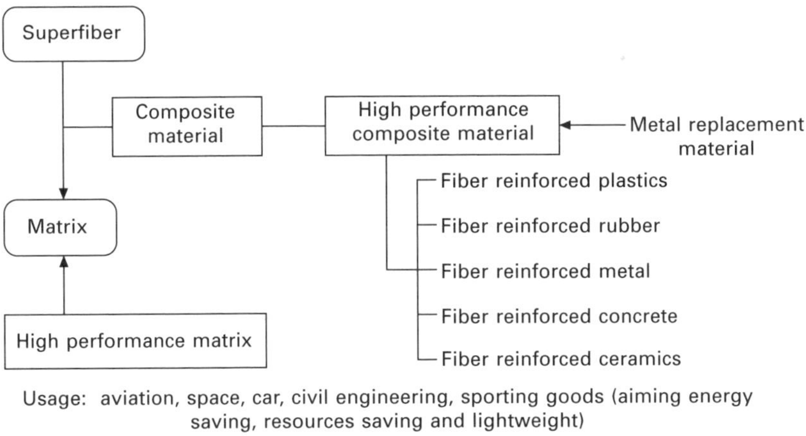


*3.4* High performance composite materials as metal replacements.

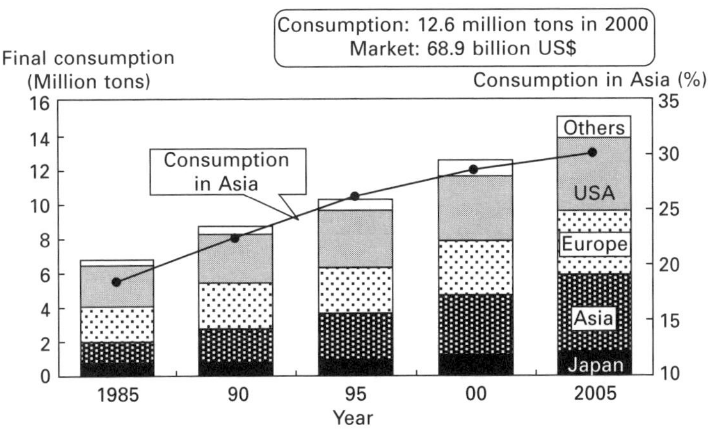


*3.5* Global consumption trend of fiber in non-clothing field (Toray).

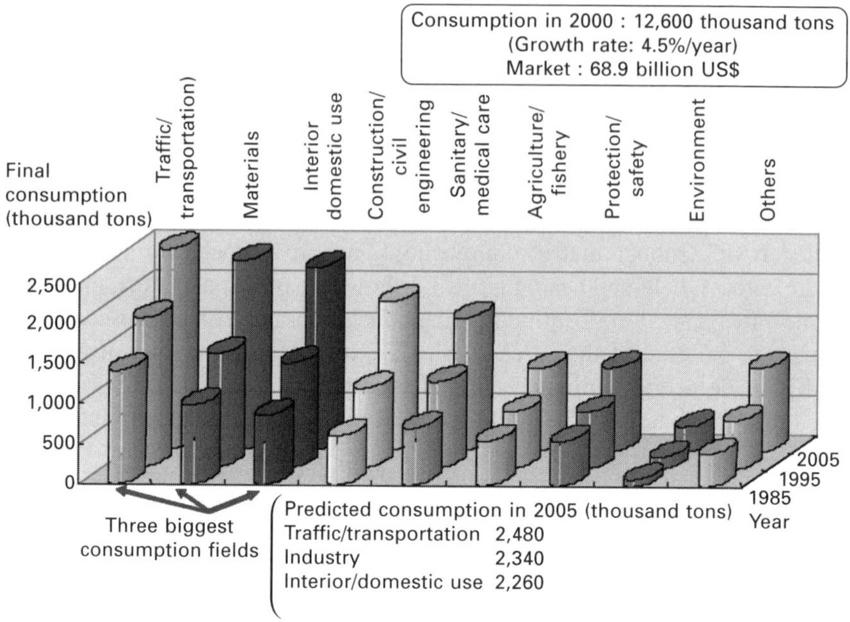


*3.6* Consumption trend of non-clothing fiber in different application fields (Toray).

*Table 3.3* Expected application areas of superfiber

| Field | Application |
|---|---|
| Aviation/space industry | Composite materials for aviation and space development |
| Transportation | Advanced composite material reinforced materials (cars, trains, ships) |
| Civil engineering/construction | Geotextiles, fiber for concrete reinforcement |
| Ocean development | Floating structure, artificial tideland, artificial ocean farm |
| Medical field | Artificial organs, fiber for health, bioreactor |
| Sports/leisure | Various goods |
| System/information | Electronics–related field, optical fiber |
| General industrial material | High performance fiber, high functionality fiber |

the radial tire becomes more popular, more aramid-type superfiber is introduced into the carcass. Nylon cord is mostly used for the cross-ply tire, is suitable for driving on a bad road at a low speed. The radial tire ensures a stable revolution and high-speed driving, but requires more suppression of friction and reduction of rolling resistance. Aramid fiber is also employed for the belt of the continuous velocity transformer (CVT) and has also replaced asbestos as the friction material for the brakes.

*3.7* Position of fiber materials in the carcass section.

### 3.3.2 Applications in sports goods

Carbon fiber is applied to the frame of a tennis racket and the shaft of a golf club. The hybrid of aramid and carbon fiber was used initially in order to complement the respective weak points of aramid (compression resistance) and carbon (impact resistance) fiber. Those fibers are compounded with epoxy resin matrix for practical use. The ski stock and ski board are also made of superfiber. The soles of ice skates also contain carbon fiber.

### 3.3.3 Aerospace technology

Lightness is a key factor in this field. Composite materials with glass or carbon fiber are commonly employed. Advanced composite materials (ACM) are becoming more and more important, where aramid or carbon fiber is applied. For example, the Boeing 777 is built with 13.5 tons of carbon fiber composites in order to reduce its weight. The composite materials and applications in the field of aerospace technology are listed in Table 3.4.

### 3.3.4 Civil engineering

The Kansai-Awaji Earthquake in January 1995 stimulated concerns about the safety of domestic infrastructure. Thus composite sheet composed of

*Table 3.4* The composite materials and applications in the field of aerospace technology

| Term | Materials | Applications |
|---|---|---|
| Short to middle | Lightweight structure materials for ACM<br>Seat molding compounds<br>Heat-resistant flame proof fiber | Aviation/space devices<br>Cars, aircraft<br>Interior material in aircraft, uniform |
| Middle to long | Lightweight structure materials for ACM (SST for next generation)<br>Oxygen enrichment membrane/fiber<br>Membrane material for airship | Space base camp, Large aircraft engine by one molding<br>Sky-net plan |

aramid or carbon fiber has been used to reinforce the concrete columns of a bridge girder because it is easy and lighter to handle.

Geotextiles are the textile products used for civil engineering. The practical application of geotextiles was advanced in Europe and the United States, but the introduction of superfiber to geotextiles was made in Japan. Aramid fiber reinforced composite is also applied in ground fill. A reinforcing iron rod is gradually replaced with aramid or carbon fiber reinforced composite material, which is free from rust. Staple-fiber reinforced concrete (FRC) is also now becoming popular.

### 3.3.5 Need for standard specifications

Composite materials reinforced by superfiber have expanded steadily into the fields of aerospace engineering, ocean engineering, civil engineering, transportation engineering, sports/leisure engineering and medical engineering. The development of fiber/textile for industrial use takes at least five to six years, since no technical data is available for new materials. The required specifications for fiber/textile for industrial use are listed in Table 3.5. When the specifications and regulations are well established, the production of fiber/textile for industrial use can be expected to expand enormously. However, each engineering field has its own regulations, and no standard specification has been established. Here the basic technical data should be gathered from the various industries and standardized in order to spread the application in various industrial fields. Close cooperation is required among government institutions, civil engineering industries and chemical fiber industries.

It must be recognized that superfiber is a new material in the industrial field in comparison with concrete or metal. Table 3.6 shows the potential superfiber applications in the field of civil engineering.

*Table 3.5* Required specifications of fiber/textile for industrial use

| Required specifications | Original fiber | | | | | Adhesion | Heat resistance | Processing durability | Flame resistance | Anti-weatherability |
|---|---|---|---|---|---|---|---|---|---|---|
| | Strength | Modulus | Toughness | Heat shrinkage High | Heat shrinkage Low | | | | | |
| Tire cord | ◎ | ○ | | | ○ | ◎ | ○ | ◎ | | |
| V-belt | ○ | ◎ | | | ◎ | ○ | | ◎ | | |
| Conveyor belt | ◎ | ○ | | | ○ | ◎ | | ○ | | |
| Ropes/net | ◎ | | ○ | | | | | | | ○ |
| Heavy cloth | ◎ | | | | ○ | | | | | |
| Bags/wrapping | | | ○ | | | | | | | |
| Sewing | ○ | | | | | | | | ○ | |
| Base for artificial leather | | | | | | | | ○ | | |
| Electric material | ○ | | | | ○ | | ◎ | | | |
| Filtration | | | ○ | | | | ○ | | ◎ | |
| Safe belt | ◎ | | ○ | | | | | | | ◎ |
| Felt | | | | | | | | ○ | | |
| Non-woven fabrics | | | | | | | | ○ | | |
| Hose | ◎ | ○ | | | ○ | ◎ | | ○ | | |
| Thread for *tatami* | ◎ | | | | ○ | | ○ | | | |
| Carpet | | | | | | | | | | |
| Curtains | | | | | | | | | | |
| Sheet for car | | | | | | | | | ○ | |
| Required specification | 10 | 4 | 4 | | 7 | 4 | 4 | 7 | 3 | 2 |

◎ Very important specification.
○ Important specification.

*Table 3.6* Potential superfiber applications in the field of civil engineering

| Term | Material | Application |
|---|---|---|
| Short to middle | Cement reinforcement fiber<br>Asbestos replacement material<br><br>Optical spectrum conversion material | Construction/civil engineering<br>Heat retention, heat insulation packings<br>Agriculture, fishery |
| Middle to long | High strength, low density, heat resistance (>500°C), durability, anti-weatherability | Anti-weatherability covering material<br>Filter (heat resistant, chemical resistant)<br>Flame proofing, heat proofing cloth (clothes, sheet)<br>Radioactive rays protection<br>Electronic base material, robot |

## 3.4 Frontiers of superfiber applications

The information/communication industries and biotechnology could be leading industries in the first half of the twenty-first century. Superfiber will be a leading technology in the fiber/textile field and the following examples demonstrate the frontiers of superfiber applications in these areas.

### 3.4.1 Kevlar, liquid crystal spun para-aramid fiber

*Ultra-thick Kevlar wire to fence the ocean culture farm from floating ice*

About 60 000 tons of scallops are produced every year and earn a total of about US$ 10 million in Lake Saroma in Hokkaido. Since Lake Saroma is directly linked to the Sea of Okhotsk, floating ice will surge into the lake in late January, and may damage the farm. In fact about US$ 22.2 million was lost by the damage caused by the floating ice surge in 1974. Global warming continues, and in recent years floating ice has often surged into the lake, even in late March.

Hokkaido Development Bureau investigated how to cope with floating ice surges, and adapted the ice boom proposed by Professor H. Saeki's group (Hokkaido University). The ice boom is a floating wire fence. A single boom is 100 m long, and consists of a main wire with 28 floats (1.2 m in diameter and 3 m in length) and 4 m wire net underneath the water. Since a large quantity of floating ice will surge into the lake, the steel wire should be thick enough to cope with an ice surge, and become too heavy for the floats of practical size to sustain. Since Kevlar is about five times lighter than steel in terms of the same tensile strength, a Kevlar ice boom can be made lighter and become practical. Kevlar has another advantage, since it does not rust in seawater. The main wire, produced by a rope manufacturer in Gamagori (Aichi Prefecture), is made of ultra-thick Kevlar fiber, 13 mm in diameter, which can stand a load of 500 tons. The bird's-eye view of this ice boom

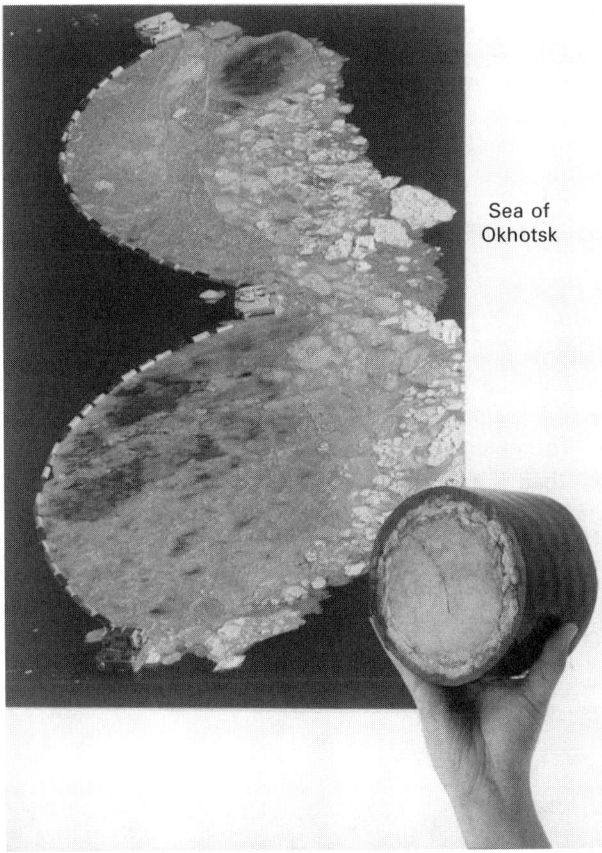


*3.8* Bird's-eye view of ice boom set in Lake Saroma to protect from floating ice, and the cross-section of Kevlar wire (Du-Pont Toray).

looks like an oil fence (see Fig. 3.8). The cross-section of Kevlar wire is also shown in Fig. 3.8.

*Kevlar tension member for optical fiber*

The twenty-first century is the era of high-speed information, so that optical fiber is indispensable. The information transmission capacity of optical fiber is about 1000 times that of conventional metal cable. Two types of optical fiber are available. One is the inorganic type of quartz fiber, and the other is made of organic polymethyl methacrylate (PMMA). The quartz optical fiber is suitable for long distance transmission, and its network extends from Hokkaido to Kyushu as well as from Japan to the United States. Superfiber supports this optical fiber network. Optical fiber is extremely thin and thus is weak against tension. When optical fiber is stretched, the optical transmission characteristics can be damaged. The high Young's modulus of Kevlar can be utilized to support such optical fibers. The optical fiber cable is composed of Kevlar fiber axis as a tension member and optical fibers arranged spirally around Kevlar as shown in Fig. 3.9.

### 3.4.2 Technora®, wet-spun para-aramid fiber

Teijin developed wet-spun Technora in 1973, and has produced almost 1600 tons per year in its Matsuyama Plant since 1987. Technora is synthesized by copolymerization of the same components as Kevlar (terephthalic acid chloride and *para*-phenylene diamine) and the third component diamine containing an ether linkage. Technora is wet-spun and drawn. The ether linkage in its molecular structure makes Technora more flexible than liquid-crystalline-spun Kevlar. Since Technora has improved characteristics with regard to solubility and drawability, the filament surface is smooth and no fibrillation takes place. Technora is used for ropes and nets for fisheries and civil engineering, and for protective clothing, including bullet-proof jackets, knife-proof clothes and gloves, where not only high-density woven fabric but also sheet-like structure is used. Technora is also used as FRP tension materials to reinforce rod-concrete construction against a big earthquake.

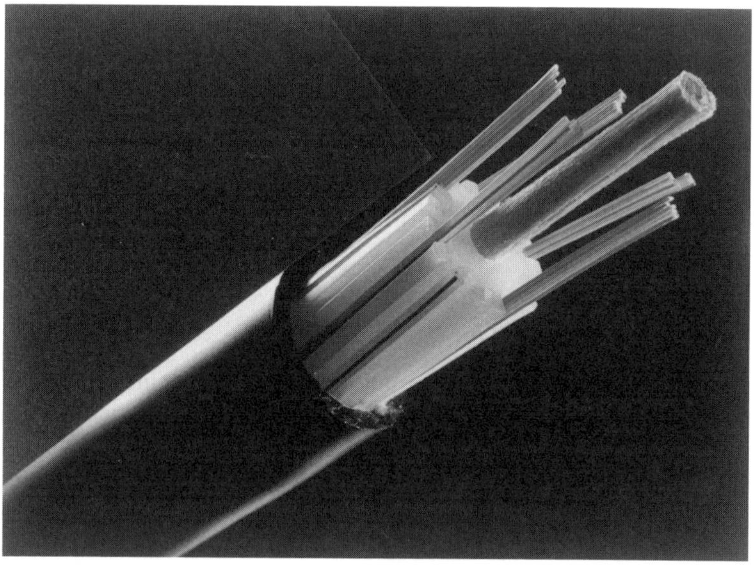

*3.9* Kevlar fiber as a tension member for optical fibers (Du Pont-Toray).

The application to printed circuit boards deserves a special mention. Conventionally, glass/epoxy resin is used for printed circuit boards for electronic equipment. Recently the light and easy-processed circuit boards from *para*-aramid paper have been developed and are replacing the conventional glass/epoxy resin boards.

We should also mention examples of high-durable aramid FRP rod applications. Aramid FRP is a composite of polyvinyl ester resin matrix reinforced by Technora-fiber. Aramid FRP rod has a rugged surface with wound aramid fiber, or by twisting in order to achieve a good adhesion to concrete. Aramid FRP rod possesses almost equal or even higher tensile strength than PC steel. Its modulus is about a quarter that of steel and has no yielding point. Its weight is about a sixth that of steel. Aramid FPR rod exhibits no deterioration in water, seawater, or alkaline solution, so no anti-rust processing is required. Figure 3.10 shows the aramid FRP rod anchor to protect the river bank in the Kyushu district, where 5160 m of aramid FRP rod of 7.4 mm in diameter was used. Figure 3.11 shows the application examples of aramid FRP.

## 3.4.3 Thermotropic liquid crystalline spun Vectran® fiber

Kuraray produce the polyarylate superfiber Vectran. Polyarylate contains aromatic aryl groups. By controlling the molecular structure, polyarylate melts at a certain temperature and forms thermotropic liquid crystals. Vectran is melt-spun from such polyarylate.

One of the most novel applications of Vectran was as an airbag for the Mars Pathfinder. NASA (National Aeronautics and Space Administration) announced a large space science project (the Origin Plan) in 1996. This project aims to explore the origins of space, and planned to launch a series of space probes. Mars Pathfinder was launched from the Kennedy Space

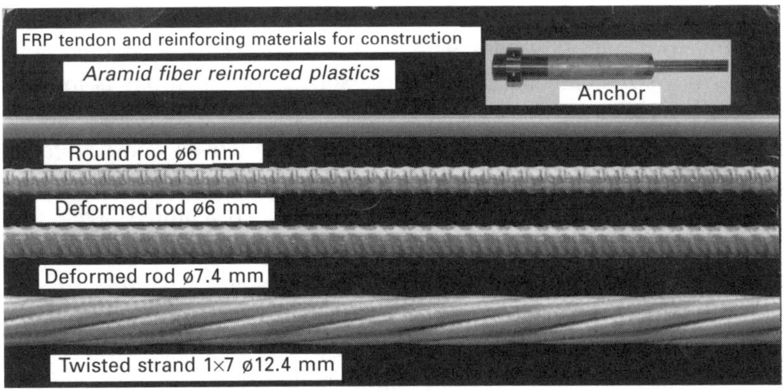


*3.10* Aramid FRP rod (Teijin).

*Figure 3.11* Application examples of aramid FRP (Teijin).

Center in Florida on 4 December 1996 in order to arrive on Mars on Independence Day (4 July 1997). The most difficult part of this project was how to soft-land on Mars. The staff at JPL (Jet Propulsion Laboratory, NASA) was asked to develop a less expensive soft-landing device quickly. They innovated and produced a new soft-landing device utilizing an airbag, which costs US$ 600 million, compared with US$ 1000 million for the conventional soft-landing device. This airbag is made of four-layered woven fabrics of Vectran fiber laminated inside with silicon polymer. Various designs were tested at NASA's Lewis Research Center, including the shape of the airbag, the fabric materials and the seam. The red surface of Mars is covered with rocks of various sizes and shapes. Thus the surface of the airbag is loosely seamed in order to adjust its position according to the external load and to absorb the impact shock. Vectran fiber of 200 d was woven into five-layer high-density fabrics laminated with silicon polymer with warp directions shifted 45° with respect to each layer. The outer two to three layers are designed to yield at certain impact strength and to absorb the impact energy. The fifth layer (innermost layer) is made completely airtight and possesses high strength. Figure 3.12 shows the space probe Mars Pathfinder covered with the Vectran airbag.

When the Mars Pathfinder (its front is protected with heat-resistant shield) is separated from the spaceship, it descends into the Martian atmosphere.

*3.12* Space probe Mars Pathfinder covered with Vectran airbag (NASA).

Several minutes before soft-landing, the heat-resistant shield is cut off and Mars Pathfinder descends by parachute. At 300 m above the surface (8 s before landing), four airbag sets (each composed of six beach-ball-like bags) inflate within 0.5 s at each side of tetrahedral Mars Pathfinder. A retro-rocket device is ignited immediately and the speed of descent is reduced. Mars Pathfinder falls on to the Martian surface, bounces up and down several times and then stops. The Vectran rope then folds the airbags to the space probe surface. The outline of this landing is shown in Figure 3.13. This Mars Pathfinder system is made up of the tetrahedral landing device (the lander) and the rover. Total weight, including airbags, must be less than 360 kg, because of the limit of load to the airbags. The landing craft opens to expose its interior when landed, and the rover is sent to explore the surface (see Fig. 3.14 showing Mars surface probing car 'Rover').

### 3.4.4 Zylon®

Zylon is a commercial name for PBO (poly-*p*-phenylene benz-*bis*-oxazole) fiber. Toyobo started Zylon production at its Tsuruga Plant in October 1998. The production started with 200 tons/year at the beginning and increased to 360 tons/year by 2002. The fiber is gold-colored and its appearance is similar to Kevlar. The characteristics of Zylon are an extremely high tensile strength, an extremely high modulus, high heat resistance, and a high flame resistance. The material is a rigid molecule, and its processing is difficult.

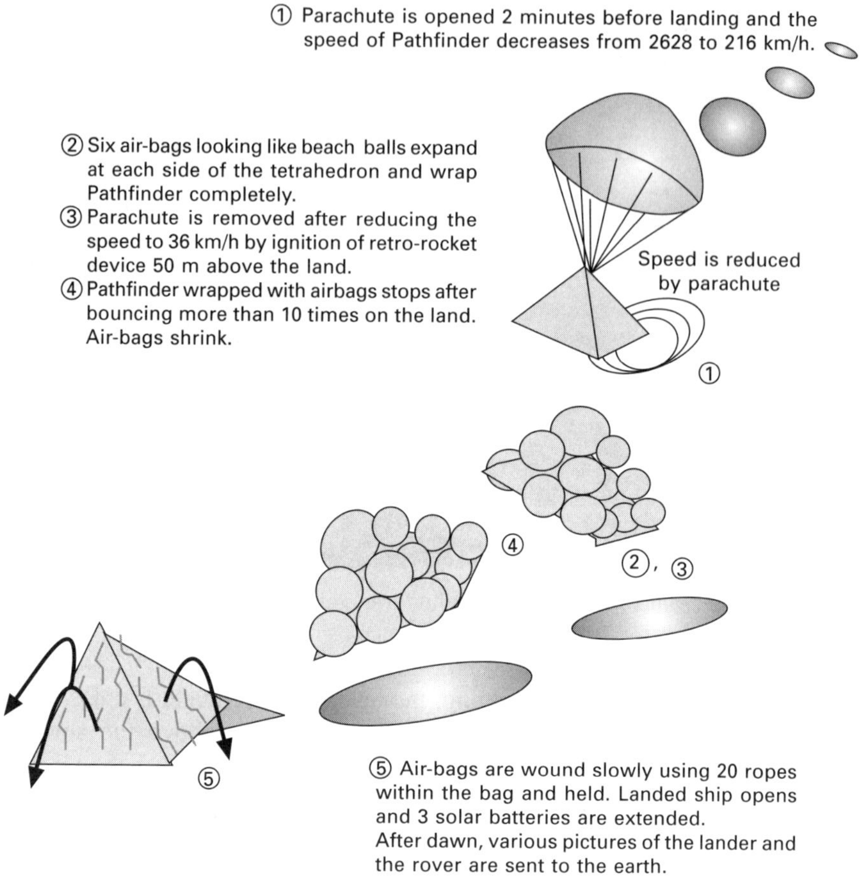


*3.13* Outline of Mars Pathfinder landing (Yomiuri Shinbun, 5/7/1997).

*PBO fiber from Toyobo*

PBO fiber had a long history before its commercialization. When *p*-aramid fiber (Kevlar) appeared in the 1970s, the researchers at Air Force Materials Research Institute (USA) started a new challenge to synthesize polymers having higher modulus and better heat-resistance than Kevlar. They published the basic results in 1979, and PBO was one of the heterocyclic polymers developed in this project. Dr Wolf patented PBO at Stanford Research Institute. Dow Chemical bought the rights of this patent and attempted to make fibers from PBO for seven years without success. Consequently, Dow Chemical was looking for a partner to develop PBO fiber. During this period, however, Dow Chemical found a new method to improve the monomer (*p*-phenylene diamine) yield and a more efficient polymerization to form PBO.

*3.14* Mars surface probing car 'Rover' (Taken at Tokyo Big Site exhibition hall).

Toyobo were interested in PBO, and proposed a joint venture with Dow Chemical to develop PBO fiber. At that time, the tenacity of PBO fiber was similar to that of Kevlar but its modulus was higher. Toyobo was confident that PBO fiber could be made stronger because of its higher modulus. Later Dow Chemical withdrew from the development of PBO fiber because of a change in the company's business strategy. Figure 3.15 shows a yacht's sail made using Zylon®.

*Tenacity of PBO fiber*

The tenacity of PBO fiber was initially 20 to 22 g/d. The tenacity was improved to over 40 g/d during the period of the joint development, and now commercial Zylon has an average tenacity of 45 g/d, twice as strong as Kevlar.

Kevlar activated further research into high-tenacity polycondensates including nylon and polyester. Conventional polyester fiber for industrial use has a tenacity of 6.5 to 10.0 g/d. The tenacity is still lower at 4.5 to 6.5 g/d for synthetic fiber use for clothing. However, the tenacity of polyester is calculated from the fully extended ideal structure as 230 g/d. Many research projects have been initiated to develop superfiber from such conventional polymers.

*3.15* Race yacht sail made by Zylon® (Toyobo).

*Unique spinning of Zylon*

How is PBO fiber spun? PBO is a heterocyclic polymer, classified as poly benzazole, polymerized from diamino and terephthal acid in polyphosphoric acid. Zylon melt is dry-wet spun into fiber. The PBO spinning is shown schematically in Figure 3.16, as presented by Ledbetter. At the initial stage of the project at the Air Force Materials Research Center, the researchers examined theoretically how to design rigid and heat-resistant polymer molecules prior to the actual experiments. However, major innovations are often done by serendipity, and we will notice that the PBO case is rare.

NASA planned ULDB (Ultra Long Duration Balloon), a program to float a pumpkin-shaped balloon (see Fig. 3.17) at an altitude of 33.5 km for up to 100 days and obtain scientific data by atmospheric observation. The balloon is composed of a layered polyester and polyethylene film about the thickness

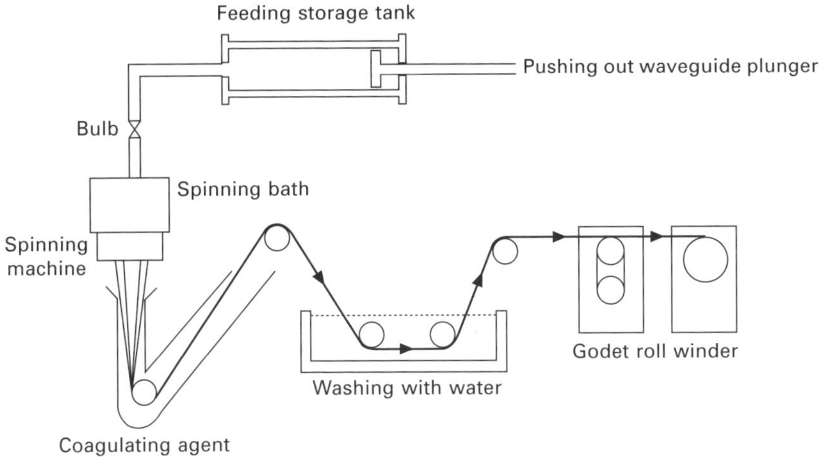


*3.16* Outline of PBO spinning (H. D. Ledbetter *et al.*, MRS **134**, 253 (1989).

*3.17* Pumpkin-shaped balloon for atmospheric observation (NASA).

of ordinary plastic food wrap (62 g/m$^2$). Two rings on the top of the balloon are pulled by Zylon pulling members to reduce the required strength of the membrane at planning.

Another application example of Zylon is a bullet-proof vest as shown in Fig. 3.18.

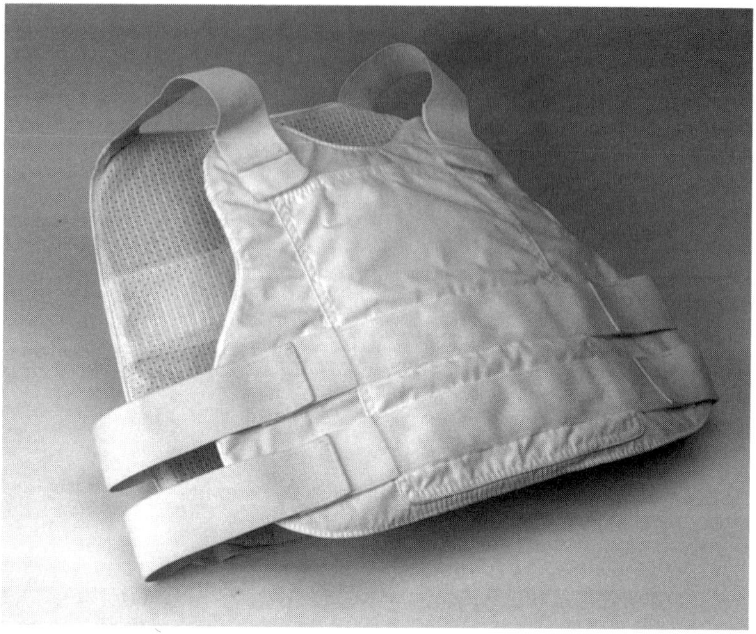

*3.18* Bullet-proof vest made of Zylon (Toyobo).

*Tension member for optical fiber*

As mentioned previously, two types of optical fibers are available commercially. In either type, plastic or quartz, the optical fiber is extremely thin, and breaks easily by bending or stretching. In practical use optical fibers are protected with a tension member. The tension member is conventionally made of steel wire, but steel is not ideal since it picks up noise from thunder and cars. Non-conductive superfiber is an ideal material for acting as the tension member.

An optical cable for long distance communication is made of a few hundreds to a few thousands of optical fibers bundled around a rod and covered with the protective material. Another advantage of using superfiber for this purpose is its weight. Since PBO fiber has twice the tenacity and modulus of other superfibers, the tension member can be made thinner and lighter. For example, a conventional optical cable of 80 mm diameter will be reduced to 50 mm by using PBO fiber.

*Balloon for Venus probe*

JPL (Jet Propulsion Laboratory, NASA) is planning to send a planetary probe to Venus in about 2006. Venus is covered with a sulfuric acid cloud, and the distance between the cloud and the surface of Venus is about 48 km. The temperature of Venus's surface is 460°C, but the temperature is as low as –10°C in the sulfuric acid cloud. The atmosphere is mainly composed of carbon dioxide, containing a small amount of sulfurous acid gas. In order to transmit information gathered on the surface of Venus back to Earth, the balloon probe needs to rise above the cloud. The balloon should go up and down several times from the surface of Venus above the sulfuric acid cloud. This up-and-down movement will be controlled by inflating or deflating the balloon. PBO fabric or PBO film, for this reason, is the only candidate for the flexible heat-resistant membrane for the balloon probe.

The balloon probe will be launched with a satellite and be parachuted down to the surface of Venus through the sulfuric acid cloud. Since PBO is weak against acid, PBO fiber will be coated with gold. Thus Zylon could help in the achievement of a considerable scientific goal.

### 3.4.5 Gel spinning of flexible polymer

*Dyneema® stands for a strong fiber*

Three major fiber industries in the world are producing high tenacity polyethylene fiber. DSM (Holland) and Toyobo (Japan) jointly developed and produce Dyneema®. Allied Signal Inc. produces Spectra®.

Dyneema® is a composite Latin word made up from 'dyne' (power) and 'neema' (fiber). Dyneema thus means 'a strong fiber'. Professor Pennings, Dr P. Smith (formerly-Professor of California Institute of Technology) and Professor Lemstra (Eindhoven Technical University) collaborated to develop gel spinning of high molecular weight polyethylene. Commercial high tenacity polyethylene fiber is based on this basic process developed by these three scientists in 1979.

Dyneema SK60 has a tenacity of 30 to 35 g/d. In 1999 Toyobo first put Dyneema SK71 on the market, having a tenacity of over 40 g/d. The basic concept of gel spinning is to reduce crystalline defects. Chain ends, entanglements and the folded chains in the amorphous region are counted as defects. The tenacity of each filament fluctuates considerably, so that the filaments should be produced without tenacity fluctuation. The tenacity fluctuation of high tenacity polyethylene fiber is shown in Fig. 3.19.

*Characteristics of Dyneema*

The density of Dyneema is less than one, and it floats on water. Commercially available Dyneema (Dyneema SK71) is made of flexible polyethylene, and its elongation at break is around 4%. High tenacity polyethylene fiber has a good balance of tenacity and elongation, so that it is easy for later processing including fabrication and knitting. Its impact strength is excellent, and is used for protecting and reinforcing materials. Polyethylene is chemically

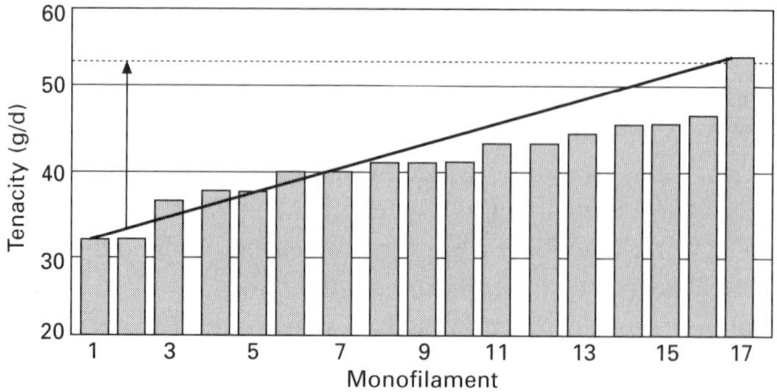


*3.19* Tenacity fluctuation of high tenacity polyethylene fiber.

stable and has a good chemical resistance in a wide range of pH except for some organic solvents. No degradation will take place in water, so that it is suitable for use in humid places exposed to the sun. The molecular weight of high tenacity polyethylene is over 1 000 000, and its shape stability is good. However, the melting point is low (150°C) and it creeps at higher temperature.

These characteristics should be taken into account in any application. Since high tenacity polyethylene is light and strong, it is mostly used for ropes. Anchoring rope and tag rope for ships made of high tenacity polyethylene is light and not water-absorbent. A fishing line is a high value-added application of high tenacity polyethylene fiber. Since Dyneema has a high sonic modulus, a bite can be detected quickly, resulting in advantageous fishing conditions.

#### *Application to superconductive materials*

Dyneema is highly crystalline and highly oriented. Thus its filament shrinks with lowering temperature. That is, its line thermal expansion coefficient is negative. By integrating with other resins of normal thermal expansion behavior, a composite can be made, which deforms in any direction as the temperature changes. A bobbin for a superconductive coil can be made from high tenacity polyethylene fiber and epoxy resin composite (see Fig. 3.20). The bobbin will expand a little in a radial direction at liquid helium temperature, so that superconductive coil will not become loose even when the superconductive coil stretches by Lorenz forces at liquid helium temperature. A superconductive energy storage system for electric power peak saving is shown schematically in Fig. 3.21.

### 3.4.6 Prospects for superfiber

Superfiber applications are capable of serving a great need within our society. Energy and resource conservation will require lighter materials of higher

*3.20* A bobbin for a superconductive coil made from high tenacity polyethylene fiber and epoxy resin composite (Toyobo and Kyushu University).

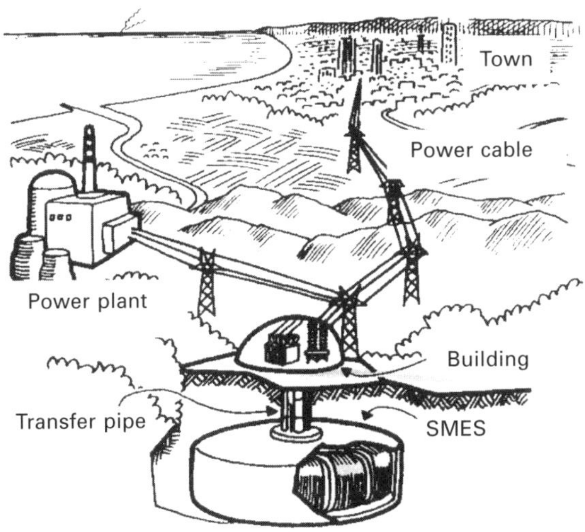


*3.21* Superconductive energy storage system for electric power peak saving (Toyobo).

performance. The prospect of such superfiber needs in non-clothing fields is summarized in Fig. 3.22.

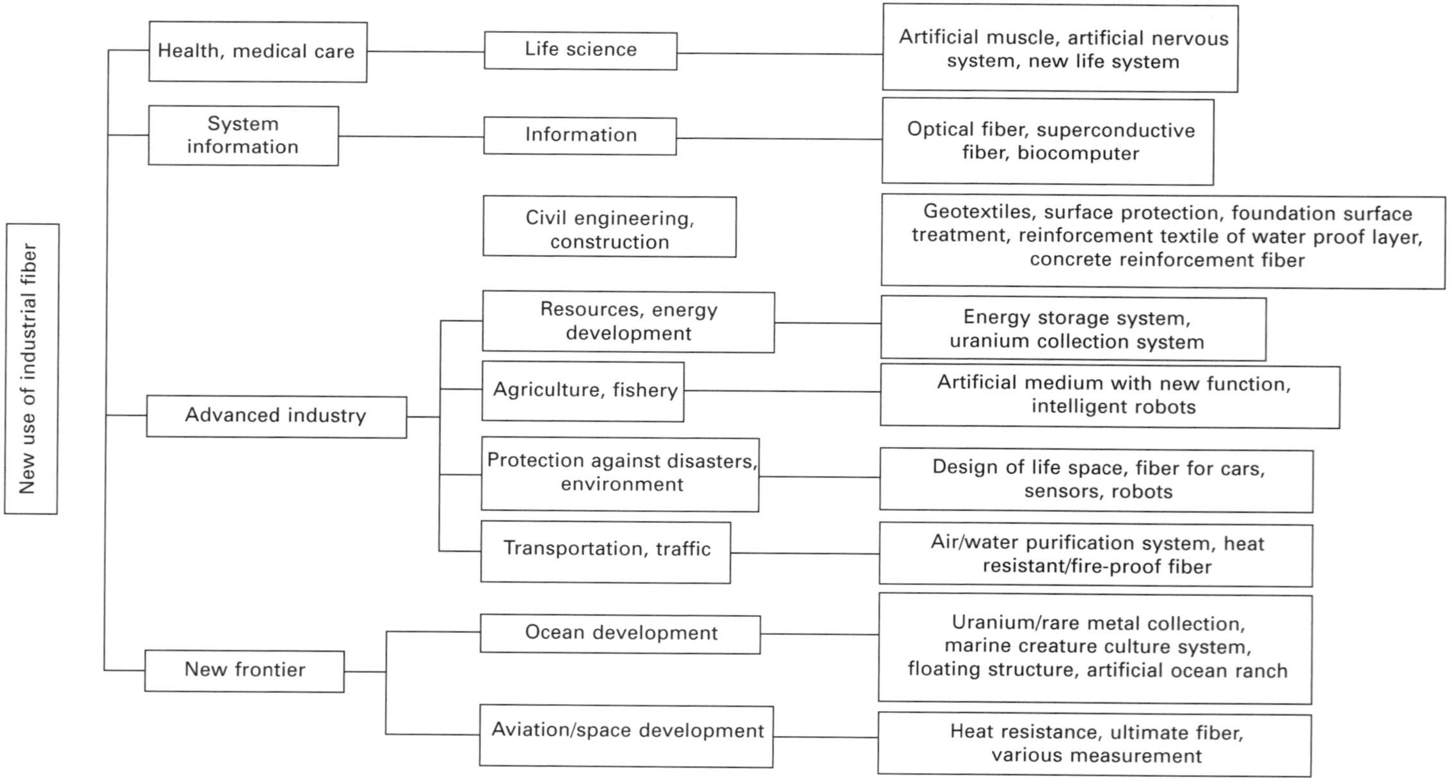


*3.22* The prospect of superfiber needs in non-clothing fields.

## 3.5 Nanofiber (carbon nanotube)

While nanofiber refers to DNA (deoxyribonucleic acid) in the field of biology, it stands for a carbon nanotube (CNT) in the field of fibers. CNT consists of carbon atoms connected in a cylindrical manner with a diameter of several nm to several tens nm and a length of several μm. Aspect ratio (ratio of length to diameter) is in the range of 1000 to 10 000, and the strength of the nanotube is expected to be 40 times of that of carbon fiber, judging from the crystal structure. It belongs to a new group of molecular fibers. CNT is classified into two types according to the thickness of the wall of the tube: single-walled CNT (SWCNT) and multi-walled CNT (MWCNT). The former attracts attention from the academic point of view.

The method of synthesis of CNT includes arc discharge, laser ablation method and chemical vapor deposition methods. Professor M. Endo (Shinshu University) first discovered the carbon nanofiber. The fiber formation system using the floating catalyst method is shown in Figure 3.23. Dr S. Iijima (NEC) discovered the carbon nanotube and he showed transmittance electron micrographs of MWCNT and SWCNT in 1991 and 1993, respectively. A schematic model of the carbon nanotube is shown in Fig. 3.24.

SWCNT consists of a cylindrical graphite sheet with a diameter of 0.6–1.8 nm and a length of several μm. MWCNT is composed of several concentric

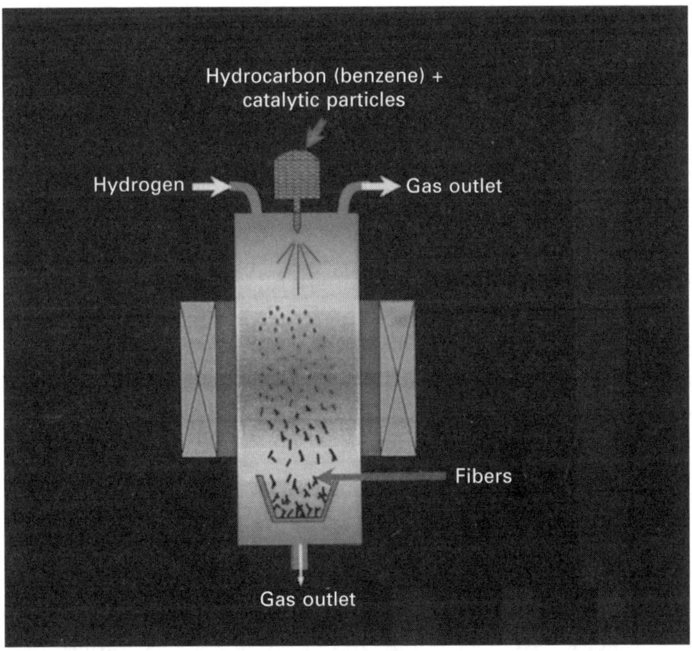


*3.23* Fiber formation system using floating catalyst method (Professor M. Endo, Shinshu University).

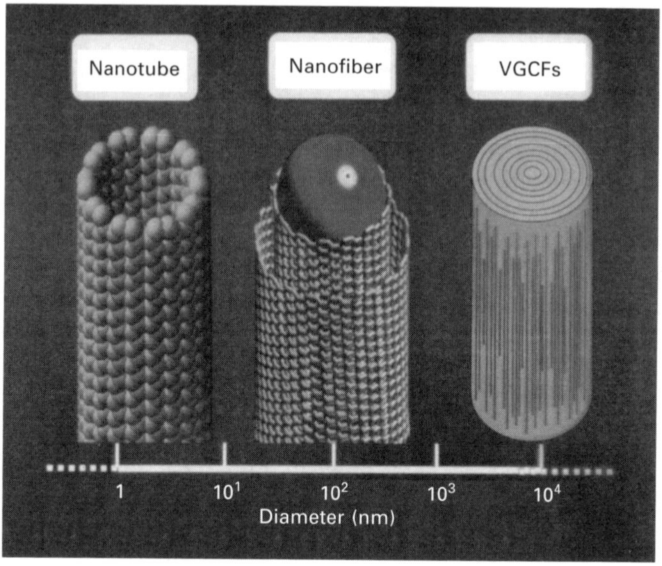


*3.24* Schematic model of the carbon nanotube (Professor M. Endo, Shinshu University).

cylinders with diameters of 5–30 nm. CNT did not attract attention at first, but has attracted considerable attention subsequently, particularly when the electric conductivity of CNT was measured successfully. As a material, it has very fine structure, high tenacity strength and modulus, and specific properties in heat resistance, electric conduction and heat conduction. As a result, it is expected to be a material to drive science and technology in the twenty-first century. It is already used as the cathode plate of lithium ion batteries, hydrogen-occlusion material, electro-conductive filler in resin and paint, field emission display material (FED), in atomic force microscopes (AFM) and nano-tweezers.

Carbon fiber with carbon nanotube as a core and optional diameters in the range of several tens to several hundreds nanometers is now almost ready for manufacture. Nano composite fiber, composed of nylon and CNT as the carbon fiber, is used for the smallest gear of the second hand in a watch. Future possibilities look almost endless. The technology fields shown in Fig. 3.25 will be opened by nanotechnology.

## 3.6 High polyketone fiber

Asahi Kasei got off to a bad start in the high strength fiber field. But now it regards polyketone fiber as a high strength fiber such as aramid fiber and has a project to develop it with the support of NEDO. The polyketone fiber could be a masterpiece of new millennium fiber.

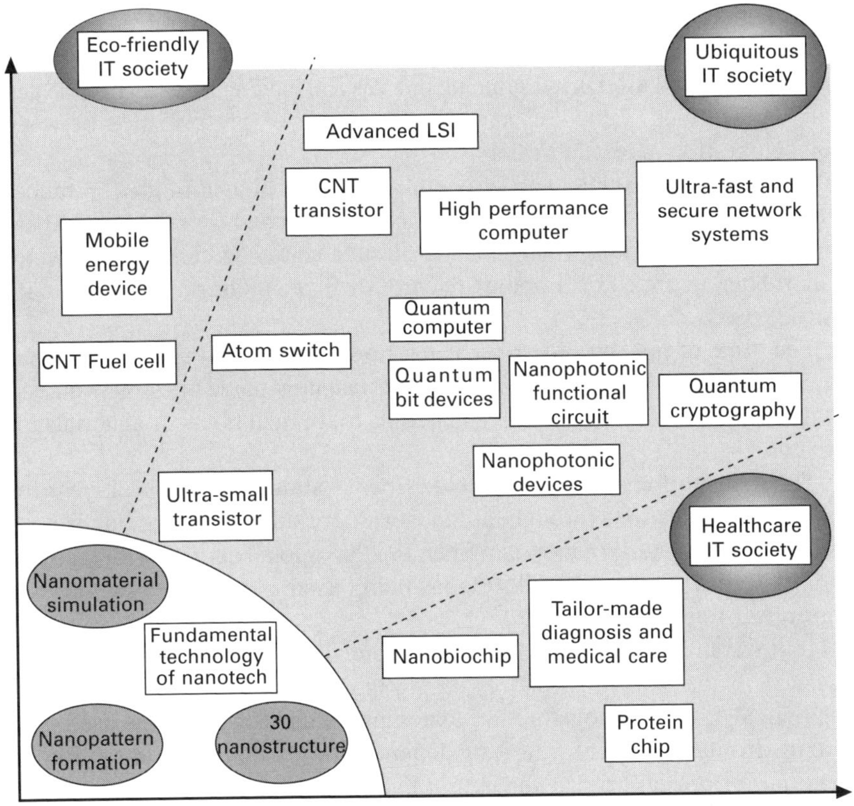


*3.25* Technology fields that will be opened by nanotechnology (NEC).

High costs have held back multipurpose development of aramid type materials. The high price is due to the special raw materials used and the special and complex equipment and methods necessary. Asahi Kasei, however, has now developed low price high strength fiber of polyketone fiber with strength and durability as good as aramide fiber. According to the company announcement and patents, this fiber has a molecular structure which includes carbon monoxide and is also composed of ethylene. Thus it contains only carbon, oxygen and hydrogen, so it has a low manufacturing cost compared to other high strength fiber containing other atoms such as nitrogen. Figure 3.26 shows reactions of the polyketone polymer. The low price raw materials and the simple structure of this polymer have reduced the price by 40%.

$$CO + CH_2{=}CH_2 \longrightarrow \left( CH_2CH_2\overset{\overset{\displaystyle O}{\|}}{C} \right)_n$$

*3.26* Reaction of polyketone polymer.

So far there has been only one polyketone polymer produced, but it is difficult to spin to form fiber. Only Asahi Kasei has developed a low cost inorganic solvent, used in the spinning process and have successfully fabricated polyketone fiber. Table 3.7 shows comparison of performance between polyketone fiber and other fibers.

Polyketone is excellent in strength and has a high affinity for rubber within raw fibers. This feature has opened up a demand for reinforced fibers in composite materials which include plastics and concrete, and especially with rubber, in tire cords. For the tire cord, steel and nylon and polyesters are usually used.

The price of polyaramide fiber is ten times that of polyester fiber. It has a market share of less than 1%. Asahi Kasei can now move into a new market for low cost new fiber to reinforce materials for optical fibers or materials for tire cord.

The demand for *para*-type aramide fiber is about 36000 tons a year for industrial applications throughout the world. 10000 tons of aramide is used as reinforcing material for optical fiber. In 2005, global demand will increase to 50000 tons a year, with 20000 tons being used as a reinforcing material for optical fiber.

Teijin aramide fiber, which ranks second in global market share, will increase its turnover to 35 000 000 000 yen a year as a reinforcing material for optical fiber and construction materials. Asahi Kasei makes use of its strong affinity to rubber for tapping demands for tire cord and for reinforced material for optical fiber as the largest market in high strength fibers. Asahi Kasei had not previously produced a high strength fiber, but hereafter rallied with the development of the polyketone fiber.

Toyobo developed a high strength and high heat proof fiber (Zylon) in 1998 and now manufactures 200 tons a year for firemen's uniforms and bullet-proof vests.

*Table 3.7* Comparison of performance between polyketone fiber and other fibers*

| | Polyketone fiber | Ester | Rayon | Aramid |
|---|---|---|---|---|
| Tenacity (g/d) | 20 | 3 | 6 | 23 |
| Stretch (%) | 5 | 13 | 11 | 4 |
| Elasticity (g/d) | 400 | 120 | 130 | 490 |
| Heat contraction at 150°C (%) | 0.5 | 3.9 | 1.7 | 0.5 |
| Density | 1.5 | 1.4 | 1.5 | 1.4 |

*Nikkei Industry Newspaper on 25 July 2002.

## 3.7 Bibliography

### *3.1 Description of superfibers*

Hongu T. and Phillips G.O., *New Fibers*, 2nd edn, p. 6, Woodhead Publishing, Cambridge, 1997.

Kunugi T., Ohta T. and Yabuki K., *Ko-Kyoudo Ko-Dannseiritsu Sen'i (High-Strength High-Tenacity Fibers)*, p. 58, Kyoritsu Shuppan, Tokyo, 1988.

### *3.3 Superfiber as a reinforcing material*

#### Carbon Fiber – aerospace (CFRP)

Brewer D., *Advanced Composites*, p. 58 (Jan./Feb., 1987).

Boeing, *7J7 Program Review*, Oct. 1987.

Hardness J.T., *14th National SAMPE Tech. Conf., 26* (1962); *Performance Materials*, 1987-10-19, p. 3

Williams J.G., O' Brine T.K. and Chapman III A.J., *NASA CP2321*, p. 51 (1984).

Masters J.E., Courtier J.L. and Evans R.E., *31st Int. SAMPE Symp.*, p. 844 (1986).

Smith B.D., *Aerospace Composites Materials*, 61 p. 20, Shephard Press, 1992.

*Aircraft Engineering and Aerospace Technolology*, p. 29, MCB University Press Ltd, 1989.

Stover D., *Advanced Composites Sept./Oct.*, p. 30 (1991).

Mendon H., *13th SAMPE Europe Int. Symp.*, p. 13 (1992).

#### Carbon fiber – vessel (CFRP)

Momoshita S., *Proc. of ICCM IV*, p. 1710 (1982).

#### Carbon fiber – technical textile (vehicles)

Siejak V., *Conference of 'Bag & Belt 2000', 6th Int. Symp. on Automotive Occupant Restraint Systems*, Cologne, Germany, March 2000, p. 287.

Kitamura M. and Siejak V., *Conference of 'Bag & Belt 2000', 6th Int. Symp. on Automotive Occupant Restraint Systems*, Cologne, Germany, March 2000.

Tanaka K. and Isoda H., *Function and Materials* (Japanese edn), Feb. 1997, p. 31.

Oka T. and Ohta Y., *The 53th Annual Symp. of Tex. Mach. Soc. Jpn.*, Osaka, June 2000, p. 80.

#### Geosynthetics

Koener R.M., *Designing with Geosynthetics,* 3rd edn, Prentice-Hall, 1994.

Ochiai H., *et al., Proc. of Int. Symp. on Earth Reinforcement (IS Kyushu 96)*, Vol. 2, 1997.

Giround J.P., *Proc of the 5th Int. Conf. on Geotextiles, Geomembranes and Related Products, Special Lecture & Keynote Lectures*, p. 3, 1998.

Knipschild F.W., *Proc. of Int. Conf. on Geomembrance*, **2**, 739 (1984).

Imaizumi S., Yokoyama Y., Takahashi S. and Tsuboi M., *Proc. of the 12th Southeast Asian Geotechnical Conference,* **1**, 57 (1996).

Imaizumi S., Futami T. and Nomoto T., *Proc. of the 12th Southeast Asian Geotechnical Conference*, **1**, 325 (1998).
Koerner R.M., *Designing with Geosynthetics*, 4th edn, Prentice-Hall Inc., 1998.
Janson L.E., *et al., ASCE Int. Conf. on Underground Plastic Pipe*, 1981.
Kimura K., *Engineering Materials*, **48** (10), 49 (2000).

## *3.4 Frontiers of superfiber application*

### *para*-oriented aramid fibers – Kevlar

Cooper J.L. and Sakai H., *Sen'i Gakkaishi*, **43** (4), 125 (1987).
Kwolek S.L., *Liquid Crystalline Polyamides*, 179th Am. Chem. Soc. Natl. Meet, Houston (1980).
Du Pont, Japanese Patent (Examined) 1980-14170, 1975-8474.
Tanner D., *Proc. of the Int. Sym. on Fiber Sci. and Tech.*, Hakone, 1985.
Dobb M.G., *et al., Polymer*, **22** (7), 960 (1981).
Koralek A.S., *et al., 19th Annual OTC*, Houston, 1987.
Ghosh T.K. and Barker R.L., *J. Ind Fabrics*, **4**, 20 (1986).
Hoiness D.E. and Frances A., *32nd Int. SAMPE Sym. & Ex.*, 1987.

### *para*-oriented aramid fibers – applications of aramid fibers to digital communication device and cable

Sakamoto K., *et al., The Evolution on Continuing Development of ALIVH High-Density Printed Wiring Board*, IPC Printed Circuits EXPO Program Committee, April 2000.
Kato S., *Sen-i Gakkaishi*, **43** (4), 130 (1987).

### *para*-oriented aramid fibers – Technora

Teijin, Japanese Patent (Examined) 1977–39719.
Hiratsuka S., *TEIJIN TECHNOLOGY REPORT*, 1985.
Murayama S., *et al., Sen'i Gakka-shi*, **13** (1), 58 (1998).
Noma T., *Sen'i Gakkaishi*, **56** (8), 245 (2000).
Fujita K., *NIKKEI ELECTRONICS*, **1** (13), 15 (1997).

### *para*-oriented aramid fibers – Towaron

Picken S.J., *Macromolecules*, **22**(4), 1766 (1989).
Picken S.J., Aerts J., Visser R. and Northolt M.G., *Macromolecules*, **23** 3849 (1990).
US Patent 3,767,756 (1973), US Patent 3,869,429 (1975).
Yang H.H., *Aromatic High-Strength Fibers*, John Wiley & Sons, 1989.
Aramid Profile: U.S., R.M. Kossoff & Assoc., Inc.

### Polyarylate fiber (Vectron)

Ueda K., *Sen'i–Gakkaishi*, **40**, 135 (1987).
Nakagawa J., *Proc. of Int. Conf. on Advanced Fiber Materials*, p. 31, 1999 Ueda, Japan.

Nakajima T., *Advanced Fiber Spinning Technology*, pp. 161–171, Woodhead Publishing Ltd., Cambridge, 1994.
Nobile M.R., Amendola E., Nicolais L., Acierno D. and Carfagna C., *Polym. Eng. Sci.*, **29**, 244 (1989).
Blizard K.J. and Baird D.G., *Polym. Eng. Sci*, **27**, 653 (1987).
Mithal A.K., Tayebi A. and Lin C.H., *Polym. Eng. Sci*, **31**, 1533 (1991).
Kiss C.G., *Polym. Eng. Sci*, **27**, 410 (1987).
Crevecoeur G. and Groeninkx G., *Polym. Eng. Sci*, **30**, 532 (1990).
Dornheim M.A., *Aviation Week & Space Technology*, 14 July 1997.

## PBO Fiber (Zylon)

Yabuki K., *Proc. of Int. Conf. on Advanced Fiber Materials*, p. 27, 1999 Ueda, Japan.
Kato K., *Kobunshi*, **48**, 20 (1999).
SRI, US Patent 4, 225, 700; 4, 533, 692; 4, 533, 693.
Teramoto Y. and Yabuki K., *Kogaku-Sochi*, **8**, 83 (1997).
Radler M.J., *et al., Polym. Mater, Sci. Eng.*, **71**, 328 (1994).
Wierschke S.G., *Mater. Res. Soc. Symp*, **134**, 313 (1989).
Wierschke S.G., *et al., Polymer*, **33**, 3357 (1992).
Kuroki T., *et al., J. Appl. Polym. Sci.*, **65**, 1031 (1997).
Fujishiro H., *et al., Jpn. J. Appl. Phys.*, **36**, 5633 (1997).
Kitagawa T., *et al., J. Polym Sci. Polym. Phys.*, **36**, 39 (1998).
Yabuki K., *Sen'i Gakkaishi*, **54**, 16 (1998).

## Polyethylene

Pennings A.J., *et al., Colloid Polym. Sci.*, **253**, 452 (1975).
Lemmstra J., *et al.,* in *Developments in Oriented Polymers: 2*, I.M. Ward (ed.), p. 39, New York, 1987.
DSM, Japanese Patent (Examined) 1985-47922, 1989-8372, 1989-4887 etc.
Ohta Y., *Sen'i Gakkaishi*, **54** (1), 8 (1998).
Ohta Y., *Series of Basic Lectures on Fiber & Textile*, p. 66, The Society of Fiber Science and Technology, Japan, 2001.
Ikaga T. and Ohkoshi Y., *et al., Sen'i Gakkaishi, **58**,* 16 (2002).
Yamazaki Y., *Proc. of the 6th Spring Seminar* p. B1, The Textile Machinery Society of Japan, 2000.

## *3.5 Nanofiber (carbon nanotube)*

Dresselhaus M.S., Dresselhaus G., Sugihara K., Spainm I.L. and Goldberg, H.A. *Graphite Fibers and Filaments*, Spring-Verlag, Berlin, Heidelberg (1988).
Oberlin M., Endo *et al., J. Cryst. Growth*, **32**, 335 (1976).
Endo M., *Chem. Tech.*, p. 568, ACS, 1988.
Speck J.S., Endo M. and Dresselhaus M.S., *J. Crystal Grouwth*, **94**, 834 (1989).
Endo M., *et al., Sen'i Gakkaishi*, **51**, 412 (1998).
Ohya A., *Engineering Materials*, **50** (4), 98 (2000).
Dresselhaus M.S., Dresselhaus G. and Eklund P.C., *Science of Fullerenes and Carbon Nanotubes*, p. 764, Academic Press, 1995.

Endo M., in *CHEMTECH*, pp. 568–576, ACS, 1988.
Iijima S., *Nature*, **354**, 56
Endo M., *Proceedings of 1st Int. Congress on Nanofiber Sci., Tech.*, 'Aim for the practical Application' p. 2 On Fiber Sci. Tech. Japan, June 28, 2004.

## *3.6 High polyketone fiber*

*Nikkei Industry Newspaper*, 24 July 2001.
Morita T. and Kato J., JPNPAT 2001-115007.
Taniguchi R. and Kato J., JPNPAT 2001-295134 and other 10 materials.

# 4 Carbon fiber expands towards the twenty-first century

## 4.1 PAN-based and pitch-based carbon fiber lead the world

Carbon fiber was developed in Japan as a new material some 30 years ago. This fiber can be distinguished as polyacrylonitrile-based (PAN-based for short) and pitch-based carbon fibers. The main chemical component of carbon fiber is carbon, but the function of carbon fiber depends on the structure of the fiber. Using their unique properties, the application of carbon fiber is most pronounced in the superfibers category. Carbon fiber is used in Europe and the United States mainly in the aviation and space industries. It is the physical and mechanical functions of carbon fiber, which are harnessed in resources saving, energy saving, and protection against disasters. However, new biological applications have enabled their expansion into new fields such as control of aqueous environments. Moves to create thin, light, and strong fibers, can offer new areas of development to meet the needs of society. The changes in the global carbon fiber market are illustrated in Fig. 4.1.

### 4.1.1 Carbon fiber in high technology

PAN-based carbon and pitch-based carbon fiber, derived from pitch or tar as a byproduct of the carbonization of coal or in refining petroleum, have different applications. Pitch-based carbon fiber can be used for general-purpose products (isotropic) and high performance products (anisotropy). Carbon fiber is also classified according to the properties, whether high strength or high modulus fiber.

Generally carbon fiber is a conflux of thin fibers (filaments) with diameters of 5–8 microns. The conflux is called tow. According to the number of filaments in tows, it is classified 3K or 6K. K stands for 1000, 3K and 6K, therefore, mean 3000 and 6000 filaments, respectively. 3K and 6K tows are called small tows. On the other hand, large tows, with in excess of 50 000 filaments, have been produced recently to reduce the cost of production. The

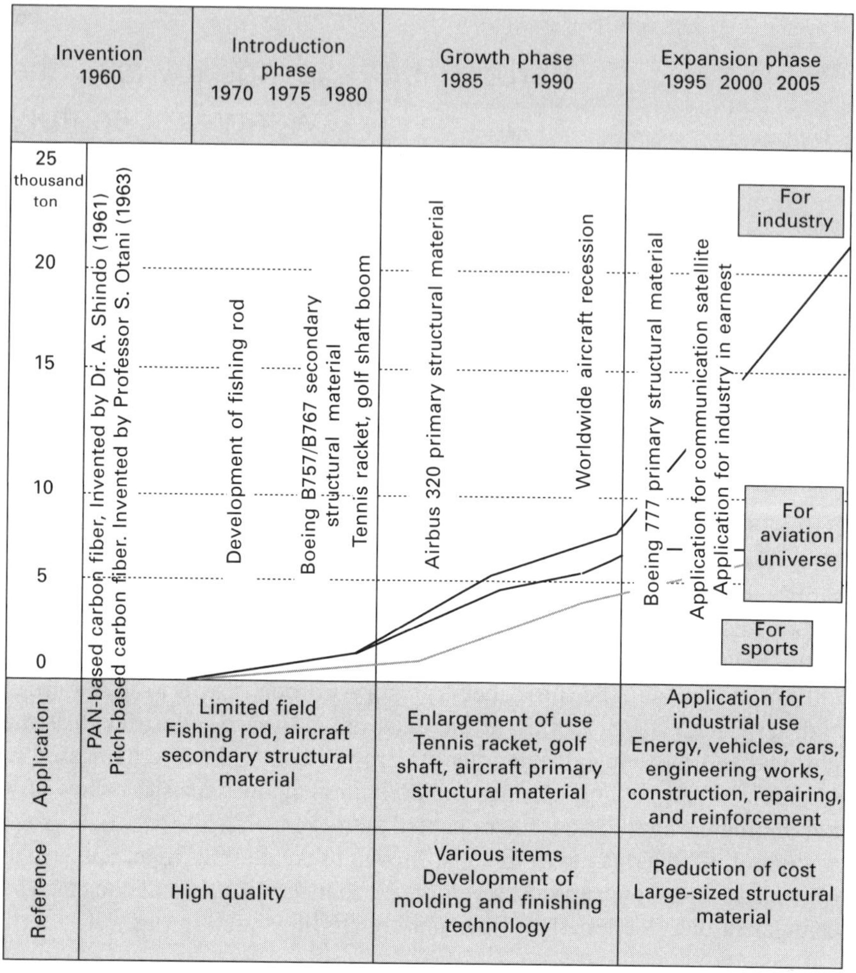


*4.1* Changes in the global carbon fiber market.

large tow is cut into 6 mm lengths, mixed with thermoplastic resin for reinforcement, and thus substitutes for 12K tow.

The production of carbon fiber can be compared to 'charcoal making', but of course it is not so simple. Wood is baked to charcoal over an intercepting air supply. However, fibrous PAN or pitch is processed at high temperature to carbonize. The difference between charcoal making and carbon fiber manufacture is that carbon fiber with different strength and moduli can be produced by changing tow materials and/or giving different heat treatment of the materials. The classification of carbon fibers is shown in Table 4.1.

Carbon fiber is not used on its own, but mixed with resin as a fiber reinforced material. This is called a composite material and is now one of the most important structural and heat resistant materials. The composite material

*Table 4.1* Classification of carbon fiber

| Classification | Species | Main application |
|---|---|---|
| PAN-based<br>Pitch-based | High-performance carbon fiber | Advanced composite material |
| | Carbon fiber for general purposes | Insulating material<br>Electro-conductive material |
| | Activated carbon fiber | Absorbent<br>Electrode material for battery |
| Pitch-based | Carbon fiber developed in vapor phase | Electro-conductive material<br>Electrode material for battery |

made of glass fiber and plastic is distinguished as glass fiber reinforced plastic (G-FRP) from that made of carbon fiber and plastic (carbon fiber reinforced plastic, C-FRP). The first FRP, appearing in the 1940s, was G-FRP, and it was used in familiar household articles. In the 1960s, glass fiber and polyester resin were replaced by carbon fiber and epoxy resin, because of the higher performance and higher heat resistance, respectively. Currently it is a high-tech material, and is used in the field of electricity/electronics, medical care/welfare/care equipment, sports, leisure, aviation/space. More than 10 000 and 2000 tons/year PAN-based and pitch-based carbon fibers are produced, respectively. PAN-based carbon fiber is now established as a structural material.

It is noteworthy that both PAN-based and pitch-based carbon fibers were originally produced in Japan. PAN-based carbon fiber was invented in 1959 by Dr A. Shindo of Government Industrial Research Institute, Osaka (presently National Institute of Advanced Industrial Science and Technology (AIST)). It was first industrialized in 1971 by Toray. Mitsubishi Rayon and Toho Rayon (presently Toho Tenax) started manufacturing in 1983 and 1973, respectively. Pitch-based carbon fiber was invented in 1963 by Professor S. Otani of Gunma University (now Emeritus Professor of Gunma University), and it was industrialized as a general-purpose staple fiber in 1970 by Kureha Chemical Industry. This Japan-based technology is now used world-wide.

### 4.1.2 PAN-based carbon fiber goes around the world

In 1957–1958 the first rayon–based carbon fiber was produced in the United States, mainly from rayon. The Cold War which existed between the United States and the Soviet Union at that time drove the United States on to develop space technology and as a result there was a flourishing collaboration among industry, university and government bodies on carbon fiber research.

Barnebey-Cheney made carbon fiber derived from rayon fiber in 1957. Furthermore, National Carbon (an associated company of UCC) delivered carbon cloth to US air-force material research laboratory in 1958. In May 1959, this news was published by Nikkan Kogyo Shinbun, and Dr A. Shindo read this article and started work on making carbon fiber. In this way the PAN-based carbon fiber was manufactured. In autumn 1996, 37 years after the invention, Dr Shindo was decorated. He used many kinds of fibers to make carbon fiber, and finally he used 'Orlon' (a trade name for acrylic fiber made by US Du Pont) to demonstrate that PAN-based carbon fiber could provide high performance. It took only three months' research after reading the article for Dr Shindo to identify three important heat treatment steps necessary: flame-proofing (heat stabilization), carbonization at 1000°C, and graphitization at 2500°C. In 1959, a patent for 'method to produce carbon manufacture from polyacrylonitrile-based synthetic polymer' was applied for a patent, notified in 1962 and registered in 1963.

A report, 'study of carbon fiber' in the 317th report of Government Industrial Research Institute, Osaka was published in English in March 1961. As a result, a group at the Royal Aircraft Establishment (RAE) in Britain, initiated investigation into the application of composite materials in aircraft structures.

### 4.1.3 Carbon fiber used for impeller blade of jet engines

The National Research and Development Corporation (NRDC) was in charge of making a commercial scale production system using this new material. Courtaulds, the fiber company in the United Kingdom, Morganite, a research and development company in heat resistant materials, the Rolls-Royce Company, the manufacturer of jet engines and cars and the Atomic Energy Research Establishment (AERE), Harwell, UK joined to develop the material. All the British carbon fiber composite material in 1967 was made by batch cycle and thus the propeller blade of an engine was produced.

This propeller blade is blade of the rotor which is at the air intake of the aircraft engine. Unfortunately a bird flew into an engine made of carbon fiber composite material (RB-211) during the test, caused damage and as a result made the blade unusable in aircraft at that time. Subsequently, the design philosophy changed. Epoxy-resin replaced unsaturated polyester, and glass fiber was replaced by carbon fiber, and eventually in the latter half of 1960, a new type of engine (GE-90) was developed by American General Electronics. The engine, which mounted this blade, was made of carbon fiber that entered this field from 1967 to 1970. The news galvanized the acrylic fiber manufacturers.

Because the engine was considered as made of metal only until then, researchers working for acrylic fiber manufacturers were encouraged by the chance arrival of carbon fiber composite, and at one sweep, their entry into the field of carbon fiber began. Dr J. Matsui in charge of development of composite fiber (formerly of Toray) describes in detail the situation in those days in *Reinforcement Plastics*, vol. 43.

## 4.2 A step of development of carbon fiber

The nature of the demand for carbon fiber today has changed greatly from that of the days when carbon fiber was first developed. The full-scale global manufacture of carbon fiber started in 1971 by Toray, in 1973 by Toho Rayon (now Toho Tenax), and in 1983 by Mitsubishi Rayon. All the companies were acrylic fiber manufacturers. About 30 years passed already after Toray began manufacturing. The quantity of current world demand increased to about 20000 tons in 2005. The demand suddenly grew about 1980, but weakened just before 1990, but again began to increase suddenly around 1994.

In this chapter, some representative PAN-based carbon fibers which could be used in the twenty-first century are described. PAN-based carbon fiber is merely described as 'carbon fiber' hereafter. Pitch-based carbon fibers are quite different and will not be discussed here.

### 4.2.1 Advanced composite material (ACM)

Carbon fiber is more often used as composite material than used alone. Composite material, in which carbon fiber is used as reinforcement fiber, is classified according to the associated material used in the matrix, whether resin, rubber, metal, concrete, or ceramic. Here, resin composite material which is much in demand is mainly described.

Matrix resin is divided into so-called thermosetting resin, which is stiffened when treated with heat, and thermoplastic resin. Epoxy resin is a representative thermosetting resin, and it is a main matrix material of C-FRP at present. The reason why thermosetting resin is used is that it is superior in specific strength and specific Young's modulus. Thermoplastic resin is used when keeping the main characteristic of the fiber at a maximum is the principal objective. The so-called super engineering plastics, PEEK (poly ether ether ketone) or PPS (polyphenylene sulfide) are used for this purpose. This material shows superior properties in reducing weight and in electroconductivity. Composite material is used to enhance the characteristics of the component materials, and so making up for their individual defects.

From ancient times reinforced composite materials could make a wall from reinforced plaster, with straw, wood and bamboo providing the fiber. Composite material currently has progressed from a simple combination to a high performance material using glass fiber reinforced plastic (G-FRP) in the first generation (1940 to 1960) and carbon-fiber reinforced plastic (C-FRP) in the second generation (1960 to 1980s). G-FRP was used for personal applications such as helmets and bathtubs; C-FRP was used for aviation/space material, and the use developed rapidly.

Application examples are shown in Table 4.2 in the fields of sports, leisure, aviation/space and general industry. The resin-based composite materials

*Table 4.2* Application of carbon fiber in various fields

| Field | Species of resin | Application | Properties | Application field |
|---|---|---|---|---|
| Sports/ leisure | CFRP | Fishing goods | Lightweight, rigidity, sensitivity | Fishing rod, reel |
| | | Golf | Lightweight, rigidity, sensitivity | Shaft, head, face |
| | | Racket | Lightweight, rigidity, sensitivity | Tennis, badminton, squash |
| | | Ocean | Lightweight, rigidity, vibration retardation | Yacht, cruiser, boat for race |
| | | Others | Lightweight, rigidity, vibration retardation | Bat, ski, ski pole, sword for kendo, bow, radio-controlled car, ping-pong table, billiards table |
| | CFRTP | Fishing goods | Lightweight, rigidity, corrosion resistance | Reel |
| | | Golf | Lightweight, rigidity, design | Head, shoes |
| | | Racket | Lightweight, rigidity, design | Racket |
| Aviation Space | CFRP | Airplane | Lightweight, fatigue resistance, heat resistance | Primary structural material: wings, tail assembly, main body |
| | | | Lightweight, fatigue resistance, heat resistance | Secondary structural material; aileron, vertical rudder, elevator |
| | | | Lightweight, fatigue resistance, heat resistance | Interior material; Floor panel, beams, lavatory, seat |
| | | Rocket Artificial satellite | Lightweight, fatigue resistance, | Nozzle cone, motor case |
| | | | Lightweight, fatigue resistance, dimension stability | Antenna, solar battery panel, tube truss structural material |
| General industry | CFRP | Car | Lightweight, high-speed, fatigue resistance | Propeller shaft, racing car, CNG tank |
| | | Bicycle | Lightweight, high strength | Frame, wheel, handlebars |
| | | Vehicle | Lightweight, rigidity | Rolling stock, linear motor car rolling stock, seat |

*Table 4.2* (Continued)

| Field | Species of resin | Application | Properties | Application field |
|---|---|---|---|---|
| | | High-speed solid of revolution | Fatigue resistance, rigidity corrosion resistance, light-weight, high-speed | Centrifuge rotor, roller for industrial use, shaft |
| | | Parts of electrical equipment | Lightweight, rigidity, vibration absorber | Parabolic antenna, speaker |
| | | Pressure-resistant container | | |
| | | Medical instrument | Lightweight, high strength | Hydraulic cylinder, cylinder |
| General industry | | Engineering works/construction | Lightweight, X-ray penetration | X-ray grid, surgical instrument, wheel-chair |
| | | Others | Lightweight, corrosion resistance | Cable, concrete-reinforcing material |
| | | | Lightweight, dimension stability, electroconductivity | Mold for resin, umbrella, helmet, plane heating element |
| | CFRTP | OA/office instrument | Rigidity, electroconductivity | Bearing of printer, cam, housing stand |
| | | Parts for electronic/electrical instrument | Lightweight, rigidity, electroconductivity, high accuracy | VCR parts, CD parts, IT applications |
| | | Parts of machine | Lightweight, rigidity, abrasion resistance | Bearing, gear, cam, bearing retainer |
| | | Precision instrument | High strength, dimension stability | Camera parts, plant parts |
| | | Others | High strength, vibration absorption | Sound speaker cone, glasses frame |

CFRP: Carbon fiber reinforced plastics
Carbon fiber reinforced thermosetting plastics: Mainly used for materials requiring high tenacity and modulus
CFRTP: Carbon fiber reinforced thermoplastics
Carbon fiber reinforced thermoplastics: Mainly used for materials requiring lightweight and electroconductivity

need to have enough strength and dimension stability when they are used in daily life. However, heat resistance and abrasion resistance of these materials are not sufficient to be used in the next generation of supersonic passenger aircraft. But in recent carbon fiber advanced composite materials (ACM), it is their strength, rigidity, modulus of elasticity raised rapidly, which are in the forefront. When an aircraft is lightened using ACM, flight performance and fuel efficiency increase. Inevitably then, an increasing quantity of carbon fiber will be used in aircraft serving to increase size of future aircraft. The same thing will occur in the space sector, since reducing the weight leads to energy saving. On this account, aluminum alloys or light metals are reinforced with carbon, boron, silicon carbide fiber, and thus metal-based composite material can be developed to make lightweight, heat resistant, and abrasion resistant material, possessing superior properties. Boron fiber reinforced aluminum which first appeared around 1980 is a metal-based composite developed as a third generation material. It was used as a structural brace for the fuselage of the space shuttle and as a car component plunger because of its superior heat resistance. It is also going to be used for the development of the Orient Express 'super sonic transport' linking New York and Tokyo in the future. In this way, the development of composite materials from the first generation to the third generation is accompanied by a high performance and a high price. However, a price will be naturally reduced if applications spread, and it can be mass-produced.

Advanced composite material (ACM) such as (1) resin composite material, (2) metal composite material, and (3) ceramic composite material are currently the main products. Resin-based advanced composite material can be made by autoclaving and molding the layered intermediate material (prepreg) by impregnating carbon fiber with high strength and high modulus epoxy resin. This material is lightweight, has strength and resilience which exceeds that of aluminum alloys, but remains more expensive than aluminum alloys.

### 4.2.2 Applications in fishing rod, golf and tennis in Japan, and aircraft/space in Europe and the United States

Originally the golf shaft was made of steel. Then came glass fiber. Carbon fiber is superior to glass fiber in its vibration damping. For this reason glass fiber was replaced by carbon fiber comparatively early. The technique of making a fishing rod with carbon fiber was applied to a golf shaft, because the properties required of golf shafts were to be light, thin, minimum torsion, durability, feel of the driven ball, and a long carry. The properties of golf shafts thus improved quickly and the so-called 'black shaft revolution' started. The demand grew rapidly in the early 1970s.

In 1976 application to the tennis racket started and Kawasaki Racket produced rackets in Japan. At that time the share of goods using carbon fiber rose to 70% in the field of sports and recreation in Japan. Among them, fishing rods, golf shafts and tennis rackets consumed 90% of the production of carbon fiber (see Fig. 4.2).

As the material has high physical properties such as specific strength, specific Young's modulus and high reliability, freedom of design made it possible to develop materials with new performance which was impossible with conventional materials.

In the later half of 1970s, carbon fiber begun to be used in airplanes and for munitions in Europe and the United States. Export of munitions was not allowed in Japan at that time. Accordingly the demand for carbon fiber changed from military jets in the latter half of the 1970s, to private passenger aircraft in the 1980s.

### 4.2.3 Application to commercial aircraft cabin

The development of a commercial aircraft cabin takes about ten years. Boeing began to use carbon fiber in its 757 and 767 type planes during 1975 to 1980. Lightening of an aircraft by using composite material is shown in Fig. 4.3. There are two ways of using carbon fiber in aircraft. One is in the primary structure, and another is as secondary structural material. Primary structural materials are used for parts such as main wing or tail assembly. The secondary structural materials are used for parts such as floor and other internal sections and operating systems.

Subsequently the European airbus (Airbus Industries) adopted C-FRP for their primary structural materials (vertical tail). Ironically, a much higher technology is demanded for private passenger aircraft than for military aircraft. Construction of private machines comes close and here the demands are severe in achieving the necessary safety factors.

The United States increased armaments expenditure after the attack on Afghanistan by the Soviet Union in 1982. Then there was a great increase in use of carbon fiber which spread swiftly into the sports market. Demand forecasting of PAN-based carbon fiber is shown in Fig. 4.4.

### 4.2.4 CF composite material specification announcement for 'Boeing 777 type'

Boeing, in particular, decided to use carbon fiber composite material in the primary structure of passenger aircraft, and 'a specification for carbon fiber composite material to use to Boeing 777 mode' was announced to material manufacturers throughout the world. They identified the characteristic material necessary for aircraft, and progress of the carbon fiber composite material which was usable as primary structural material was achieved.

Fishing rod appeared in the first half of 1970

Tennis racket appeared in the latter part of 1970

Black golf shaft appeared around 1973

*4.2* Three main uses of carbon fiber (Daiwaseiko).

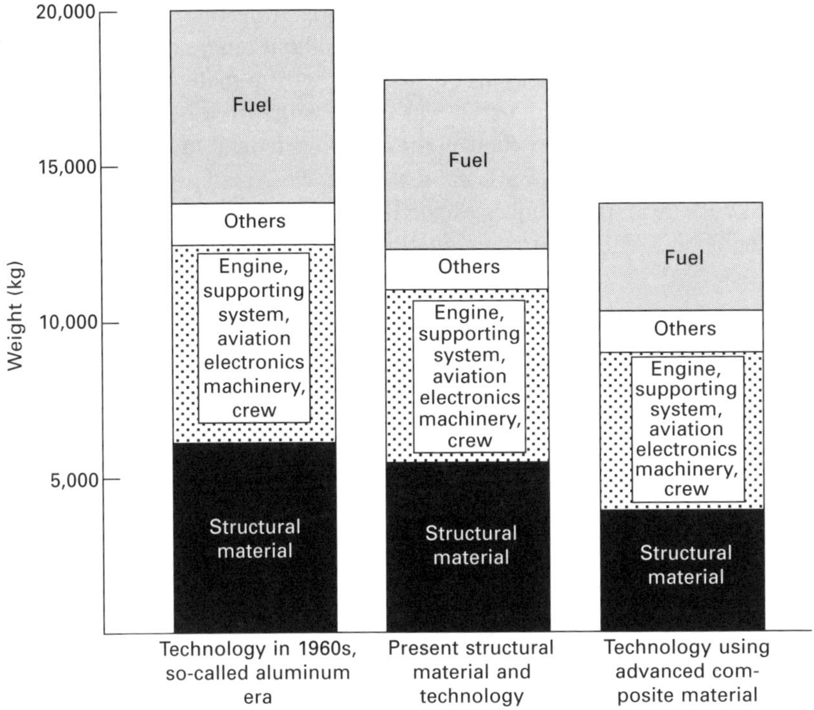


*4.3* Lightweight aircraft made of composite material (New material user's book).

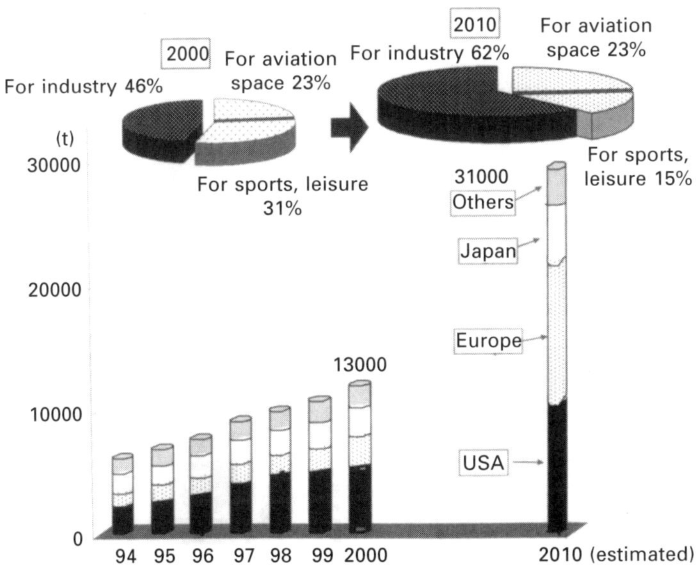


*4.4* Demand prediction of PAN-based carbon fiber.

Carbon fiber composite material prepreg 'P-2302' from Toray was authorized for this work in 1989. This development took about ten years. In 1995, 13 tons, at a use rate of 15%, of carbon fiber composite material was used to make one Boeing 777. The material was used for the floor beam and part of the tail assembly. In this way the weight of the horizontal tail of the Boeing 777 type plane decreased by 25% and that of the whole tail assembly by 15% compared with the corresponding use of aluminum alloy. The assembly plant of the Boeing 777 and the implementation are shown in Fig. 4.5.

### 4.2.5 The American 'black shaft' main jib

In 1990 an entirely new quality of carbon fibre was developed, meeting, for example, the demands of supersonic aircraft. The development work took place from 1986 to 1987 as the reduction in the tension in the Cold War reduced the demand for carbon fiber in ammunitions. This did not affect Japanese manufacturers directly. However, European and American manufacturers faced big problems, despite the growth in sports and private plane applications for they had greatly increased their production capability. Therefore, the relationship between supply and demand grew out of balance. There was a large recession in airplane construction.

Thus the demand for carbon fiber as a high technology material decreased sharply. But this was only temporary and shortly demand increased again. It was carbon fiber used for the black golf shaft in about 1992 which started the

*4.5* Carbon fiber composite material used for floor girder material of Boeing 777 (Boeing).

rejuvenation. The black shaft boom started in Japan, but was slower in the United States. There are reasons for the development of the market.

*One professional golfer offered an opportunity*

In 1992, 20 years later than in Japan, the carbon fiber golf shaft came into the mainstream in the United States. It started with the professional golfer, Davis Love III, who used a golf shaft made of carbon fiber when he won the US Open Championship. It was Gay Bruier, who used the black shaft when he won the championship at Pacific Ocean Club Masters in 1972 that started the fashion in Japan. It helped also that the price of carbon fiber had fallen dramatically in the latter half of 1980s to 1990. It remained extremely expensive in Japan, so when the carbon fiber golf shaft boom came to the United States, the lower price had an adverse effect on prices in Japan. The black CF shaft was easy to use and as the price became comparable to conventional shafts their popularity increased. It became the savior of the carbon fiber manufacturers at that time. It was expected by the Japanese carbon fiber industry in 1993–1994 that equipment for Boeing 777s would be made using carbon fiber. However, the recession in the aircraft industry lasted into 1995 and 1996.

*Industry applications begin to spread*

In 1990 applications, for aviation/space and sports were about 40% each, and, industry generally, it was only around 20%. Carbon fiber composite material used in sports and recreation is shown in Fig. 4.6. The share of C-FRP was 50%, 27% and 23% for industry, sports, and aviation/space.

New industrial applications began to develop in about 1992–1993, particularly in the field of engineering works/construction. Its use as a repairing material also began to grow.

### 4.2.6 Two major characteristics of carbon fiber

Development of a 'strong fiber' was the dream which fiber engineers pursued. Carbon fiber can be classified by its mechanical properties. There are two representative characteristics: tensile-strength and elastic modulus. Tensile strength is expressed by grams needed to run out 1 denier fiber pulled from end points. Centi Newton per Deci tex (CN/dtex) is used internationally following the International System of Units decision in October 1999. Grams per denier unit was used previously. Elastic modulus indicates degree of hardness. The larger the value, the harder the material. What was not expected was the added strength which carbon fiber provided to fiber reinforced composite material. The nature of this effect depends on the type of carbon fiber. So every company developed various kinds of carbon fiber.

*4.6* Carbon fiber composite material used in ski and ski-poles.

In the mid 1980s Toray developed 'T-1000' which has a maximum strength of 630–640 Gpa. For parts of an airplane, a higher strength is demanded than the high elastic modulus. Development emphasis was on 'how to increase strength'. To achieve this, it is necessary to make a fiber with less solid state defects in the graphite structure. Generally, to produce carbon fiber, the process involves baking of feedstock fiber under strain at 200–300% in air for about ten minutes. Then the temperature is increased to 2000°C at the rate of 1000°C over several minutes in an inert atmosphere. When fiber with high elastic modulus is burnt at about 3000°C, in nitrogen or oxygen the acrylic fiber can be converted to a carbon fiber having a graphite structure without defects and hence is very strong. This is treated to have properties like C-FRP.

### 4.2.7 Japanese enterprises started from acrylic fiber

The reason why three carbon fiber manufacturers in Japan (Toray, Toho Rayon (now Toho Tenax), and Mitsubishi Rayon) are strong in carbon fiber production is that they produce acrylic fiber as a starting material. Courtaulds in the United Kingdom and BASF in the United States withdrew from carbon fiber manufacturing in 1990 and 1993, respectively. Amoco and Hexcell in the United States are companies who are good at carbon fiber production technology. Therefore, they produce or buy acrylic fiber only in small amounts. In contrast Japanese manufacturers developed their manufacturing from acrylic fiber and so achieved an overwhelming share of the world market.

### 4.2.8 Material development

Since carbon fiber seemed to have little practical use initially, universities did not have it in their materials laboratories. It was always compared to metals and its value and strength in composites was not recognized. So it was left to business to undertake the fundamental developments.

Superfiber is currently superior to metal and within a multi-component composite material could be leading the material revolution in the twenty-first century. Such materials are worthy of study by universities.

### 4.2.9 Space developments are connected directly with weight and price

In the field of space exploration the weight of the material is related directly to the price of the operation. For example, an artificial satellite is launched using a Saturn rocket, so if the weight can be reduced there is a price and engineering benefit. Japan entered commercial space exploration with H II type rocket. Since 19 billion yen is required to launch one satellite, the target is to reduce the price by about one-third. When composite material is selected, one of the main constraints is service temperature. According to the Ministry of Trade and Industry (MITI), the body surface temperature increases to around 250°C by aerodynamic heating during supersonic transport (SST) at Mach 2.5, to fly Tokyo to Los Angeles in four hours. The National Space Development Agency (NASDA) failed in its attempt to launch the DH II rocket (No. 8) in November 1999, and Institute of Space and Astronautical Science (ISAS) of the Ministry of Education and Science failed to launch M – V rocket (No. 1) in February 2000, because of abnormality of the number one stage engines. The price of the operations ranged up to 10 billion yen.

Carbon fiber composite material with high modulus has good thermal conduction, and thermal expansion is almost zero. Therefore it is used in radio antennae to contact ground masts. In addition, carbon fiber is used within storage sections of the space shuttle and in the manipulator for its remote control.

### 4.2.10 Development of supersonic transport (SST)

Work in Japan has started on the next generation of supersonic transport. Manufacturers will work on the body and the National Aerospace Laboratory will collaborate with research organizations for the propulsion systems. It is anticipated that the SST which incorporates carbon fiber composite material can overcome the problems which have been associated with the supersonic plane Concorde, and coexist with other aircraft and the global environment. The hope is that the advantages to the life of the nation and the economy will spread globally and contribute to the same extent that new trunk rail transport of Japan has. The planned SST has a capacity of 250–300 seats, cruising speed of Mach 2.0–2.4, and a flying range of 9300–11 000 km. Development research is ongoing, and flight model developments are expected around 2005.

The development of next generation supersonic transport (SST) is shown in Fig. 4.7, and the one-day travel zone in the world is shown in Fig. 4.8 when SST is finally developed. The vision is to travel from Tokyo to New York in two hours.

The airplane would take off from a normal runway, and rise into the sky. The Pacific would be crossed at Mach 8 (8 times the speed of sound, 9000 miles per hour). The race to journey from Tokyo to New York in 7 hours, half of the present 14 hours, has started. Sixty per cent of material used for such a supersonic transport will be composite material, including both organic and inorganic materials, and the base material is carbon fiber.

The development of a new type of engine is being pushed forward separately. The materials used must have exceptional heat resistance, since supersonic transport lifts off vertically and enters into the stratosphere immediately. In

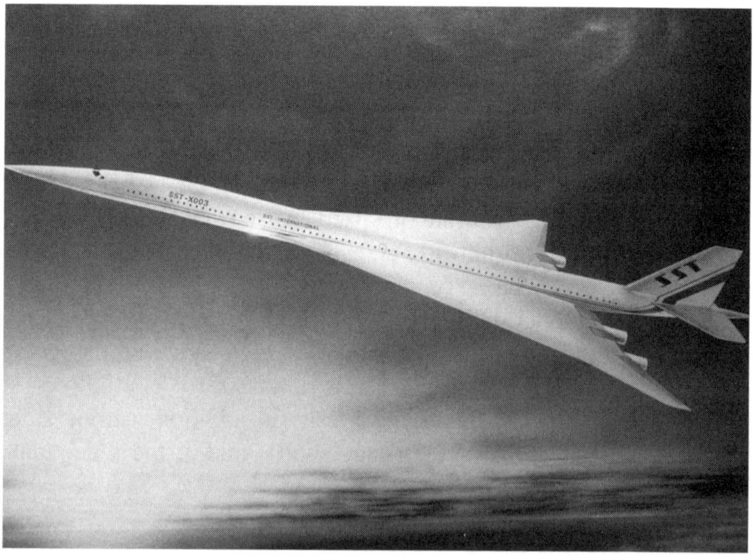


*4.7* Development of supersonic transport for next generation (The Japan Society for Aeronautical and Space Sciences).

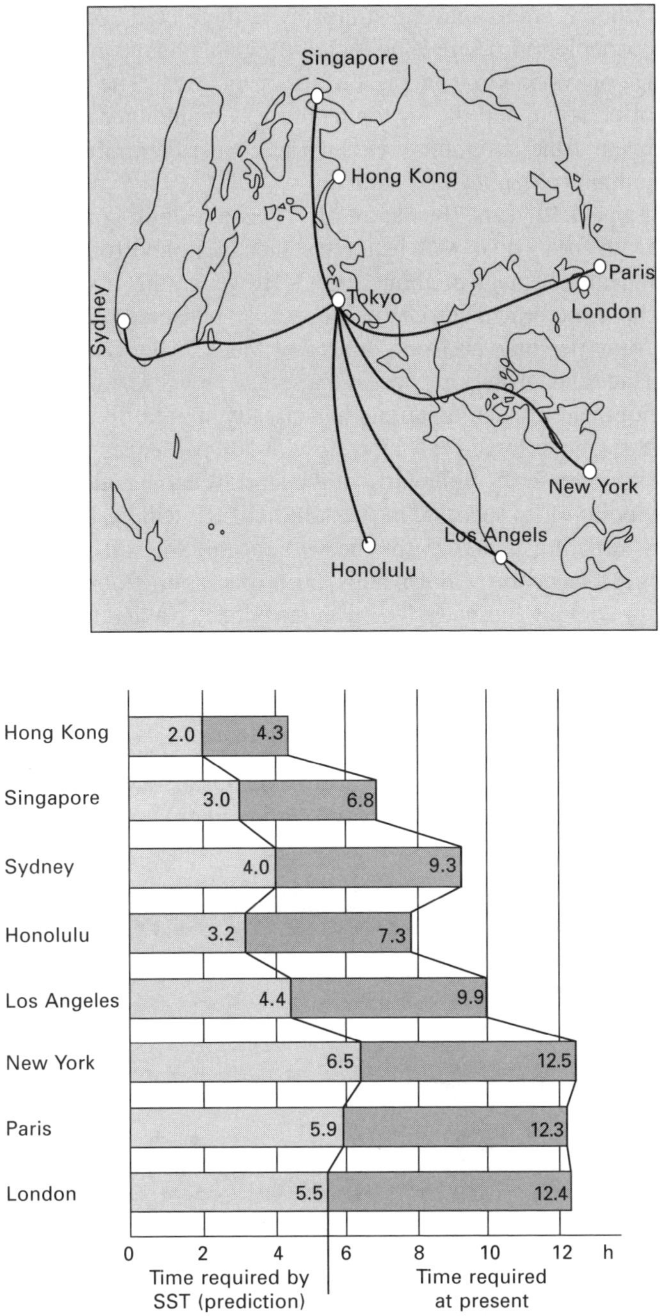


*4.8* One-day air travel ranges from Japan using new generation SST (Japan Aviation Space Engineering Society).

addition, when it enters into the stratosphere there could be the problem of ozone layer depletion. There is no meteorological change in the stratosphere in the range of 10–55 km from the surface of the earth. The wind blows with uniform orientation, and the air temperature is maintained almost constant. On this account the automotive exhaust gas must be minimized so that an environmental problem can be settled.

Over the past 30 years, the nature of the base resin of composite material changed from epoxy resin with heat resistance of around 150°C to polyimide resin with heat resistance of about 300°C. However, this is not enough to be used as the structural material, because heat resistance of 500–900°C is required for hypersonic transport at around Mach 5. For SST materials high heat resistance and tenacity are necessary, so that new resin must be developed. The development of such materials has already started. In the development of the A380, a super large-scale aircraft with 500 passengers to be developed by Aerospacial by 2005, lightening of the aircraft using composite materials is the key point of its success. In the A380 C-FRP will be used as the main secondary structural material for the vertical tail fin, tail plane, tail cone, main wing, floor beams, floor panels, strut to support floor, and composite material for cockpit and body. The trial to reduce cost and to introduce new molding technology is now ongoing. The A380 is shown in Fig. 4.9.

### 4.2.11 Expansion to geotextiles

After the enormous earthquakes in California, Osaka-Kobe-Awaji in Japan, and Taiwan, the demands for PAN-based carbon fiber for industry application,

*4.9* A380, a super large-sized passenger plane with 500 seats, to be developed by 2005 (By computer graphics, Airbus Japan).

especially in engineering works and construction, have grown quickly. The necessity of strengthening commercial buildings and bridge piers has increased the demand for carbon sheet and cloth. Recently, the market for industrial applications, especially in engineering works and construction, accounts for 50% of carbon fiber production.

Nowadays high performance fiber, including carbon fiber, is used as a replacement for iron materials, particularly where rust and etching are a problem. Using this material also allows a reduction in manufacturing and maintenance repair energy. While with iron, the material is cheap, labour costs are high because of the difficulties in fabricating iron plates at the site and the heavy weights involved.

However, carbon fiber is soft, light and strong in composite with resin. After incorporating carbon fiber into a bridge pier, for example, and fixing with resin, the pier becomes very strong and able to withstand earthquakes. The material may be expensive, but the labour costs are less. Moreover, carbon fiber is light so that it does not overload the original pier. So this construction method now attracts attention. Therefore the quantity used for repair/strengthening in engineering works has increased very much. Examples of its execution and mechanism used are shown in Figs 4.10 to 4.14. It is the reason why the demand for carbon fiber in engineering and construction grew so quickly around 1992–1993.

### 4.2.12 The first building in the world to use a space truss made of C-FRP

In the construction of the dining room of the Toray Ehime factory, the roof space truss made of C-FRP which can be lifted in one piece. The space truss

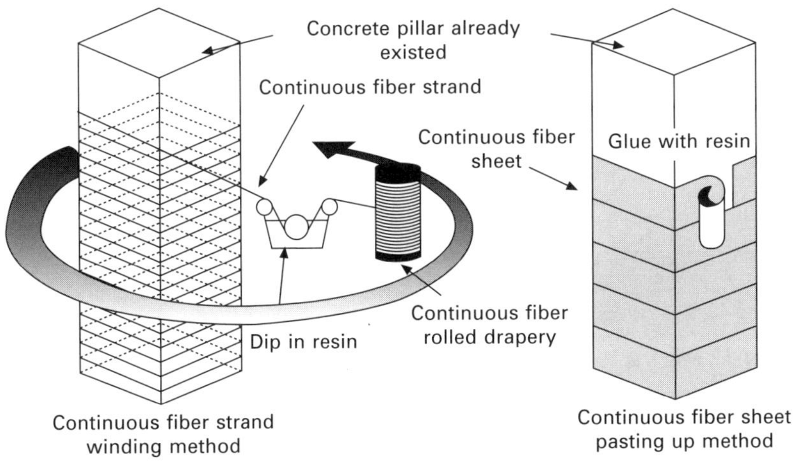


*4.10* A continuous fiber sheet pasting up method and continuous fiber strand winding method.

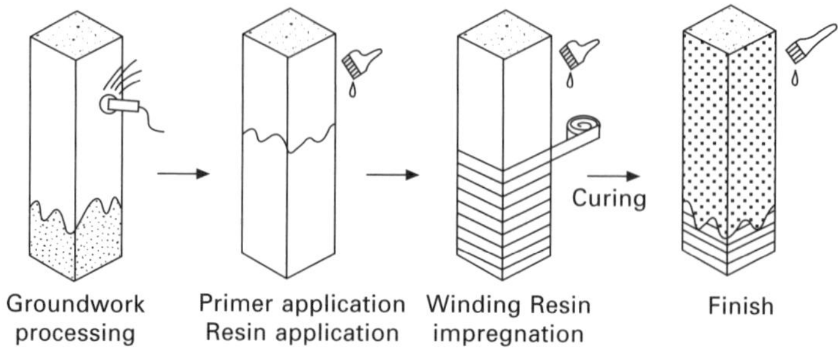


*4.11* An execution procedure of a continuous fiber sheet and tape pasting up method.

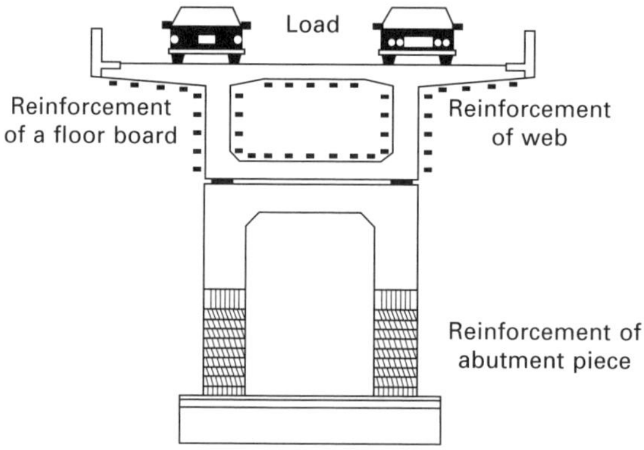


*4.12* Reinforcement of a bridge.

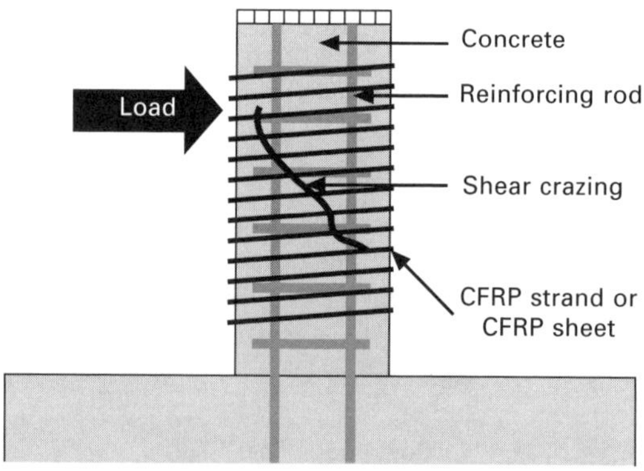


*4.13* Mechanism of shear strength reinforcement.

roof is shown in Fig. 4.15. This truss is composed of a triangle and tetragon made of C-FRP. The truss is 1/30 000 in weight of the structure, superior in corrosion resistance, and the same or greater in strength compared to that

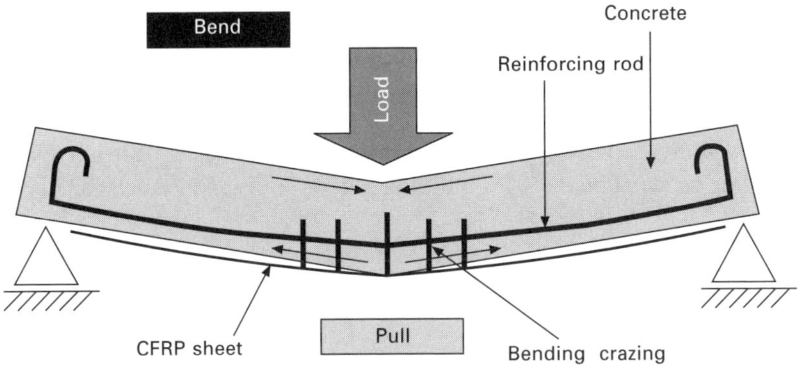


*4.14* Mechanism of bending strength reinforcement.

*4.15* A view of Toray Ehime factory dining room where a solid truss roof made of CFRP was used for the first time (Toray).

made of iron or steel. In addition, it could be built effectively and safely in a short time and the design is geometrically beautiful. Such a space truss made of C-FRP was developed jointly by Toray, Shimizu Construction, and Nippon Aluminum after getting authorization from the Minister for Construction. Six workers made the space truss roof with manual operation in four days, assembling 800 lightweight main parts.

Two 50 ton cranes used in usual construction work were used, and they lifted the parts to about six meters from the ground, and it was installed in units of ten steel columns. The duration of the operation was about 30 minutes. If the same scale roof were constructed using an iron space truss, two 100 ton cranes with a gross weight of 18 tons would be necessary. Such a space truss will be used extensively for building indoor pools and gymnasia, etc. in the future.

### 4.2.13 Application for clean energy

*Wing for wind power generation*

In the United States and Europe, there are many wind power generation facilities which harness clean energy. At the end of 1999 there were more than 200 power generating windmills in Japan. However, there are few places where the wind blows uniformly all through the year. Because there are many islands in Japan, wind power generation is attractive and is being examined for regions where transmission of electricity is not easy. Table 4.3 shows an international comparison of the quantities of wind power/solar power generation. In wind power generation, a propeller is moved by the wind. Thus toxic waste gases such as nitrogen and sulfur oxides are not produced as in thermal power generation by combustion of coal or petroleum. An example of wind power generation is shown in Fig. 4.16.

Only a few places are suitable for power generation because of the need for maximum wind power. Tomamai and Shiribeshi towns in Hokkaido and Tappimisaki in Aomori prefecture are such locations and attention to this form of energy is now being given by local government and private companies.

The CRC in Tokyo, using meteorological data from the Meteorological Agency, has developed a system for predicting wind velocity in every direction

*Table 4.3* Comparison of wind and solar power generation (capacity of facility (1000 kW))

| | Wind power | Solar power |
|---|---|---|
| Japan | 38 | 133 |
| Germany | 2579 | 54 |
| USA | 2055 | 100 |
| UK | 474 | 1 |
| Netherlands | 425 | 6 |
| China | 294 | – |
| Italy | 155 | 18 |
| Switzerland | – | 12 |
| World total | 9841 | 392 |

NEDO (New Energy and Industrial Technology Development Organization)

*4.16* Wind power generation.

and consequently calculating the potential electric power generation. A great deal depends upon the generating efficiency of a blade itself. For this reason, lightweight wings are requested. Thus wings made previously of G-FRP are now being replaced by C-FRP.

### *Strengthening of natural gas storage tank*

Each State in the United States can have its own regulatory systems. Exhaust gas regulation is one example. As a result there are States where exhaust gas regulation is so strict that cars using gasoline are severely hampered. There cars using natural gas became popular. There are many advantages. Natural gas is cheaper and reduces the quantity of harmful exhaust gases. The situation is similar in Canada. In Japan liquefied natural gas is imported from the Near and Middle East and delivered to each house through a pipe in the form of gas. In the United States, it is produced in large quantities and there is no need to liquefy, so that it is comparatively cheap. To transport in gaseous form a strong gas tank is required. An iron gas tank reinforced by carbon fiber is used for this purpose, and is shown in Fig. 4.17.

If all cars in the United States and Canada are eventually converted to use natural gas, the market will become very big.

Natural gas cylinder on the roof

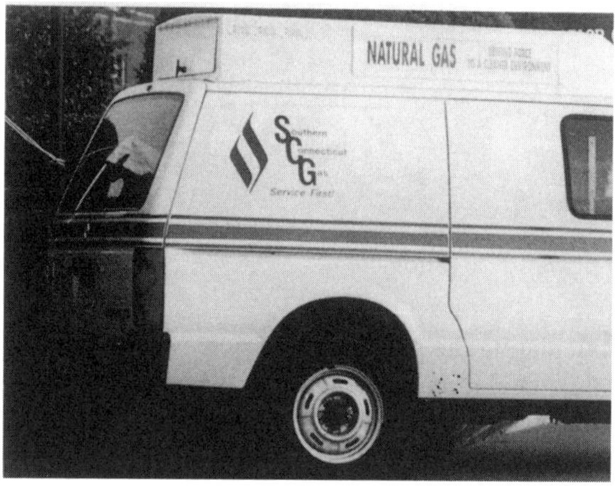


*4.17* Natural gas car.

### 4.2.14 Safety-related fields

Firemen carry a storage tank on their backs for breathing air when fighting a fire, in the way a diver carries an aqualung. Thus the tank for a fireman should be strong and light, and for this purpose carbon fiber is used. To reinforce such a tank about 1 million tons/year CF is now used globally.

Teijin uses lightweight FRP composite for the container tank for firemen and patients at home who need an oxygen cylinder. The company extended its use of natural gas container tanks to small-sized commercial vehicles in 1994, and to low-level Tokyo buses in spring 2000.

Lightweight 'Ultressa' which can lifted by a child is shown in Fig. 4.18. 'Ultressa' is made from a seamless, high strong corrosion resistant aluminum alloy reinforced with epoxy-resin impregnated glass-fiber, and is half the

*4.18* Lightweight air-breathing device made of CFRP (Teijin).

weight of an ordinary iron container. The container, in which glass fiber is replaced with carbon fiber, is 30–40% lighter.

## 4.2.15 Application in other fields

*Beds used for CT (computed tomography) scanning*

Recent medical technology has witnessed a marked enhancement, not only in diagnosis techniques, but also in the equipment used. The CT scanning bed using carbon fiber is one example. Previously, the bed was made of wood or plywood, covered with thick, white cloth to cushion the skin.

Carbon fiber is used instead of wood because of its better X-ray permeation than wood. Polystyrene foam covered with carbon fiber is used and is much less a burden for the human body and clearer X-ray photographs are obtained. The X-ray transparency of carbon fiber is about 10 times greater than aluminum and wood, and on this account carbon fiber is now introduced in such medical X-ray equipment. Carbon fiber is used for most CT scan beds now, and the quantity of carbon fiber used in this field is 100–200 tons per year.

*Lightening of various rollers*

Roller used for printing machines used to be made of metal. The use of carbon fiber has revolutionized the production of the roller in these operations such that they can be managed by one or two people and the danger of handling heavy rollers has decreased. Carbon fiber is also used for rollers in other fields.

*Challenge for America's Cup (2000)*

The yacht used in the America's Cup races had a coxswain of 24 m, a maximum width of 4.5 m, a displacement of 23 t, and huge mast of 34 m in length. The development of new craft was designed to give a lightweight and well-balanced hull and mast. The planning was undertaken by Professor H. Miyata of Tokyo University with a team of 30 researchers from the best marine engineers from five universities and five companies. They spent four years and one billion yen to develop the new craft. They combined leading shipbuilding techniques with seeking new lightweight materials. The result was JPN44 (pet name of Ashura) which performs well in strong wind and the lighter JPN 52 (pet name of Idaten), a narrow-width craft which can move fast in light breeze conditions. Yachts which challenged in the Americas' Cup are shown in Fig. 4.19.

CFRP aramid honeycomb sandwich structure materials were used for the hull, and CFRP and aramid composite used for the mast, boom, and spin pole. The manufacture of the hull was undertaken by Nippon Challenge Building, made up of invited public participants. The Japanese team finished fourth in the preliminary race, and the CFRP technology was proven in a global arena.

### 4.2.16 New applications in the field of aqueous environment

Ecology, aiming at harmonization with nature will be important in the twenty-first century. Applications of carbon fiber to aqueous environments can assist here. For this reason the Ministry of International Trade and Industry (MITI) started their carbon fiber-aqueous environment project in this field. Carbon fiber in water can collect large quantities of microorganisms which can decompose soap (carbon source) and domestic sewage (nitrogen source). In this way microorganisms grow to form large colonies, which can purify

*4.19* Yachts challenging for America's Cup (Mitsubishi Rayon).

wastewater quickly. A direct purification experiment in a river is shown in Fig. 4.20.

Professor A. Kojima of Gunma National College of Technology first observed this phenomenon which led him to utilize resources and techniques of the textile industry to conserve the aqueous environment. How does carbon operate in this way? Professor M. Matsuhashi of Tokai University discovered that an extremely weak signal is given by carbon, which revitalizes a nearby cell which then grows. The subsequent colony of grown cells send stronger

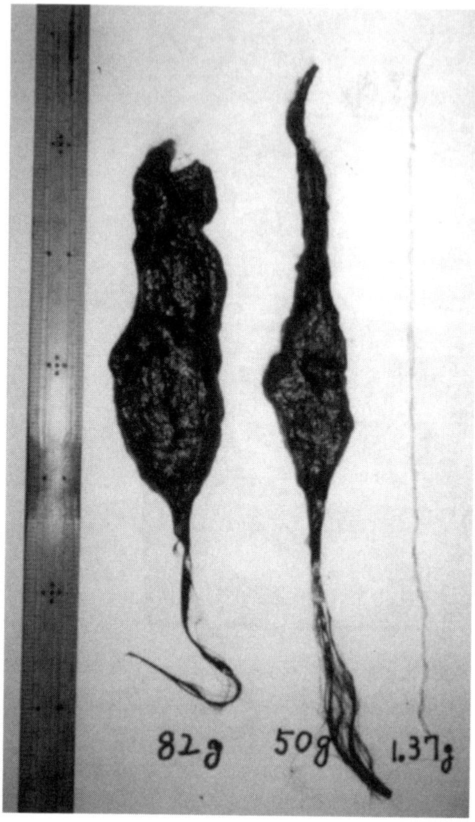


*4.20* Purification of river water by microbe adhered on carbon fiber (Professor A. Kojima of Gunma National College of Technology).

signals which proliferate even more cells. By using his study, relations between microorganism anchoring, propagation and good algae field formation were understood. The surface of carbon fiber becomes covered with a gel-like material which is secreted by the microorganisms or protozoa. Carbon-loving microorganisms anchor and grow. This anchoring and proliferation are promoted by electromagnetic wave promoted by the graphite structure of the carbon fiber.

Algae form within the field of carbon fiber as microorganisms anchor to the monofilament of carbon fiber in large quantities. As carbon fiber moves like waterweed because of its high modulus and large elastic recovery, fish are convinced that carbon fiber is a waterweed. So fish lay their eggs willingly, and a living circulatory system is produced in a naturally occurring form. This phenomenon has never been observed by any other traditional fiber materials, but is well known for natural waterweed. Even when large quantities of microorganisms anchor on to the carbon fiber, water from outside can penetrate into the carbon fiber and microorganisms can continue growing due to the movement of carbon fiber. Professor S. Ohtani, the project leader called this phenomenon the 'net pump action'.

This finding is useful for the purification of water and replanting in desert regions, leading to effective resource and energy saving. Comparing this method to a chemical purification method demonstrates that it is much more eco-friendly. Microorganisms anchoring to carbon fiber have now been observed in rivers, marshes, ponds and domestic sewage. The experiments have also progressed in desert replanting and in the gathering places of fish in the sea. Tests have been repeated in Australia, China, India, and Thailand.

Textile companies in the Kiryu district produce carbon fiber and cloth, knit, and their combination for these developments. The market is big and carbon fiber is set to replace ordinary water purification systems.

## 4.3 The future of PAN-based carbon fiber

Table 4.4 shows the application of carbon fiber in sports in Japan and the US market, but developing more slowly in Asia. For aviation/space, expansion along the Asian route and ultra high capacity passenger transport airplanes is

*Table 4.4* Development of carbon fiber application in 21st century (Toray)

| Usage | Item | Example | Growth rate/year (%) 10 20 30 |
|---|---|---|---|
| Industry | Container for high pressure gas | CNG tank, air-breathing device | |
| | Engineering works/construction | Reinforcement of abutment piece, construction material | |
| | Transportation device | Boat, truck, car, train | |
| | Energy | Flywheel, windmill | |
| | Parts of machine etc. | Roll, pipe, container | |
| | | Compound, medical instrument etc. | |
| | Others | | |
| Sports | Golf | Carbon shaft share 60% → | |
| | Racket | 80% | |
| | Fishing | Developed in Japan and USA | |
| | | Developing in Asian countries | |
| Aviation/ space | Aircraft space satellite | Super jumbo aircraft HSCT (high speed commercial transportation) | |

expected. Most growth is expected in industrial fields such as high-pressure vessels, engineering works/construction, transportation and machine parts. The application developed by NASA and DALPA will be used in energy-related fields. The application of carbon fiber to cars, boats and vehicles will have enough advantage because carbon fiber is light. In France, they experiment to replace iron as the core with carbon fiber to make a lightweight electrical wire. Furthermore, carbon fiber can show the same electroconductivity as silver, when a chemical compound of chloride is introduced inside. It can therefore contribute to energy saving.

Carbon fiber utilized as reinforcement of concrete poles and durability strengthening of buildings following the Osaka, Kobe and Awaji great earthquake disaster have contributed greatly to energy saving. The centrifugal separation device for enriching uranium (revolution drum) for nuclear applications is now made of composite material in Europe, and concentration of the fuel can now be carried faster. The polar plate of fuel cell is now also made of carbon fiber.

In addition, resource mapping satellites are made of carbon fiber composite material as in windmills and tendons in oil rigs to probe oil fields. Secondary lithium cell cathode materials, super-high-speed energy storage equipment will also use carbon fiber. Japanese manufacturers have a 70% and 90% share of PAN and pitch, respectively, and could be the leading producers.

## 4.4 Bibliography

### *General*

Shindo A., *Ceramics Japan* **21**(10), 941 (1986).

Matsui J., *Reinforced Plastics*, **43**(5), 29, (6), 20, (8), 32, (9), 37, (12), 25 (1997), **44**(1), 25 (1998).

Kobatake K., Proceedings of 9th composite material seminar by carbon fiber meeting, 29 (1996).

Sato T., *Journal of High Pressure Institute of Japan*, **35**(39), 11 (1997).

Nakama K., *Polymer Applications*, **47**(7), 19 (1998).

Kojima A., *Engineering Materials*, **47**(47), 52 (1999).

Otani S., Okuda K. and Matsuda S., *Carbon fiber, Kindai Henshuusha* (1983).

Otani S., 'Wonder's Carbon'. *Diamond* (1978).

*The 13th Composite material seminar, 'Carbon fiber – Challenge to Change How to answer the variety of needs'*, The Japan Carbon Fiber Manufacturing Association (2000).

Carbon Fibers for Aerospace Engineering, 2004 EXPO/Fibers for New Era, *Proceeding of Exhibition Symposia at National Museum of Emerging Science and Innovation,* 23 August 2004.

### *PAN-based carbon fiber*

Yamane S., Hiramatsu T. and Higuchi T., *32nd Int. SAMPE Symp.*, 298 (1987).

Noguchi K., Hiramatsu T., Higuchi T. and Murayama K., *Int. Carbon Conf.*, 178 (1984).
Simida A., Ono K. and Kawazu Y., *34th Int. SAMPE Symp.*, 2597 (1989).
Odagiri N., Kishi H. and Nakae T., *Proc. 6th Technical Conf. Of Am. Soc. For Composites*, 43 (1991).
Sato T., *Series of Basic Lectures on Fiber & Textile II*, p. 114, The Society of Fiber Science and Technology, Japan, 2001.

## *Pitch-based carbon fiber*

Otani S., *Carbon*, **3**, 31 (1965).
Otani S., *Mol. Cryst. Liq. Cryst.*, **63**, 249 (1981).
Weinberg V.L., *et al., 15th Conf. on Carbon*, p. 144 (1981).
Ali M.A., Majumder A.J. and Rayment D.L., *Carbon Fiber Reinforced Cement*, **2**, 210 (1972).
Endo M., *J. Mater. Sci.*, **23**, 598 (1988).
Mochida I. and Yanagida K., *et al,, Proc. of Int. Symp. On Carbon*, Tsukuba, p. 50 (1990).
Tomonoh S. and Sawanobori T., *4th Int. Conf. On Composite Interfaces* (1992).
Yamamoto I., *et al., SAMPE Int. Symp.*, **37**, 820 (1992).
*Mitsubishi Kasei R & D Review*, **177**, 198 (1993).

# 5
# High function fiber

## 5.1 Prospects for high function fiber development

Japan is a forerunner in high function fiber development, although in terms of total output of synthetic fiber, Japan is ranked only fifth in the world. The main direction in the high function fiber development is to utilize biomimetics, namely mimicking the high-order structure of natural fibers such as cotton, wool, and silk. Efforts have been focused on enriching human life by pursuing high function, aesthetic appreciation, and human sensitivity. Such fiber is being applied, not only to clothing, but also to industrial applications. In this chapter, the expected developments for high function fibers and associate textiles are summarized and classified.

### 5.1.1 Present status of high function fiber

Functional fibers are developed according to social needs. The function can include health, comfort, and mental satisfaction as summarized in Fig. 5.1.

### 5.1.2 The concept of function will change according to consumer needs

The nature of fiber function can vary in different circumstances and so requires a variety of mechanical, physical, chemical, and biological characteristics, as shown in Fig. 5.2. The broad function can be classified into four categories according to the level of design:

- Basic function fiber (the first order function)
- Higher function fiber (the second order function)
- Super high function fiber (the third order function)
- Intelligent function fiber (the fourth order function)

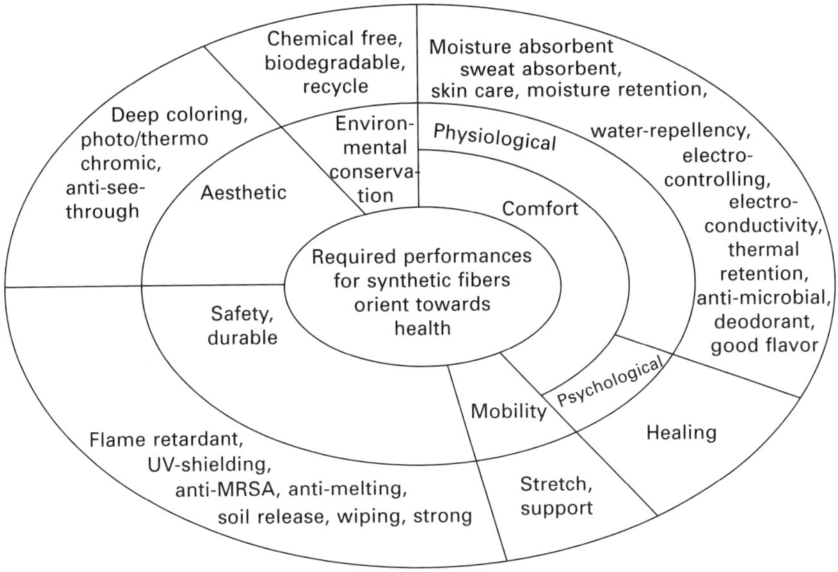


*5.1* Present status of high functional fibers.

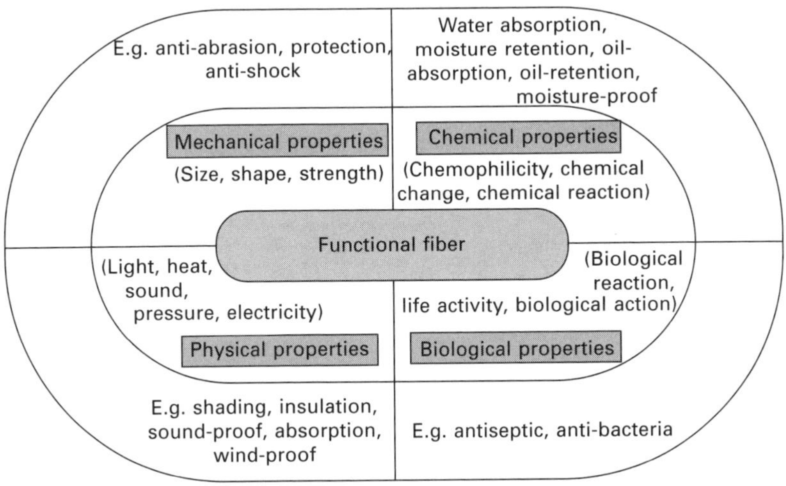


*5.2* Materials for function fibers classified by properties.

Overlapping in multiple fields will be required, as a higher degree of function is structured. The development of fiber functions is summarized in Fig. 5.3 according to these various stages.

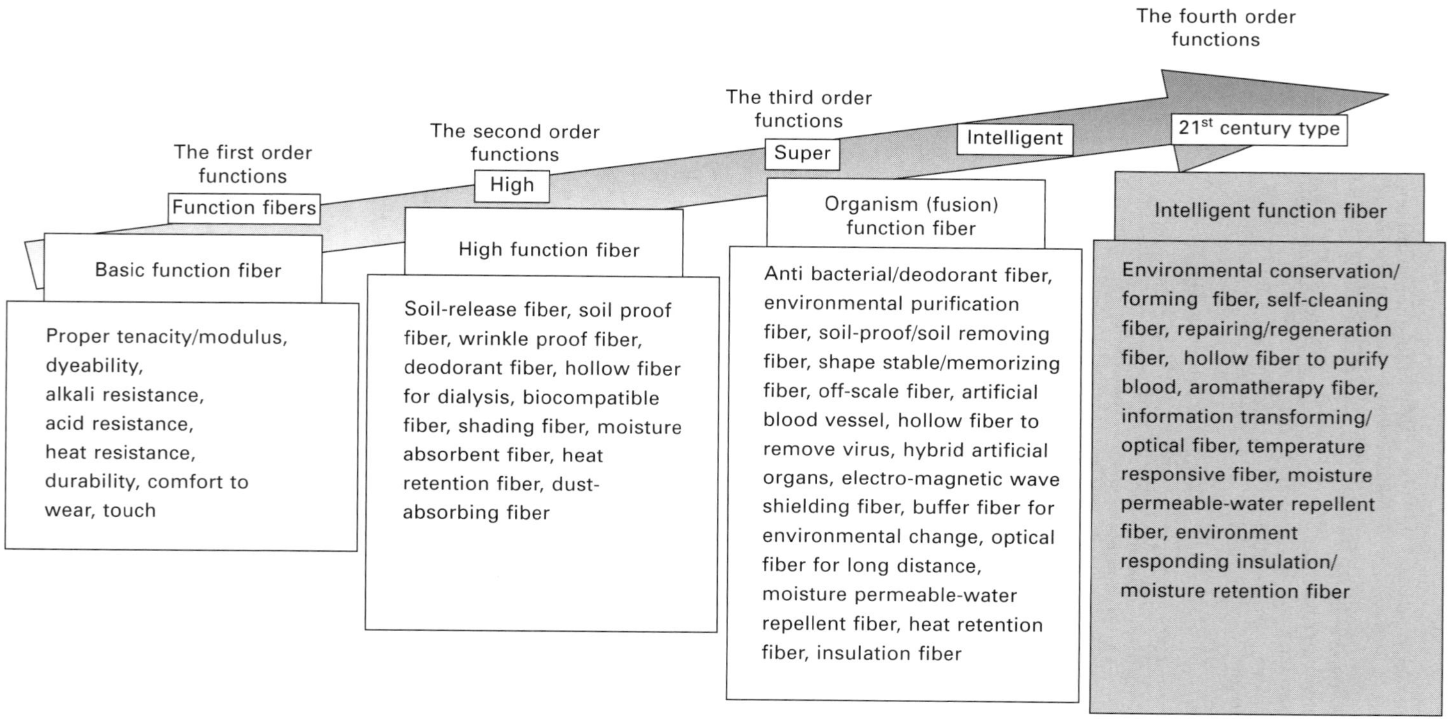


*5.3* Examples of development of fiber functions.

*Basic function fiber (the first order function)*

The basic function of fiber is dependent on the long and thin shape of fiber. Thus the basic function is already present, does not require further functionalization, and is thus referred to as a first order function.

- *Mechanical function*: Typical examples of the basic functions are the tensile strength of fiber, the pressure relaxation by bulkiness, the abrasion resistance and the flexibility. These functions emerge from the chain orientation and crystallization along the fiber axis.
- *Physical function*: The physical function includes the heat resistance, moisture retention, the anti-electrostatic effect, and the transparency. The physical function utilizes the fiber characteristics for thermal, electric, and light stimuli. Conductive fiber is also considered a physical function of fiber. More and more attention is being given to static electricity, which damages computers and contaminates pharmaceutical/food/IC production by collecting dust. Conductive fiber was developed for carpets some twenty years ago in the United States by introducing a metal coating. However, this coating-type is not suitable for clothing and now the carbon composite type is produced by blending high conductive carbon with polyester or nylon.
- *Chemical function*: Chemical functions of natural fibers, such as high moisture absorbency, provide a comfortable in-cloth climate for apparel. Those natural fibers possess high density functional groups on the fiber surface.
- *Biological function*: A weak anti-bacteria effect of silk and wool is a typical example of the biological first order function. Figure 5.4 shows the flow chart of the development of the function fiber towards higher order functions.

*Higher function fiber (the second order function)*

A new function can be added intentionally by utilizing the characteristic shape of fiber according to the needs. Such second order function for clothing is mainly added by processing, to achieve a comfortable in-cloth climate. However, the second order function for industrial use is introduced into the polymer raw materials by molecular design and can involve mechanical, physical, chemical, and biological functions of fiber.

A second order mechanical function is closely related to the shape of fiber. Examples are suede-type artificial leather with ultra-fine napping and silk-like fiber with a triangular cross-section. Biomimetics is a key concept in such function fibers, where the relatively organized structure of biofibers is mimicked by precision spinning and processing.

A second order physical function is exemplified by a light-absorbing fiber, a moisture-retaining fiber, an UV-cut fiber, or a far infrared radiating

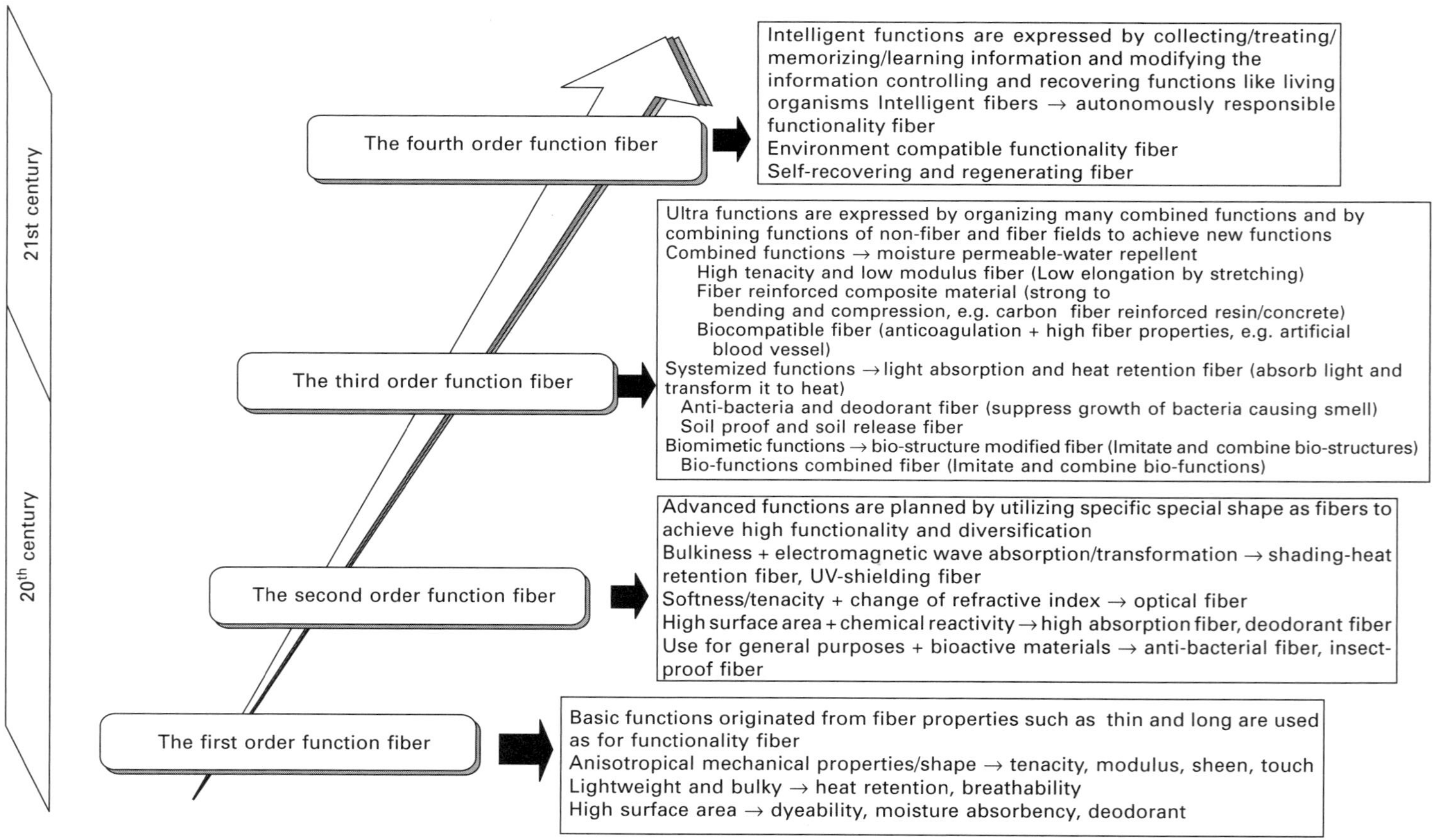


*5.4* Flowchart of technical development for function fiber.

fiber. Those functions are incorporated into fiber at the raw polymer stage.

A second order chemical function fiber includes a hollow fiber for artificial dialysis, a membrane fiber for selective permeation, or an active carbon fiber for selective adsorption of toxic substances. Among these chemical function fibers, a hollow fiber is widely applied for clothes as well as for other goods. A catalytic fiber is also available for chemical reactions, where a catalyst is fixed onto fiber.

A second order biological function fiber is mainly developed for medical uses, such as biocompatible fibers for artificial organs and cell-adhesion/cell-supporting fibers for cell cultivation. Biodegradable fiber and biomimetic anti-bacteria fiber from chitosan or hinokitiol are also included in this category.

There are many examples of water treatments using fiber technologies. Here representative water treatments using fiber technologies are given as examples of second order functional fibers. Today, fiber technologies are applied to water treatments such as separation membranes. Traditional water treatments have used activated dirt, coherent precipitation, and sand percolation; however, new techniques are using filters and separation membranes. The new techniques are applied to preparation of useful water, treatment of wastewater, and cleaning of the environment.

High technology has led to fibers, such as percolation textile made of ultra-fine fibers, ion-exchange fiber using fiber high functionality, hollow fiber membrane technology, dry-and-wet film processes, many hole opening technology using wet film process, and aromatic polyamides as reverse osmosis membrane compound. These have many application fields such as large-scale fresh water preparation from sea water, medium scale of cleaning swimming pool water and small-scale home water purification. Figure 5.5 shows the example of water treatment technologies.

Kinds of separation membranes: Separation membranes are generally categorized by size with the pore diameter determining the distribution achieved. These filters are made from textiles using plane membranes with separation functionalized layers and hollow fiber membrane combined with hole-rich membrane. Separated matter and other water treating technologies are compared in Fig. 5.6.

Decrease in filter hole size leads to microfiltration membrane (MF), ultrafiltration membrane (UF), and reverse osmosis membranes (RO), but pressure required for filtration increases with size decrease and operation costs increase. Therefore the appropriate membrane must be selected according to the nature of the separation necessary.

### Ultra-fine fiber membrane cloth, textile filter cloth

Ultra-fine fibers were developed as suede-touch artificial leather. Now a new function of the fiber is its application as a filter cloth for cleaning dirty water.

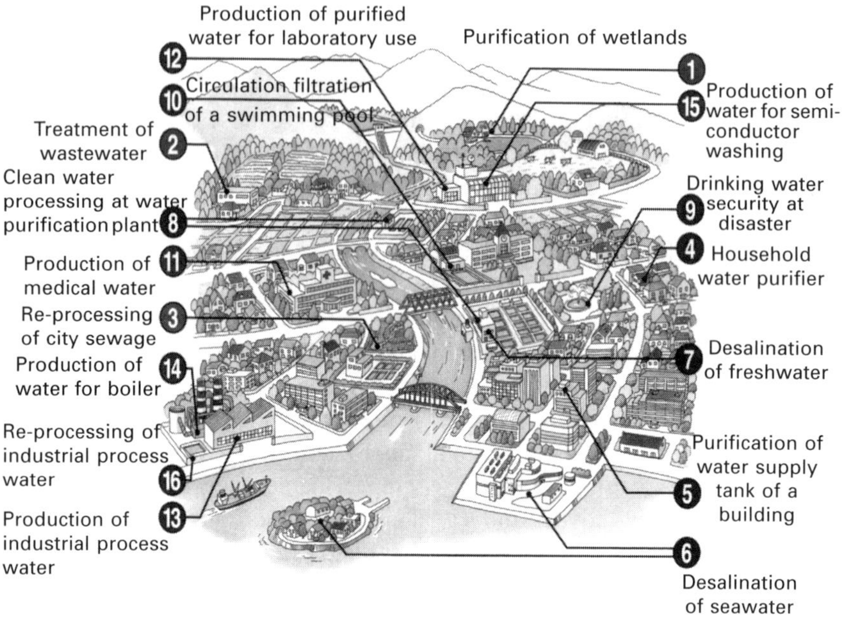


*5.5* Water treating technologies (Toray).

Toray provides the filter cloth Toralome® using ultra-fine polyester fiber.
Requirements for the textile filter cloth are:

- effective filtration of solid particles
- stability in filtration speed
- ease of peel of the cake
- high resistance to tension force (shape stability)
- chemical resistance.

High specification filtration systems require:

- high strength
- long-permanence
- anti-choking
- ability to filter particles with diameter less than a few micrometers.

The cleaned water is recycled possibly as fountains in public locations in towns.

### Fine filtration membrane, hollow fiber membrane (microfiltration membrane)

Hollow fiber membrane is the membrane made from macaroni-shaped fibers whose surface has many ultra-fine holes. When water passes on the outside

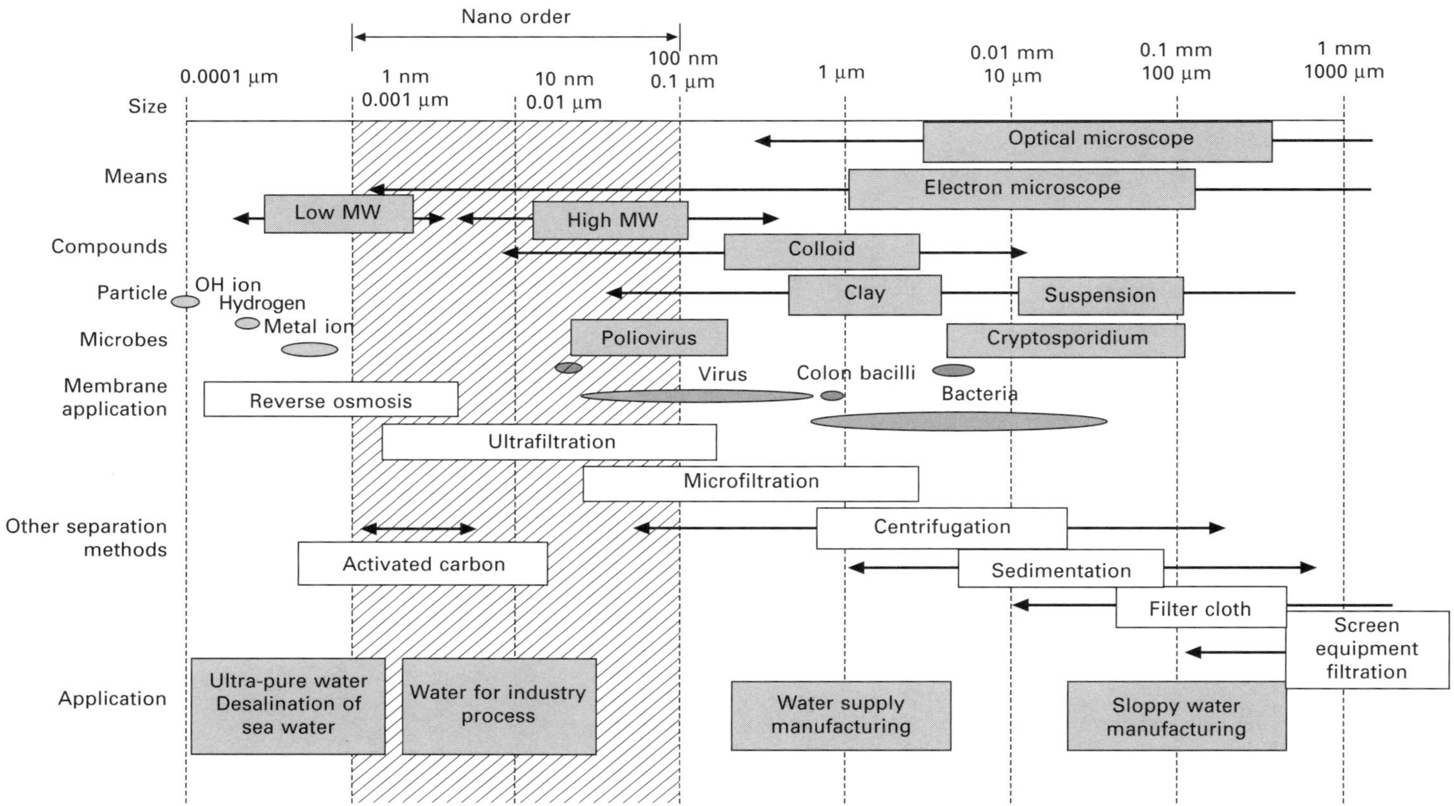


*5.6* Water treatment technologies and separated matters.

of the fiber, impurity included in the water is filtered and the filtered water comes inside the fiber. Thus the diameter of the hole determines the effectiveness of the filtration.

Pore size in the fine filtration membrane is in the order of sub-micro meter. Thus it can filtrate extremely small impurities such as cloud of water, red rust, and bacteria. On the other hand, mineral ions in water such as calcium and magnesium pass through the filter, so the filtered water remains tasty. This fine filtration membrane can be used as a home water purifier, which uses active carbon with the fine filtration membrane. The material of the hollow fiber membrane is polyestersulphone (PSF) which has high strength and high heat resistance.

#### Ultra-fine filtration membrane

As with microfiltration membrane, the ultrafiltration membrane has ultra-fine holes in the walls. The pore size is 0.01 μm less than the size of germs and viruses. Industrial application began with the filtration of germs from beer in 1968. It has been applied subsequently for small particle filtration, purification of liquids, and now the cleaning of water.

The material used for the ultra-fine filtration membrane requires good longevity for use in the water supply system. Therefore, high molecular weight polyacrylonitrile is used. A capsule packing a bundle of hollow fibers called 'Module' is widely used in water supply systems, and used for cleaning water associated with chemical products. Moreover drink water can be prepared by filtration of industrial water.

#### Reverse osmosis membrane

A semi-permeable membrane allows water to pass but does not pass salts. When sea water is treated, osmosis induces the sea water to move to the pure water side without the salts, so pure water can be obtained from the sea water. This is reverse osmosis and the membrane used is called the 'reverse osmosis membrane (RO membrane)'.

Spiral type RO membrane, prepared by rolling the seat-shaped RO membrane into a swirl-shape, has good dirt resistance and produces large quantities of pure water. This is the main type of RO membrane for producing drinking water from well-water and sea water and the preparation of ultra-fine clean water used for the semiconductor industry. Aromatic polyamide and amides are used for the RO membranes.

### *Super high function fiber (the third order function)*

A third order function fiber is defined as a super high function fiber designed to possess high multidisciplinary functions. The third function emerged from an unexpected combination of fiber science with electric/electromagnetic science, machinery/structural material engineering or cell biology. The systematic hybridization leads to a multiplicity of effects and functions. Thus super high function fiber can be regarded as a 'multi-function fiber' or 'hybrid-function fiber'. Organic optical fiber is an example of a super high function fiber in the non-clothing field, which emerged by building into the fiber a gradation of refractive index in a radial direction.

In practice, the function can be classified as multiple function, systematized function and biomimetic function. For example, the multiple functions include the water-repellent/vapor permeable fabric (that repels water but allows vapor to permeate) and the high tenacity/low-modulus elastic fiber (that does not stretch). A systematized function is represented by the heat storage fiber (that absorbs light and converts it to heat) and the anti-bacteria/odor-killing fiber (that suppresses bacterial growth and removes bad smells). Morphotex® (developed by Nissan Motor, Teijin and Tanaka Kikinzoku) is a good example of a biomimetic fiber and is composed of multi-layers of polyester and polyamide, and produces color by the interference effect of light (see Fig. 5.7) like the Morpho found in the upper reaches of the Amazon in Brazil. Biomimetic fibers include fiber whose structure is copied from the bioorganisms and whose function mimics that of bioorganisms.

Toray developed a column to treat blood poisoning with endotoxin-absorbing fiber. Infection by bacteria can cause blood poisoning, because endotoxin is produced by the bacteria. Fever, a barrier to blood circulation and a drop in blood pressure can result. Even death can occur.

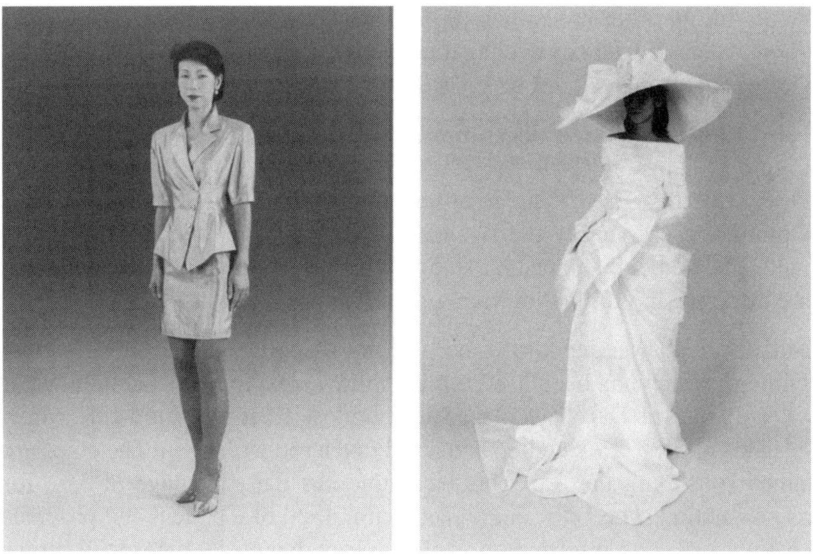

*5.7* Applications of structurally colored fiber, Morphotex®. (Nissan Motor, Teijin and Tanaka Kikinzoku).

The number of patients suffering from blood poisoning is increasing year by year. In the United States, the number of such patients is 430 000 per year. Blood poisoning is the main cause of death in intensive care units. Endotoxin is also known as lipopolysaccharide.

Blood poisoning was once treated by blocking the action of endotoxin. Medicines which combine with endotoxin were given to patients after successful animal trials. However, the medicines did not act efficiently in human clinical trials. So medicines are not now used to remove the toxicity of endotoxin.

Toray developed a treatment to remove the endotoxin by circulating the blood outside of the body using artificial kidney dialysis. In artificial dialysis, hollow fiber is used to remove low molecular weight compounds in the blood and endotoxin can be absorbed.

Endotoxin exists in the blood mixed with other essential components, which must not be removed. Only the compounds causing the disease should be absorbed and removed, so that the purified blood can be fed back into the body. The main difference from dialysis is that compounds with high molecular weighs can be removed by absorption. The compound which combines the causative is called a ligand.

Polymyxin, an antibiotic, was selected as the ligand to remove endotoxin as the result of collaborative research with Shiga University of Medical Science. Although it was known that polymyxin combines endotoxin specifically, it can also cause renal dysfunction if too large a dose is used. The solution was to immobilize the polymyxin on to fiber and the fiber packed in a column through which blood circulating outside of the body could flow. The structure of polymyxin-immobilized fiber and a column packed with the fiber are shown in Fig. 5.8.

As new outside the body cycling treatments for other diseases are developed the following points should be considered:

- Need to identify/determine filtration targets in the blood such as toxin, excess proteins and cells.
- Devise the ligands which can absorb the targets.
- Optimize a carrier for the ligand. In this way, new medical treatment methods can be developed by studying materials, including fibers, which are effective in eliminating target materials.

Mitsubishi Rayon developed the ‘fibrous-shape DNA chip’ utilizing the technology which had been used for the hollow fiber in the household water purifier. Pieces of DNA are placed in a hollow fiber-like ultra-fine straw. Thousands of these are bundled and fixed into a rod with resin. The chips are obtained by slicing the rod. This technique has the advantage of ease for mass production. The DNA chip reacts with blood of a patient and provides the patient’s genetic information, which assists diagnosis. Figure 5.9 shows the production scheme for the ‘fibrous DNA chip’.

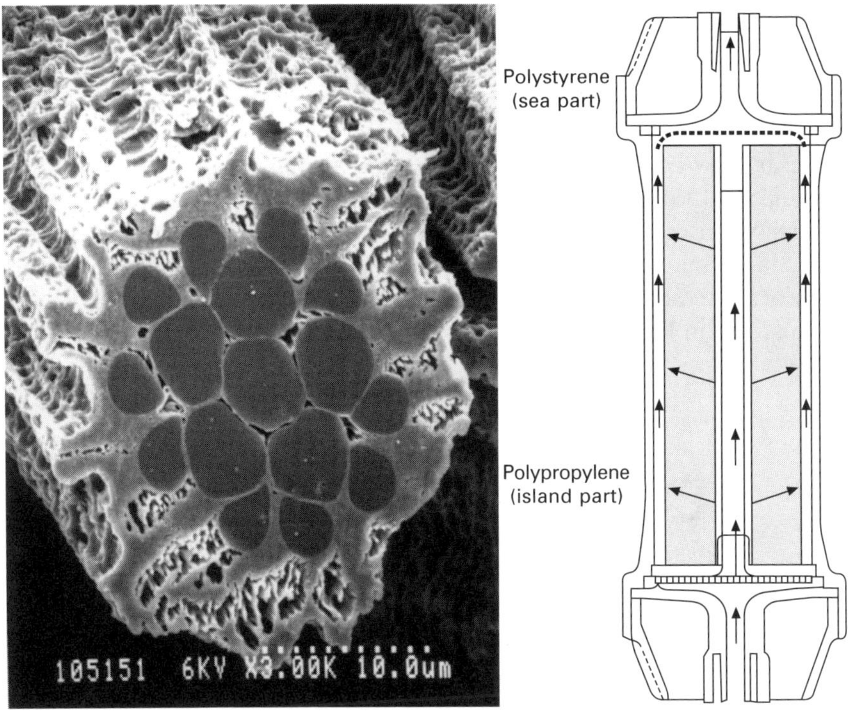


*5.8* Structure of polymyxin immobilized fiber and column packed with the fiber (Toray).

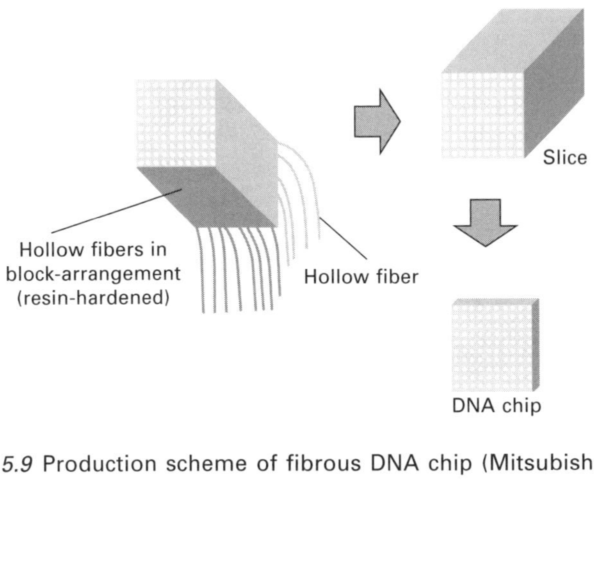


*5.9* Production scheme of fibrous DNA chip (Mitsubishi Rayon).

*Intelligent function fiber (the fourth order function fiber)*

The intelligent fiber can be classified as a fourth order function fiber. The basic concept is to provide the intelligence to reveal a specific function under a specific condition by processing the input information according to the prescribed program. For example, the heat-responsive heat-storage fiber is developed to store heat by converting sunlight, when the light absorption efficiency varies according to the ambient temperature. A typical intelligent fabric in the market is the fabric that opens or closes the texture when hot or cold, respectively. The evolution of the functions in high function fiber is summarized in Figs 5.10 and 5.11.

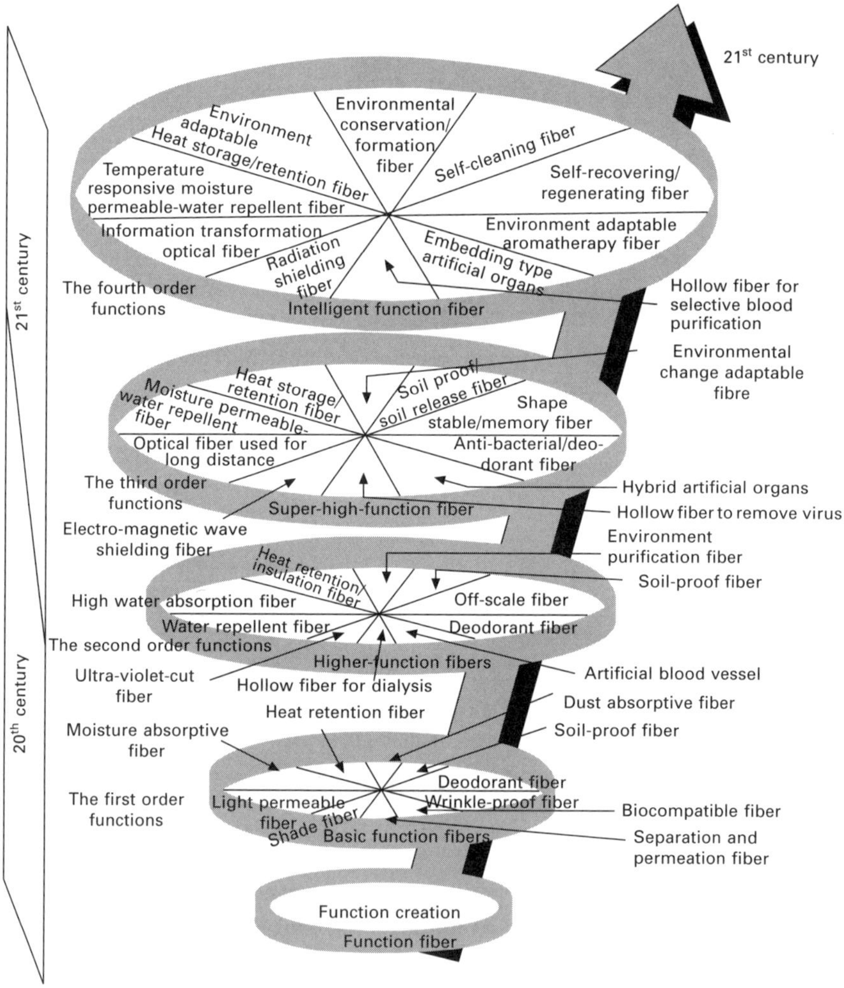


*5.10* Improvement of functions in function fibers (Professor T. Koyama, Shinshu University).

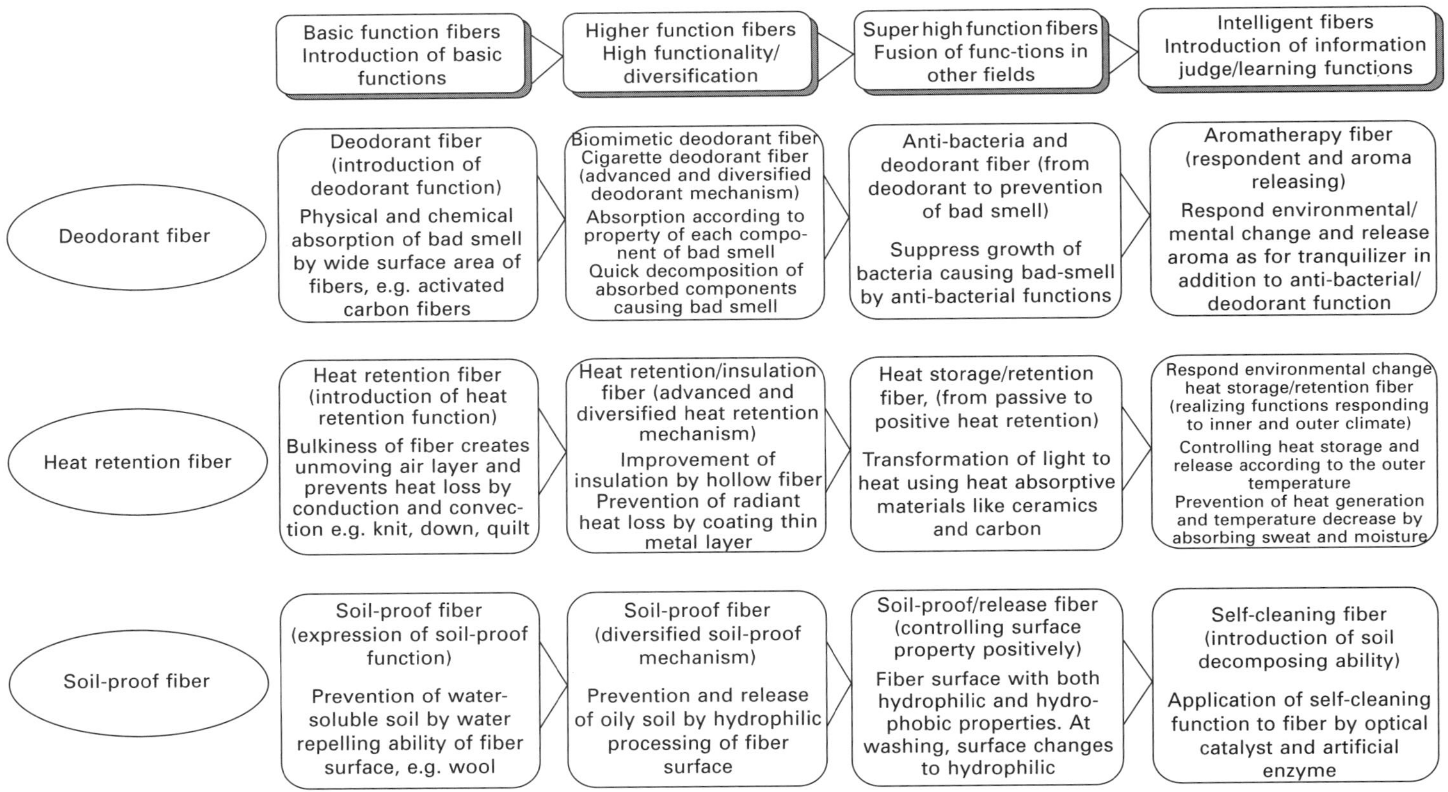


*5.11* Examples of improved functions in function fibers (Professor T. Koyama, Shinshu University).

## 5.2 Sportswear using the high function fiber

Figure 5.12 summarizes the functions and performance required for sportswear, and the required elements are summarized in Table 5.1.

### 5.2.1 High function sportswear

Although Japan imports increasing amounts of textile products, the export of the fabrics for sportswear has been steadily increasing since 1995. Sportswear needs to have more specialized functions than ordinary clothing. Thus the trends in sportswear will give an indication of new fiber material developments. The recent trend is not towards high performance but comfort. Figure 5.13 shows the position of high touch sportswear.

There is a great demand in high function sports goods. If the performance and function satisfy the consumer, the cost is inconsequential. Golf clubs and fishing rods are two examples of high function sports goods where

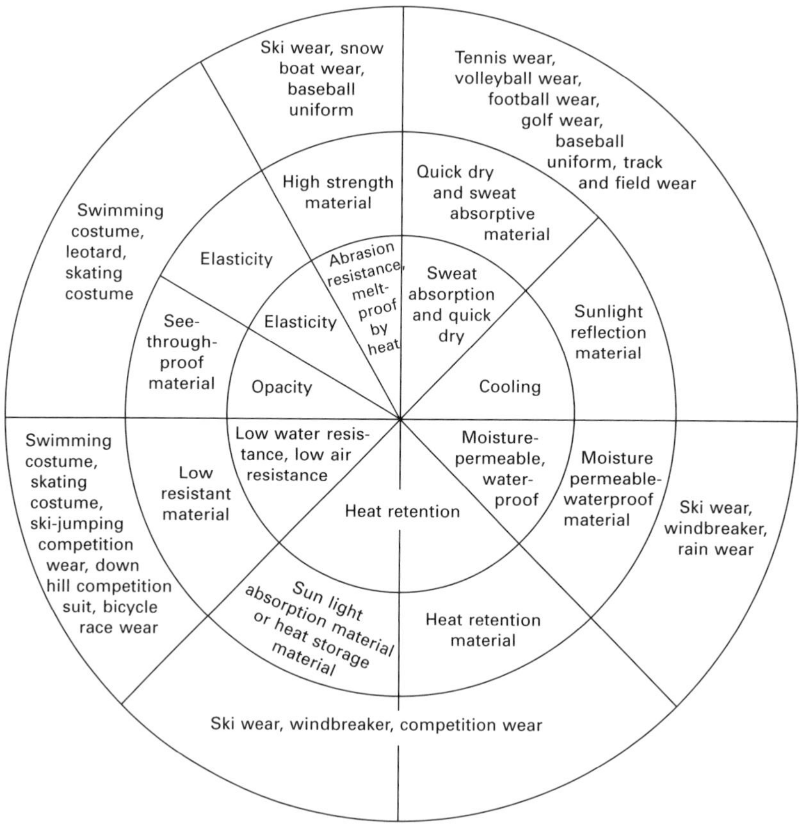


*5.12* Functions and performances required for sportswear.

*Table 5.1* Purpose of sports and required elements for sportswear

| Class of people who enjoy sports | Purposes | Targets | Characteristics | Factors required for sportswear |
|---|---|---|---|---|
| General people enjoying sports | Hobby<br>Leisure | Keeping and improvement of mental and physical health | Not stick to record<br>Achieve targets enjoying sports | Basic function<br>Kansei is regarded as important |
| | Health | Exchange between areas, family and friends | High school, university and business company clubs | Variety of prices in sportswear |
| Sportsman<br>Semi-professional | Winning<br>Challenge to record<br>Physical and mental training | Realization of purposes<br>Entry to competition<br>Compete at higher level | Preliminary state of professional activity<br>Training of body and spirit during the youth<br>Professional to sports<br>Sports clubs of high school, university and business company | Functions and functional beauty are regarded as important<br>Balance of function and price of wear<br>Special wear for each sport |
| Professional<br>Athlete | Winning<br>Challenge to record<br>Occupation | Achievement of purpose<br>Entry to international competition<br>Pride of a professional | Sports occupy big part of life<br>Top of sports<br>Very limited number of people | Function and beauty arising from function are regarded as the most important<br>Professional needs attractive and eye-catching costumes |

Source: (S. Kagechi, Soen Eye (1993), modified).

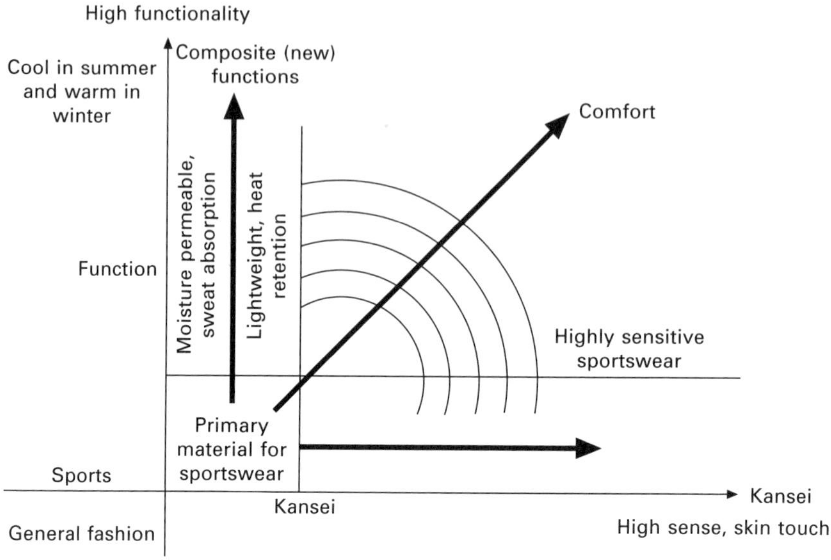


*5.13* Evaluation of highly sensitive sportswear.

carbon fiber is applied. Carbon fiber has a high Young's modulus and is used in airplanes and space shuttles in the aerospace industry. Although expensive, golf clubs, fishing rods and tennis rackets made of the carbon fiber composite attract sports lovers, since the performance satisfies their requirements.

### *Necessary performance for sportswear*

The various factors of clothing function seem to appear in more emphasized form in sportswear. Sportswear should be light and fit the body for easy movement. Durability and abrasion resistance are also required. The protective function is, of course, a prerequisite. Since heavy sweating during exercise is unpleasant, sportswear should absorb/evaporate sweat quickly and keep the body dry. The body should be kept warm in the case of winter sports. All these functions are also required for conventional clothing, hence the general value of experience in sportswear.

### *Functions from the type of sports*

An elastic response is required for swimming, aerobics and figure-skating. Transparency or non-transparency may be important in some cases. Skiing costumes require both heat insulation and physical mobility. Sportswear functions vary, as each sport requires different functions (see Fig. 5.14). The technology of fiber materials in clothing design is available to cope with the varied required functions.

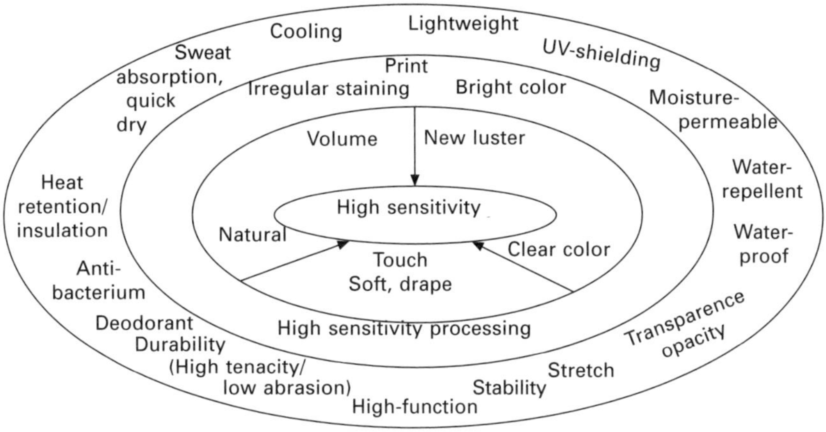


*5.14* Sportswear requires different functions.

The water-repellent/vapor-permeable textile is a typical material for sportswear. The coating/laminate type and the high-density woven fabric are commercially available for water-repellent/vapor-permeable textiles. Goretex® (Goretex Japan) and Microft Lectus® (Teijin) are two commercial examples corresponding to respective types. Commercially available quick-dry textiles are made either from water-absorbing fiber material or with a specially designed fabric structure utilizing a capillary effect to absorb sweat. Fieldsensor® (Toray) has a three-layer woven structure, which prevents absorbed sweat from flowing reversibly. Lightweight fabric is composed of hollow fiber.

Making the fabric surface as smooth as possible can reduce the surface friction of the fabric. Plain-woven high-density fabric of ultra-fine denier fiber has a smooth surface. Teflon-laminate surface is water-repellent, and produces even less friction. Since water- or airflow at the surface will be disturbed less with small surface dimples (a golf ball is a good example), such dimples are provided on the surface of swimming costumes and ski-jump coats. Garments made of heat-storage fiber were used in the Winter Olympic Games in Japan in 1999. Heat-storage fiber is a hollow fiber filled with zirconium carbide. Zirconium carbide absorbs visible light and converts it into far-infrared radiation. Transparent or non-transparent fiber materials are developed to respond to the demands from ladies wishing to wear a white swimming costume. Bodyshell® (Toray) is such a non-transparent white fabric. A conventional white fabric becomes transparent when wet. Bodyshell, however, has a star-shaped core of a different refractive index, which scatters light randomly.

Ultra high strength has been one of the major goals in the fiber/textile field and useful in sportswear. Today gel spinning and liquid crystal spinning are two industrial processes to produce ultra high tenacity fiber, represented by polyethylene and Kevlar, respectively. High tenacity polyethylene is light and strong. Scissors cannot cut aramid fiber such as Technola and Kevlar. However, the tenacity of those ultra high tenacity fibers is about 20% of the tensile strength calculated theoretically from the ideally extended chain model. Conventional polyester or nylon fiber achieves only about 5% of the theoretical tensile strength.

*Scaled swimming costume*

At the World Swimming Championships in Perth, Australia, 85% (338) of swimmers out of 400 participants used swimming costumes made of a new material Speed® developed by Toray and Mizuno. A water-repellent coating is applied to the surface of Speed. A scale-like pattern of water-repellent sheet is printed as shown in Fig. 5.15. The water-repellent scales produce

*5.15* Swimming costume, made of Speed®, with water-repellent coating (Mizuno).

the water flow speed different from that at other parts, and result in vortex in a longitudinal direction to suppress turbulence.

## 5.2.2 Sportswear for speed competition

*Utilizing the characteristics of an uneven surface*

*Stars and Stripes* from the United States won the 1987 Americas Cup. The yacht had an uneven bottom to reduce water resistance. This uneven surface is also applied to NASA rockets and the pantographs of the bullet trains. Descente applied this uneven surface principle to swimming costumes used at the Barcelona Olympic Games in 1992. This swimming costume has an uneven surface at the bottom and breast (see Fig. 5.16) by pasting on silicone sheets. Cycling wear was also developed with the same principle for the American team (see Fig. 5.17). With this uneven surface, water resistance was reduced by about 12% in the swimming costume, and air resistance by about 3% in cycling wear.

*Another use of an uneven surface*

The principle of the uneven surface was adapted to reduce the air resistance at the Nagano Winter Olympic Games in 1999. The wear for speed skating had an uneven surface, with a silicon sheet at the back, arms and legs. The athletes from the Netherlands had fringed fins at both sides of their heads, shoulders and legs to reduce air resistance.

*5.16* Swimming costume having uneven surface (Descente).

*5.17* Cycling wear (Descente).

Descente applied the same principle to the downhill costumes. The International Ski Federation (FIS; *Fédération internationale de Ski*), however, prohibits covering costume surfaces with plastics. The FIS rules specify that the air ventilation must be over 30 liters per second through an area of 1 $m^2$. Silicon sheets cannot be used since silicon is considered a plastic material. Thus Descente invented the production of dimples by utilizing a special knitting structure, and developed a new fabric with built in dimples – Dimplex®. Karl Mayer (Switzerland) cooperated with Descente, who modified their knitting machines to produce such dimples.

The dimples are applied to the ski jump suit. A major consideration in ski jumping is how to handle air resistance. Speed is the most important factor in an approach run, where the air resistance is small. The second step is the takeoff. After a takeoff, a jumper needs to be lifted up in the air. To obtain enough lift, the air resistance at the front and at the back should be significantly different. Descente developed a jump suit with Dimplex®. Here the outer side (Dimplex®) and the inner (two-way tricot) sandwiches polyurethane foam. Since FIS specifies that the thickness of the suit must be less than 5 mm, the outer and inner fabric must be as thin as possible in order to maintain some rigidity by the insulation material as fluttering causes vortex and loses lift.

### *Hollow fiber for light swimming costume*

Lightweight and heat-insulating materials have been developed by using hollow fibers for the past ten years. Hollow fibers have now been applied to swimming costumes to provide lift for the body in water and prevent it from being cooled. A hollow fiber is produced by a specially designed spinneret as shown in Fig. 5.18, where respective fiber cross-sections correspond to the shape of a spinneret. Teijin Ltd. produces the polyester hollow fiber Aerocapsule®, and the companies Unitika, Kanebo Gohsen and Toyobo have developed the nylon hollow fibers Microart®, Lightron® and Aircube®, respectively. Because of the hollow, the apparent density of a hollow fiber is less than one, and its heat conductance is small. Consequently, fabric made of a hollow fiber floats on water and insulates heat. A hollow fiber for a swimming costume is shown in Fig. 5.19, and that for clothes is summarized in Table 5.2. Table 5.3 lists the characteristics of a hollow fiber and its applications.

## 5.3 Comfort function fiber

### 5.3.1 Cooler feeling Eval fiber Sophista®

Kuraray developed an ethylene-vinyl alcohol copolymer (Eval) fiber, 'Sophista®'. Sophista is produced through the mixing of raw materials (Eval

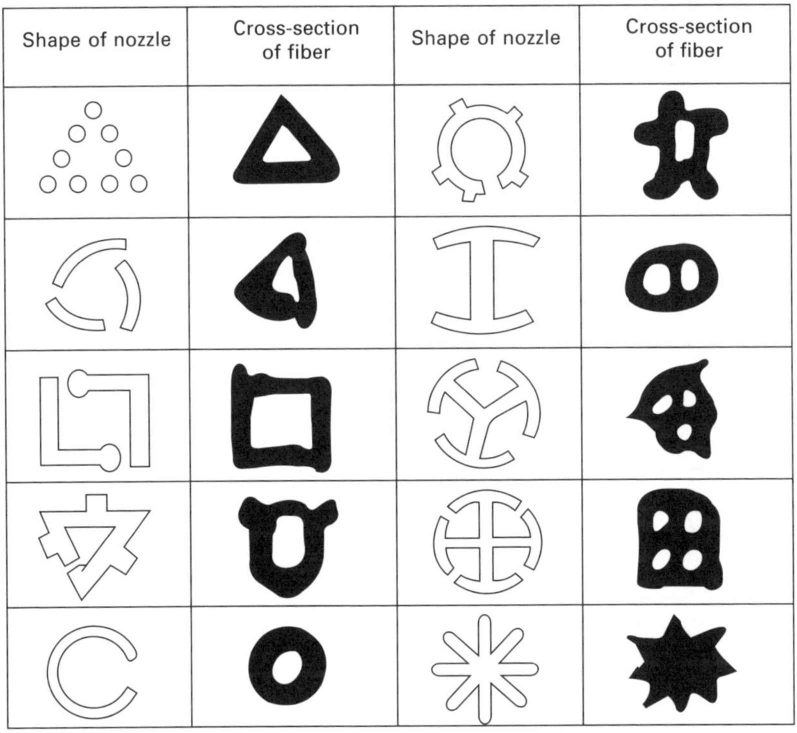


*5.18* Shape of spinneret and cross-section of fiber.

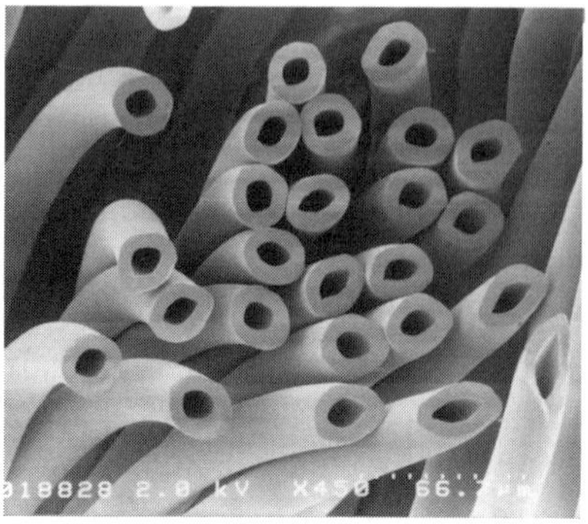

Lightron made of nylon (Kanebo)

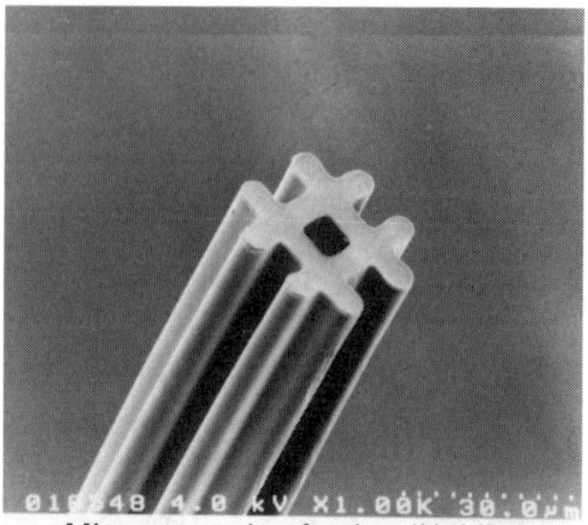

Microart made of nylon (Unitika)

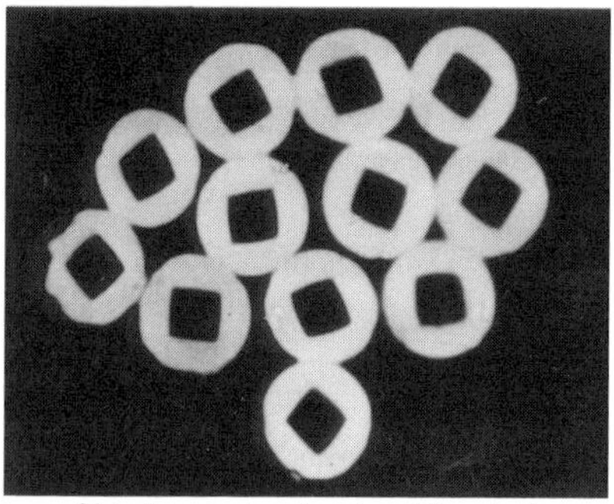

Aircube made of nylon (Toyobo)

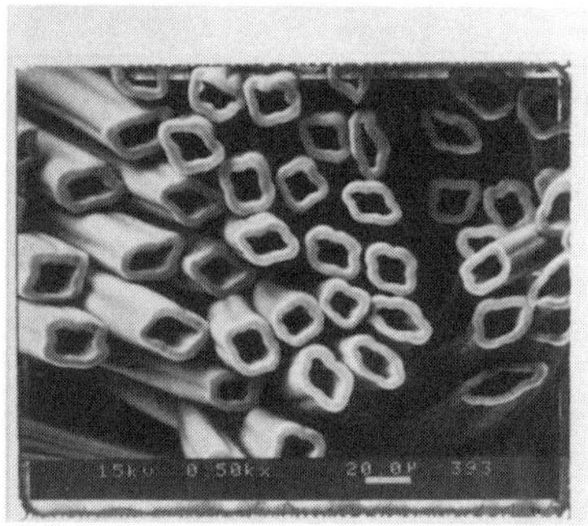

Airocapsule made of polyester (Teijin)

*5.19* Hollow fiber for swimming costume.

*Table 5.2* Hollow fiber for clothes

| Material | Manufacturer | Hollow ratio (%) |
|---|---|---|
| Killatt-P | Kanebo Gohsen | >30 |
| Aerocapsule® | Teijin | 35–40 |
| Viareggio® | Toray | 20–30 |
| NEATY® | Asahi Kasei | 20–30 |
| Airfro® | Unitika | Around 40 |
| GARLAND® | Kuraray | About 20 |

*Source:* T. Suzuki, Series of Basic Lectures on Fiber and Textiles I, The Society of Fiber Science and Technology, Japan, May, 1999.

and polyester pellets), melt-spinning and conjugate spinning. Hydroxyl groups in Eval are hydrophilic and enable Sophista to absorb and diffuse water. Sophista can be called aqua-fiber, and feels gentle and comfortable to the touch. Besides, hydroxyl groups can react with a wide range of compounds to give extra performance.

Sophista is available in three different structures (see Fig. 5.20):

*Table 5.3* Properties and usage of hollow fibers

| Properties of hollow fiber | Usage |
|---|---|
| Bulkiness, lightweight | Cotton for padding |
| Superior in heat retention | |
| Firm and strong | General clothes |
| Soil proof | Carpet |
| Water absorption and water permeation | Underwear, sportswear |
| Can be applied to filtration and dialysis | Materials for industry and medical care |

(1) Sheath-core type: core: polyester 45%, sheath: ethylene-vinyl alcohol 55%
(2) Multi-layers type: polyester 67%, ethylene-vinyl alcohol 33%
(3) Multi hollows: ethylene-vinyl alcohol 100%.

As the melting point of Eval is 160–170°C and it is not resistant to water below 100°C, Eval is not used for clothing. Kuraray introduced a cross-linker into Eval, processed and achieved characteristics such as stainability at 120°C, dry heat set at 160°C and iron-proof.

The basic physical properties of Sophista and comparisons with other fibers are shown in Table 5.4. It shows that Sophista has basic characteristics to be a fiber for clothes.

During the rainy season in Japan, 'cool' and 'fresh' characteristics are important for comfort. Cool feeling is the most important character for sports clothes. We can feel 'cool' and 'fresh' when there is no soggy and uncomfortable liquid layer between the fiber and skin. Moreover, less contact resistance, and active ventilation and heat transfer between the inside and the outside of the clothes are important. The material which gives cool feeling should have functions to absorb water/moisture quickly and let them evaporate quickly. A cool feeling cannot be obtained when the modulus decreases with water absorption.

As shown in Table 5.4, Sophista has moisture absorption of 1.5%. Sophista can absorb sweat and water on the fiber surface and evaporate the moisture move quickly than cotton and polyester. As the core of Sophista is polyester, there is no reduction of modulus when wet. It feels smooth and comfortable to touch, not soggy and sticky. During working, Sophista can maintain the body temperature at a comfortable level by latent heat caused by quick absorption and diffusion of water. It also prevents a quick drop of body temperature which occurs mostly after stopping working or exercise. Such properties give comfort to sportswear.

Compounds bonded to hydroxyl groups of Sophista may develop new application fields such as deodorization, waterproofing, and flame resistance.

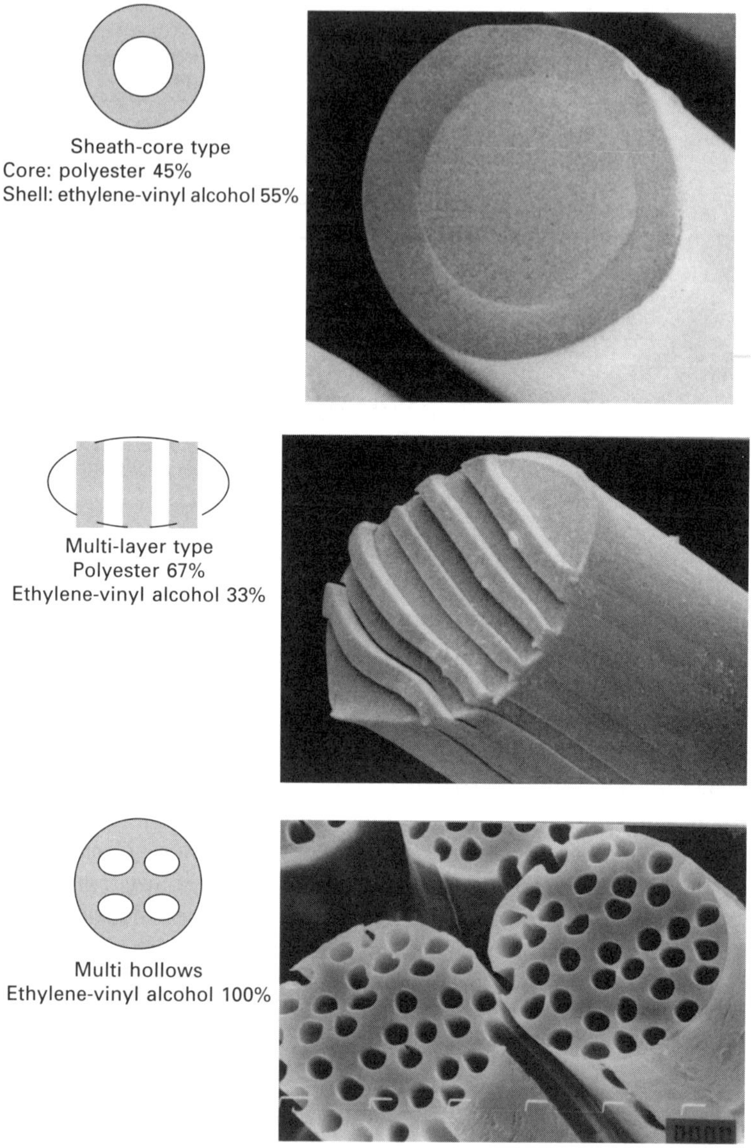


*5.20* Cool-feeling fiber, Sophista (Kuraray).

## 5.3.2 Comfort feeling materials, CORTICO®

Teijin developed comfort cool-feeling materials composed only of polyester. Using polymer and spinning technology, Teijin developed the polyester material, CORTICO, which is dry and absorbs sweat materials giving aesthetic sense and comfort functions.

*Table 5.4* Basic physical properties of various fibers

| Material | Tenacity (g/d) | Density | Refractive index | Moisture absorption (%) |
|---|---|---|---|---|
| Sophista | 3–4 | 1.23 | 1.52 | 1.5 |
| Polyester | 4–5 | 1.39 | 1.72 | 0.4 |
| Nylon | 4–6 | 1.14 | 1.53 | 4.5 |
| Rayon | 1.7–2.5 | 1.50 | 1.50 | 11 |

*Source:* Kuraray

### *Materials of CORTICO*

CORTICO is polyester filament (long fiber), fabricated using the following technology (see Table 5.5).

- *Polymer technology*: Special particles introduced into polyesters enlarge differences between thicknesses during elongation on spinning. Such particles are easily dissolved in alkaline, so alkaline processing after closing can form minute voids.
- *Spinning technology*: Special elongation technology disperses the elongation point unhomogeneously, so that the fiber axis direction and cross-section of fiber are unhomogeneous in thickness, and therefore have an anisotropic cross-section Fig. 5.21.

### *Character of cloths composed of CORTICO*

Cloths composed of CORTICO have feel and comfort functions as clothing textiles shown in Table 5.6. These characteristics are displayed in the material by stain dressing processing technology.

*Table 5.5* Character of raw fiber of CORTICO and included technologies

| Technology | Contents | Action | Character of fibers |
|---|---|---|---|
| Polymer | Special particle method | Depress crystallization orientation<br>Alkaline solubility | Single fiber<br>Maximize differences between structures in thick and thin parts<br>Minute void on fiber surface |
| Spinning | Special elongation | Unhomogeneous dispersal of elongation point | Unhomogeneous thick-thin marks in fiber axis and cross-section |
| | Anisotropic spinning hole | Anisotropic shape in fiber cross-section | Anisotropic cross-section |

*Source:* Teijin

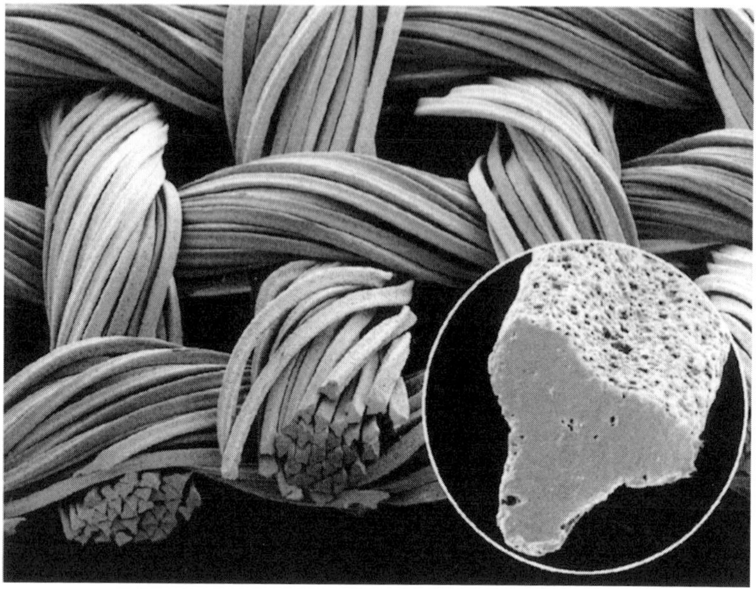

*5.21* CORTICO having anisotropic cross-section (Teijin).

*Table 5.6* Character of CORTICO textiles

| Division | Characters |
|---|---|
| 1. Aesthetic sense | Natural appearance, spun yarn feeling<br>Dry touch, mild luster |
| 2. Comfort function | Sweat absorbance, quick dry<br>Remove soggy and sticky touch |

*Source:* Teijin.

## Feeling

*Natural outside*: Natural outside is defined as the outer surface of the cloth where a difference in thickness can be recognized by eye, but no difference in color depth is recognized. CORTICO is a thick-thin fiber where the thick and thin parts are dispersed unhomogeneously so that its thick-thin morphology can be recognized but with no difference in color depth. Conventional polyester thick-thin fiber has a deep outside at the thick part and shallow at the thin part. But CORTICO has a maximum difference in inside structure of both in thick and thin parts, so appropriate stain conditions can be selected to make stains leave more easily from the thick parts. Therefore depth of colour is even throughout.

*Dry touch*: Alkaline processing in the stain perfection process dissolves special particles and makes minute voids. These random voids between fibers result in inequality at the surface which minimizes the contact area between skin and cloth, producing a dry feel.

*Mild gloss*: Random voids between fibers and inequality in the surface reflects light randomly. It displays high-class mild gloss.

Comfort function

*Sweat absorbency and rapid dry*: Minute voids maximize the surface area of the fiber. Thus the surface can get wet more easily than cotton.

*Without heavy-touch feeling*: Material which get wet easily, having capillary effects due to space between fibers can diffuse water rapidly. Thus sweating on clothes, sweat between the skin and clothes is diffused rapidly, so there is no heavy-touch feeling.

## 5.3.3 ARTIROSA™ having comfort function

Toray developed ARTIROSA, a fiber having comfort function. Its technology, structure and characteristics are shown in Fig. 5.22.

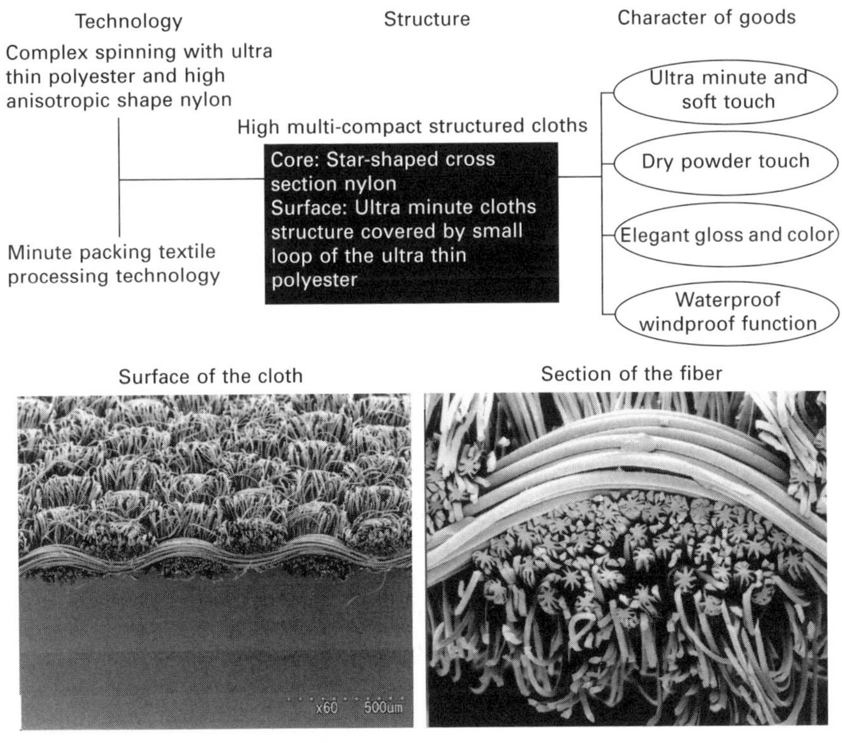


*5.22* High multi-compact materials having minute structure and flexible touch ARTIROSA (Toray).

ARTIROSA is composed of raw fibers highly composed of ultra-thin polyester fibers and star-shaped cross-section nylon fibers. Small loops of the ultra-thin polyester cover the surface of the cloths with nylon in flexible inner layers. They form an ultra-high density structure.

The soft touch of nylon and the strength and elasticity of polyester combine on a micron scale to produce a flexible feel. The structure of ultra-high density cloths make after-processing unnecessary to make the cloth waterproof and windproof. The small polyester loops give a dry and comfortable micro powder touch. The interaction between the surface loops and the inner nylon produces a special effect such as a cool and delicate gloss with deep shade.

Trends in autumn and winter coats have moved from using animal hair mixed with wool materials such as cashmere to cotton and now to artificial fiber material type. Outerwear from artificial fiber material must have light wear feeling and excellent functions, and is thus suitable for autumn, winter, and spring.

Surface materials for such clothing needs to be waterproof and windproof, so high density cloths with a flat surface using nylon and polyester, or coating and plastic processing on the surface of the cloth are used at the moment. The wear, feeling, and touch, etc., require the development of new goods using new materials.

The diversification in consumers' lifestyles, calls for a wide variety of uses, with elegant taste. There is a great increase in casual wear. ARTIROSA has high-feel with more comfortable wear-feeling, functionality and touch, so it is a new material accommodating to consumers' needs.

## 5.4 Biomimetic and intelligent fibers

### 5.4.1 The history of synthetic fiber development is the history of biomimetics

Professor Breslow of Columbia University (USA) proposed the term 'biomimetics' for a new area of research on enzymatic functions in 1972. However, in the field of fiber/textile technology, synthetic fiber has mimicked natural fibers since the 1960s. The synthetic fiber/textile copies not only the bio-structure, but also the bio-functions. Classic examples are a crimpled fiber mimicking the conjugate structure of wool, and a hollow fiber applied for blood dialysis which mimics the lumen structure of cotton. A traditional Noh costume made of polyester with a triangular cross-section scroops like real silk (see Fig. 5.23). Other examples are the water-repellent fabric mimicking the surface structure of a lotus leaf, the chromophoric fiber mimicking the microstructure of a morpho ala, and the deep colored fiber mimicking the micro-crater structure of the cornea of a moth. The odor-killing fiber mimics the supramolecular structure of an enzyme, and artificial suede micro-denier

*5.23* A traditional Noh costume made of polyester with a triangular cross-section (Toray).

fiber copies a high-order structure of natural leather. High function fibers developed using the biomimetic approach are shown in Table 5.7. These high function fibers mimic the conjugate structure, the irregular shape, the uneven surface structure, the fine structural hierarchy, and the catalytic function of natural fibres (Table 5.8). However, the mechanism of the structure formation of natural fibers is still not well understood and could be a profitable area of research.

## 5.4.2 The challenge to harness nature

Spider silk has excellent mechanical properties with tenacity similar to Kevlar and its elongation at breaking point is over 10%. However, why and how spider silk has such an excellent mechanical property remains a mystery. New function fibers learning from bio-systems are shown in Fig. 5.24. Table 5.9 summarizes the biomimetic functions of fiber and textiles under development.

The characteristics of biomaterials can be summarized as follows:

- Biomaterials are multi-functional composites (high function and multi-function).
- Synthetic materials can surpass biomaterials, but only with respect to a single property.
- Biomaterials exhibit high functions, even when biomaterials are mechanically weak.

*Table 5.7* High function fibers developed using biomimetic approach

| Discovery/ invention | Structures mimicked | Advanced-function fibers | Discoverer/inventor |
|---|---|---|---|
| | Lumen structure of cotton | Hollow fiber | Du Pont |
| | Conjugate structure of wool | Elucidation of crimp structure → Crimpled fiber | M. Horio (Kyoto University) |
| 1964 | Silver trappings of leather | Artificial leather | O. Fukushima (Kuraray) |
| 1965 | Super-fine structure of leather | Micro-denier fiber Artificial suede | M. Okamoto (Toray) |
| 1978 | Micro-crater structure of cornea of moth | Fiber with deep colors and luster | S. Yamaguchi (Kuraray) |
| 1979 | Supramolecular structure of enzyme | Odor-killing fiber | H. Shirai (Shinshu Univ.) |
| 1980 | Triangular cross-section of silk | Shingosen with silk scrooping | Y. Sato (Toray) |
| 1980 | Capillary water absorption by tree | Water absorption porous hollow fiber | T. Suzuki (Teijin) |
| 1983 | Surface structure of lotus leaf | Water-repellent fabric | F. Shibata (Teijin) |
| 1989 | Multi-layered structure | Fiber with light interference function | K. Matsumoto (Kyoto Institute of Technology) |
| 1992 | Structure of bioorganisms | Non-transparent fiber | T. Kato (Toray) |
| 1997 | Natural color of Morpho alae | Structurally colored fiber | H. Tabata (Nissan Motor)<br>M Yoshimura and K. Iohara (Teijin)<br>S. Shimizu (Tanaka Kikinzoku) |

For example, a bamboo exhibits optimum bending rigidity with a minimum amount of the material of density-gradated structure, and a shell a composite structure of multi-layer calcium carbonate bound with protein paste.

*Iridescent insects*

The color of the ala of a jewel beetle will not fade even after a thousand years, because its color is not due to a dyestuff but to a light interference. The ala electron micrograph of an ala of a jewel beetle reveals a vertical multi-layer. This multi-layer is composed of helices, which related to a specific

*Table 5.8* Biomimetics learned from functions and form of bioorganisms

| Function | Structure | Bioorganism | Biomimetics |
|---|---|---|---|
| | Conjugate | Natural Leather | Ecsaine (Toray) |
| Form | Different shape<br>Form beauty | Silk, Wool<br>Butterfly<br>Moth | Glacem (Kanebo)<br>Diphol (Kuraray)<br>Microcrater (Kuraray) |
| | Surface | Lotus leaf | Microft rectus (Teijin) |
| | Uneven | Seed of Xanthium strumarium | Hook and loop fastener |
| Micro-structure | Fluctal surface | Feather of waterfowl<br>Water strider | Fluctal, water repellent |
| | Spiral (cholosteric liquid crystal) | Insects | Colored fiber without staining<br>Conjugate material |

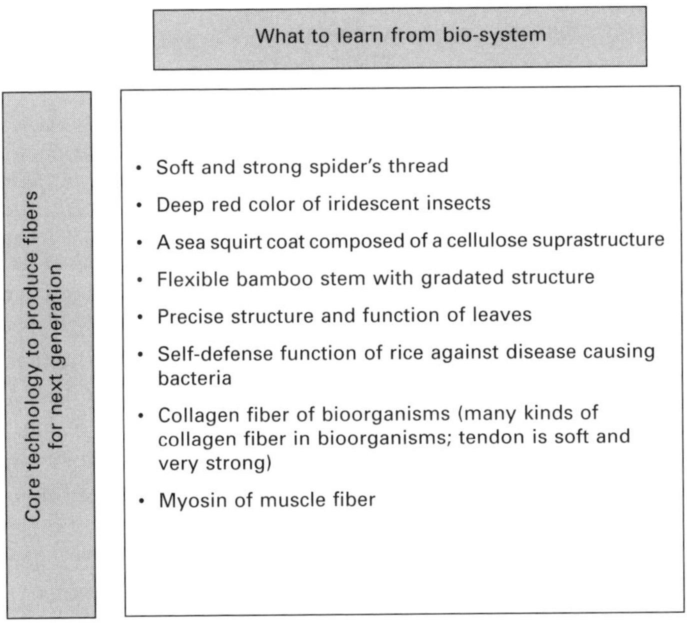


*5.24* New function fiber based on biomimetics.

liquid crystal. Such helical structures are one of the most basic structures in nature, and are found, for example, in the cellulose organization in plants, the shell of crabs and insects composed of chitin fibrous assembly, and fish scales. A part of animal bones, the Achilles tendons and eyeballs are made of

*Table 5.9* Biomimetic functions of fiber and textiles under development

| Unit to express | Structural | Basic function | Systemized function |
|---|---|---|---|
| Molecules/ assembly | Artificial silk/ wool/cotton | Dry-processed liquid crystal, Gel spinning, Highly oriented fiber Artificial spider silk | Matter recognition, Separation transmittance, Artificial enzyme immunity fiber, Suppression, Blood purification, DNA-type fibrous information storage media |
| Cell/organs | Artificial leather/ feather/bone/ teeth, Fiber reinforced ceramics | Artificial kidney/ lung/skin/muscle/ heart/nervous system | Artificial antimicrobial/ sterilization system, Artificial repairing system, Information recognition/transfer/ processing/memory |
| Bioorganisms | Marine textiles for fish gathering, Structural material, Geotextiles, Waterstorage fiber, Anchor | Fiber to reduce environmental stimuli, Air/water purification fiber | Function to assist growth and breeding, Material recycling system |

helical collagen fiber. The helical structure is energetically stable, and the helical natural polymers are concentrated in the process of growth to form a cholesteric mesophase.

The ala of an insect living in a desert reflects infrared light to survive under the strong sun. This suggests that, if we can duplicate this, we might develop a textile that cuts out the sun's rays in summer. Some synthetic polyglutamic acid esters were found to form the same cholesteric mesophase as the ala of a jewel beetle according to Professor J. Watanabe (Tokyo Institute of Technology). Cellulose derivatives also form cholesteric mesophases. The color appears in the cholesteric mesophase by light interference without dyeing. In future, it might be possible to produce a revolutionary coloring process without using a dyestuff. A similar process could be applied to cut out heat rays or UV radiation.

### *Intelligent materials*

The term 'intelligent materials' first appeared in the 1989 report of the technical council of aero- and electronic-industry in response to the inquiry of the

President of the Agency of Science and Technology of Japan. Intelligent materials are defined as structural materials with high functions. Although there are many disagreements about what exactly constitutes a high function, intelligent fiber, a possible definition is:

> 'fibrous material provided with a sensory function to sense an external stimulus and change its state, a processor function to process its state according to the change, and an actuator function to actuate itself in response to the processing.'

Biomimetic fiber can be classified as structure-mimetic (mimicking the structure of organisms) and function-mimetic (mimicking the function of organisms) fiber. The function-mimetic fiber is further classified into static-function-mimetic fiber and dynamic-function-mimetic fiber. Here the dynamic function is called 'intelligence'. The relation of the biomimetic fiber to intelligent fiber is shown in Fig. 5.25. H. Okamoto has produced a classification chart of organisms based on composite structure (Table 5.10).

Intelligent textiles are now appearing on the market. Table 5.11 shows recent intelligent textiles. To develop intelligent textiles several steps must be considered. For example, for clothing, the type and function of built-in sensor and actuator will be determined by what sort of response is expected from the environmental change by the wearers. Then the materials possessing

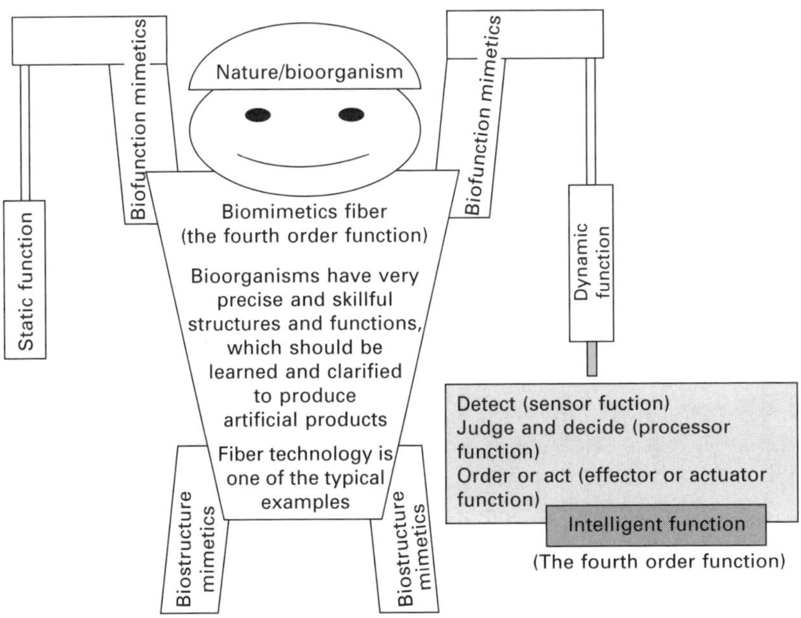


*5.25* Relation between biomimetic fiber and intelligent fiber.

*Table 5.10* Classification chart of organisms based on composite structures

| Type | Primary structure | Material | Examples of biological system | | | |
|---|---|---|---|---|---|---|
| | | | Constituent material (Matrix) | Structural element | | Structural system (composite structure) |
| Fiber-reinforced | [1-dimensional reinf.] unidirectional | Inorganic | $CaCO_3$/$CaCO_3$<br>Cellulose | Fiber/crystal (calcite)<br>Vascular tube | | Tooth of sea urchin<br>Culm part of bamboo |
| | [2-dimensional reinf.] laminates cross-ply angle-ply filament winding | Organic | Chitin/protein<br>Collagen/poly-saccharide<br>Actin-myosin<br>Collagen<br>Collagen/proteo-glycan | Cuticle<br>Parenchymal membrane<br>Muscle spindle<br>cuticle | | Crust of insect, lobster and crab<br>cornea<br>Heart, blood vessel<br>nematode<br>Intervertebral disk |
| | [3-dimensional reinf.]<br>Mixed laminate | | Cellulose/lignin<br>Actin-myosin (*)<br>* + collagen | Muscle spindle (**)<br>** + tunic | Adhesive bonding reinforcement | Wood<br>Mantle of octopus<br>Mantle of squid |
| | | Organic-Inorganic | Collagen apatite<br>Collagen calcium phosphate | Herversian lamella (crystal structure) hydroxyapatite | | Bone, tooth, ivory |
| Particle-reinforced | Prism structure<br>Nacre structure<br>Foliated structure<br>Crossed lamellar structure | | $CaCO_3$/ conchiolin<br>Silicate | (Crystal structure) calcite<br>Aragonite | | Shell (outer layer), coral<br>Shell (inner layer), pearl<br>Diatoms |

*Source*: H. Okamoto., *Biomimetics*, 2 (2) 1–13 (1994)

*Table 5.11* Intelligent textiles appearing on the market

| Uses | Function | Sensor (stimuli) | Processor (mechanism) | Actuator (response) |
|---|---|---|---|---|
| Window shade/Curtains<br>Clothes for health | Shading<br>UV-shielding | Light<br>UV | Photochromic | Color change<br>Shading |
| General clothes | Cool feeling<br>Heat retention | Temperature | Exo/endothermic | Endothermic and exothermic by increase and decrease, respectively |
| Curtains/wall paper | Soundproof | Sound wave | Piezoelectric | Absorption of sound pressure |
| Bedding | Bedsore-proof | Pressure by weight | Compression by fatigue | Decrease of contact pressure |
| General clothes | Soil-proof<br>→ No washing | Soil by organic matter | Removal by enzyme | Cleaning |

*Source:* M. Fukahara, *New Fiber Science Challenge for New Frontier.*

a required function should be selected or created. The function is revealed by mechanical, thermal, electric, magnetic, optical or chemical forces. Its mechanism is the result of structural and/or phase change of the materials, and depends on the molecular structure, cross-linking, etc. The surface property (the affinity of the boundary), the structural gradation and the distribution of physical units in terms of the size and position are also factors influencing the intelligence of the materials.

Fashion design and wear comfort are the prerequisite factors for clothing. The most difficult problem is to decide where the intelligent function should be incorporated in the process leading from polymer and spinning to a final product. The function can be incorporated in the process of dyeing and finishing by chemical reaction, coating, and/or absorption of functional groups. In the process of weaving or knitting, the fabric organization should be carefully designed. The function will be required at a particular place, but not for the whole fiber.

## 5.5 The new areas

### 5.5.1 Wearable computer

Computers in the twenty-first century could change from the portable to the wearable type. Because of the weight reduction and miniaturization of computers, it is now easy to carry a computer and various application fields have been developed on this basis. Since the first international symposium on the subject, held at the Massachusetts Institute of Technology in October 1997, wearable computers have attracted attention.

Research is active into medical-devices-wearing computers which need to have regard to fashion requirements. In the near future, fashionable wearable computers may be seen around town.

### 5.5.2 Organic electro-luminescence wearable display

Organic electro-luminescence wearable display with a thin screen has been developed and marketed by Pioneer. When a voltage is applied to specific compounds, a luminescence can be produced. Thus plastic films with a thickness of 0.2 mm can be so processed to produce a display which can be attached to clothes, bags, and the arm. The materials will be applied, not only to wearable displays, but also to mobile phones and TV and computer screens.

### 5.5.3 Mobile fuel cell using hollow fiber

The research to use a polyelectrolyte fuel cell as a power source for mobile devices is growing. The methods to supply hydrogen to fuel cells include direct supply of hydrogen to storage vessel/occlusion alloy and modification of methanol to be able to generate hydrogen. As the latter needs a specific device, there are problems in weight and size of the device.

The National Institute of Advanced Industrial Science and Technology (AIST) has now established an elemental technology to generate hydrogen from methanol, using a catalyst and hollow fiber. The technology does not need any specific devices to modify the methanol. Dr T. Okada of AIST has demonstrated the utilization of tubular polymer electrolyte membrane for direct methanol fuel cells (DMFCs). The merit of DMFCs is not only that they allow a reduction in size, but also make possible the usage of the cells by feeding methanol continuously by cartridge or syringe as shown in Fig. 5.26.

### 5.5.4 Fibrous titanium oxide optical catalyst

Ube Kosan developed a fibrous titanium oxide optical catalyst. Fibrous materials can readily be processed so that they are applied in air purifiers and smoke separators. Ube Kosan established a pilot plant in Ube Factory (Ube, Yamaguchi Prefecture) in 2002, which now markets in the form of fiber and non-woven textile. The market in optical catalysts will be worth trillion yen in 2005. Organic silica polymer and low molecular weight components such as titanium oxide crystal are combined, spun, treated in high temperature air to remove impurities, and processed to give titanium oxide with high purity. The material can hold optical catalytic ability even treated at 100°C. The diameter of the fiber is 5 μm, the core of the fiber is made of silica, and layers of titanium oxide with different crystal densities surround the core, ten to several hundred nm in thickness.

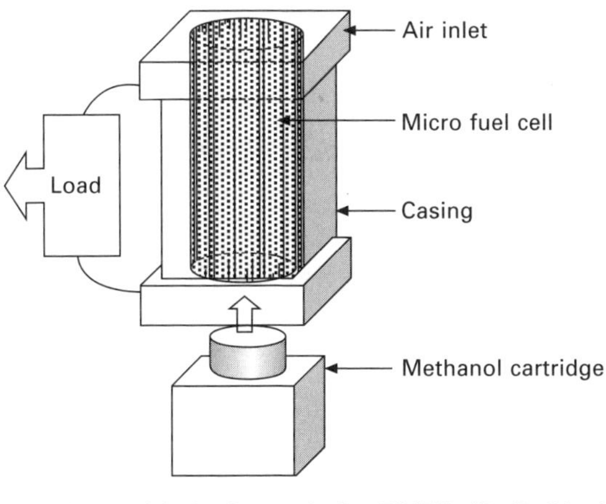


*5.26* Assembled micro-tubular DMFC. (Dr T. Okada, AIST).

Optical catalyst decomposes soil when lit and shows deodorant activity. As the development of optical catalyst that can work under visible light has now been achieved, the application fields have turned from exterior to interior use. There seems no limit to the further development of such intelligent fibers.

## 5.6 Bibliography

### *5.1 Prospects for high function fiber development*


Suzuki T., *et al.*, *Sen'i Gakkaishi*, **41**, 401 (1985).

Okamoto H., Hayashi T., Shibata F., Kawasaki S., Harada T., Sugiyama H., Hirano, Y., Kubotsu, A. and Hidaka T., 'Morphology and Function Learning from Nature', *Sen'i Gakkaishi*, **44** (3), 81–107 (1988).

Kunugi T., Sato M., Miyata S., Kaino T., Seguchi T., Santa T., Nishizawa H., Yoshioka T., Honda Y., Kai K., Shindo N., Fujiura Y., Haruna T. and Sakai K., 'Limit of Fibers: their Performance and Function', *Sen'i Gakkaishi*, **44** (9), 326–354 (1988).

Yagi K., Tsujita Y. and Ito Y., 'Surface Morphology and Performance of High-Function Fibers', *Sen'i Gakkaishi*, **45** (10), 415–432 (1989).


### *Classification according to the level of function – basic function fiber (the first order function)*


Hongu T. and Kikutani T., *Polyfile*, **34** (6), 21 (1997).


### Construction and architectures – membrane structure


Effenberger J.A., Paper 11 in *Air-Supported Structures: The State of the Art*, Institution of Structured Engineers, London, p. 144, 1980.

Motohashi K., Toyoda H. and Nireki T., *Second Int. Symp. 'Plastic and Waterproofing in Civil Engineering'* 2D, 1984.

Toyoda H., Torii T. and Taga T., *Durability of Building Materials and Components*, **6**, 93 (1993).


### *Classification according to level of function – high function fiber (the second order function)*


Yamaura K. and Kumakura R., *J. Appl. Polym. Sci.*, **77**, 2872 (2000).

Fujimatsu H., Kim Y.S., Matsuzaki H., Nakamura A., Usami H. and Ogasawara S., *Polym. J.*, **33**, 709 (2001).

Fujimatsu H., Imaizumi M., Shibutani N., Usami H. and Iijima T., *Polym. J.*, **33**, 509 (2001).

Kanekatsu R., Iizuka E., Shirai K., Kiguchi K., Abe K. and Hachimori A., *J. Seric. Sci. Jpn.*, **69**, 191 (2000).

Agnew J., Grainger K., Clark I. and Driscoll C., *Radiological Protection Bull.*, **200**, 14 (1998).

Toray Industries, MEMBRA catalogue.
Enterprise Bureau Okinawa Prefecture; Seawater Desalination Plant.
Nitto Denko Co. Ltd., MEMBRANE catalogue.

## Ultrafiltration membrane

Schmidt W., *Ann.* der Phys. und Chem., **114**, 337 (1861).
Henmi M. and Yoshida T., *J. Memb, Sci.*, **85** (1993).
Yamaura H., Fujii Y., Shiro K. and Uemura T., *Prep. of ICOM '96*, 238 (1996).
Henmi H. and Toshoka T., *J. Memb. Sci.*, **85**, 129 (1993).
Yamamura H., Fujii Y. and Uemura T., *The 1993 Int. Cong. on Membranes and Membrane Processes,* Heidelberg, Germany, 1993.

## Reverse osmosis membrane

Kurihara M., Hideshima Y. and Uemura T., *Prep. of ICOM '87*, 428 (1987).
Ikeda T., Fusaoka Y., Uemura Y., Tonouchi T. and Fujino H., *Prep. of ICOM '96*, 182 (1996).
Kurihara M., Fusaoika Y., Sasaki T., Bairinji R. and Uemura T., *Desalination*, **96**, 133 (1994).
Fusaoka Y., Kojima S., Ikeda T., Inoue T., Nakagawa K. and Kurihara M., *Prep. of ICOM '96*, 164 (1996).
Scala R.C., Ciliberti D.F. and Berg D., US Patent 3,744, 642 (1973).
Morgan P.W., 'Condensation Polymers: By Interfacial and Solution Methods', *in Polymer Reviews*, Vol. 10, Wiley, New York (1965).
Cadotte J.E., '*Materials Science of Synthetic Membranes*', ACS Symp. Series No. 269, Chap. 12, ACS (1985).
Riley R.L., *et al.*, *Desalination*, **19**, 113 (1976).
Kurihara M., *et al.*, US Patent 4,387,0224 (1983).

## Ion exchange fiber

Miyamatsu T., *Sen'i Gakkaishi*, **39**, 53 (1983).
Yoshioka T. and Shimamaura M., *Bull. Chem. Soc. Jpn.*, **57**, 334 (1984).
Yoshioka T. and Shimamaura M., *Bull. Chem. Soc. Jpn.*, **59**, 339 (1986).
Henmi M. and Toshioka T., *Desalination*, **91**, 319 (1993).
Amano O., *et al., 1991 JAIF Int. Conf. on Water Chem. in Nuclear Polwe Plants*, 85, 1991.

## Other

Robertson C.R. and Kim I.H., *Biotechnol Bioeng.*, **27**, 1012 (1985).
Stevens T.S., *et al.*, *Anal. Chem.*, **54**, 1206 (1982).
Brooks T.W., *et al.*, *ACS Symp. Ser.*, **49**, 111 (1977).
Vigo T.L. and Frost C.E., *Textile Res. J.*, **52**, 633 (1982).
Reis J.F. and Lightfot E.N., *AICHE J.*, **22**, 779 (1976).

## Wound covering materials

Hinman C.C., *Nature*, **200**, 377 (1963).
Bell E., *et al.*, *Proc. Natl. Acad. Sci. USA*, **76**, 1274 (1979).
O'Connnor N.E., *et al.*, *Lancet*, **1**, 75 (1981).
Thomas S., *et al.*, *The Pharmaceutical J.*, **241**, 806 (1988).
Ueda M., *et al.*, *Ann. Plast. Surg.*, **35**, 498 (1995).
Suzuki Y., *et al.*, *J. Biomed. Mater. Res.*, **39**, 317 (1998).

## Biodegradable fiber

Mochizuki M. and Hirami M., *Polym. Adv. Tech.*, **8**, 203 (1997).
Mochizuki M., *Int. Textil-Symp. Asia for Technical Textiles Nonwovens and Textile-Reinforced Materials*, p. 139, 1998.
Mochizuki M., *et al.*, *Macromolecules*, **30**, 7403 (1997).
Fukuzaki H., *et al.*, *Eur. Polym. J.*, **25**, 1019 (1989).
Lunt J., *Polym. Degradation and Stability*, **59**, 145 (1998).

## *Classification according to level of function – super high function fiber (the third order function)*

Koike Y. and Ishigure T., *IEICE Trans. on Communications*, **82**, 1287 (1999).
Na K., Choi H-K., Kim D-W. Akaike T. and Park K.H., *Biosic. Biotech. And Biotech.*, **65**, 2016 (2001).
Li H., Tan S. and Shen J., *Polym. Prep.*, **40**, 593 (1999).
Ruckman J.E., Murray R. and Choi H.S., *Int. J. Clothing Sci. Tech.*, **11**, 37 (1999).
Gonzales R.R., Endrusick T.L., Santee W.R., *Aviation Space and Environmental Medicine*, **69**, 1076 (1998).
Shoji H., 'Extracorporeal endotoxin removal for the treatment of sepsis: Endotoxin adsorption cartridge (Toraymixin)'. *The Apher & Dial* **7**, 108 (2003).
Shoji H., *Proceeding of Exhibition Symposia '2004 Expo Fibers for New Era'*, p. 346, The Scociety of Fiber Science and Technology, Japan, 2004.
Lewin B., *Genes V*, 109, Oxford University Press (1994).
Akita T., *Sen'i Gakkaishi*, **59**, (2003).
See also references on Morphotex® in Chapter 9.

## *Classification according to level of function – intelligent fiber (the fourth order function)*

Bocherens E., *et al.*, *Smart Materials and Structures*, **9**, 310 (2000).
Watanabe M., Shirai H. and Hirai T., *J. Appl. Phys.*, **90**, 6316 (2001).
Kwon Y.-H. and Shim S., *Clothing and Textile Res. J.*, **17**, 203 (1999).
Umeno T., Hokoi S. and Takada S., *Am. Soc of Heating Refrigerating Air-conditioning Engin.*, **107**, 71 (2001).

## Dress animation using 3D CG technology

Sakaguchi Y., *Advanced Fiber Materials Research Committee (AFMC) Preprints* No. 24, p. 31, 2000.
Carighnan Y.M. and Halmann N.M.T., *Computergraphics*, **26**, 99 (1992).
Breen D.E., House D.H. and Getto P.H., *Visual Computer*, **8**, 246 (1992).

## *5.2 Sportswear*

## High function sportswear

Kato T. and Niwa U., *Sen'i Gakkaishi*, **51**, 295 (1995).
Mori K., *The 6th Spring Seminar Proceeding*, p. A-13, The Textile Machinery Society of Japan (2000).

## *5.3 Comfort function fiber*

Nakagawa J., *The 30th Sen'i Gakkai Summer Seminar Proceeding*, p. 134 (1999).
Kamiyama N., Miyasaka N. and Mizumura T., *Sen'i-Gakkaishi*, **58**, 271 (2002).
Toray Materials of Announcement for Press on 29 July, 2002.

## *5.4 Biomimetic and intelligent fibers*

Yuan X., Mak A.F.T. and Li J., *J. Biomed. Mater. Res.*, **57**, 140 (2001).
Sarikaya M., *et al.*, *J. Mater. Res.*, **16**, 1420 (2001).
Shirai H., *et al., J. Phys. Chem.*, **95**, 417 (1991).
Shirai H., *et al., Macromol. Chem. Macromol. Symp.*, **59**, 155 (1992).
Tsuki H., *et al., Fiber*, **51**, 220 (1995).
Tsuki H., *et al., Polymer*, **51**, 220 (1995).
Tsumagari S., *Engineering Materials*, **43**, 102 (1995).
Okamoto H., *Proceedings of International Workshop on Intelligent Materials*, p. 123, March 14 (1989).

## Spider silk

Foelix R.F., 'Spider webs', in *Biology of Spiders*, Harvard University Press (1982).
Osaki S., *Acta Arachnologica*, **38**, 21 (1989).
Jackson C. and Obrien J.P., *Macromolecules*, **28**, 5975 (1995).
Osaki S., *Acta Arachnologica*, **37**, 68 (1989).
Osaki S., *et al., Polym. Prep. Jpn.*, **50**, 3493 (2001).
Osaki S., *Sen'i Gakkaishi,* **58**, 74 (2002).
Lazaris A., *et al., Science*, **295**, 472 (2002).

## Cholesteric mesophase

Neville A.C. and Caveney S., *Biol. Rev.*, **44**, 531 (1969).
Neville A.C., *Biology of Fibrous Composites*, Cambridge University Press, (1993).

## Cholesteric helical structure

Kosho H., Tanaka Y., Ichizuka T., Kawauchi S. and Watanabe J., *Polym. J.*, **31**, 199 (1999).
Watanabe J. and Nagase T., *Macromolecules*, **21**, 171 (1988).
Watanabe J., Goto M. and Nagase T., *Macromolecules*, **20**, 298 (1987).

## Intelligent materials

Fukuhara M., *New Fiber Science Challenge for New Frontier*, p. 234, Res. Institute of Economy, Trade and Industry, Tokyo (1955).
Kitagawa M., Tagami S. and Tokiwa Y., *Macromol. Rapid. Commun.*, **19**, 155 (1998).
Frisch H.L. and Wasserman E., *J. Am. Chem. Soc.*, **83**, 3789 (1961).
Dietrich-Buchecker C.O., Saubege J.P. and Kern J.M., *J. Am. Chem. Soc.*, **106**, 3043 (1984).
Ortholand J.Y., *et al.*, *Angew. Chem. Int. Ed. Engl.*, **28**, 1394 (1989).
Wulff G., *Angew. Chem. Int. Ed. Engl.*, **34**, 1812 (1995).
Bach U., *et al., Nature*, **395**, 583 (1998).

## *5.5 The new areas*

Nikkan Kogyo Shinbun-sha, Wearable Computer, TRIGGER 2000, Tokyo, July, p. 53.
Okada T. and Nakada J., Strategic Research Group for Nanofiber Technology (SRNFT), The Society of Fiber Science and Technology, Japan 2003, p. 21.
Kido J., Strategic Research Group for Nanofiber Technology (SRNFT), The Society of Fiber Science and Technology, Japan 2003, p. 18.
TRIGGER 2000, July, p. 60.
Ishikawa T., Yamaoka H., Harada Y., Fujii T. and Nagasawa T., *Nature*, **416**, 64 (2002).

# 6

# Frontier of health and comfort fibers

## 6.1 Fibers for health

### 6.1.1 Health is on the balance of physical/mental/social well-being

*Definition of health*

The World Health Organization (WHO), defined health as 'a state of complete physical, mental and social well-being and not merely the absence of disease or infirmity'. It is said that mental health, in particular, has a key control over physical, mental and social well-being. The WHO aimed at the maintenance and increase of mental health by positively preventing not only physical but also mental illness. Western medicine finds it difficult to deal with long-term recession syndrome, when a hard worker may lose motivation, or a new employee refuses to attend work.

*Homeostasis*

Figure 6.1 illustrates the extension of a healthy lifetime. The internal environment of a healthy human requires keeping a balance (equilibrium) in the face of various loads (stimulus). The reversible spring, which is within the limit of dynamic balance, stands for healthy conditions. In other words healthy condition means the condition of homeostasis (dynamic balance). If a bigger load than the limit of recovery is added to the spring maintaining homeostasis, the dynamic balance of the spring moves to the reversible area. One might say, then, that it is in a state of disease. If the spring is within resilience, the spring can return to the normal dynamic balance area by having rest or treatment. Young people can recover quickly and elderly people recover more slowly.

Suppose that a healthy spring encounters a traffic accident and is broken. If a spring is damaged badly, it is difficult for the spring to be repaired. It reaches an irreversible area, meaning death. If the damage is small, it can be

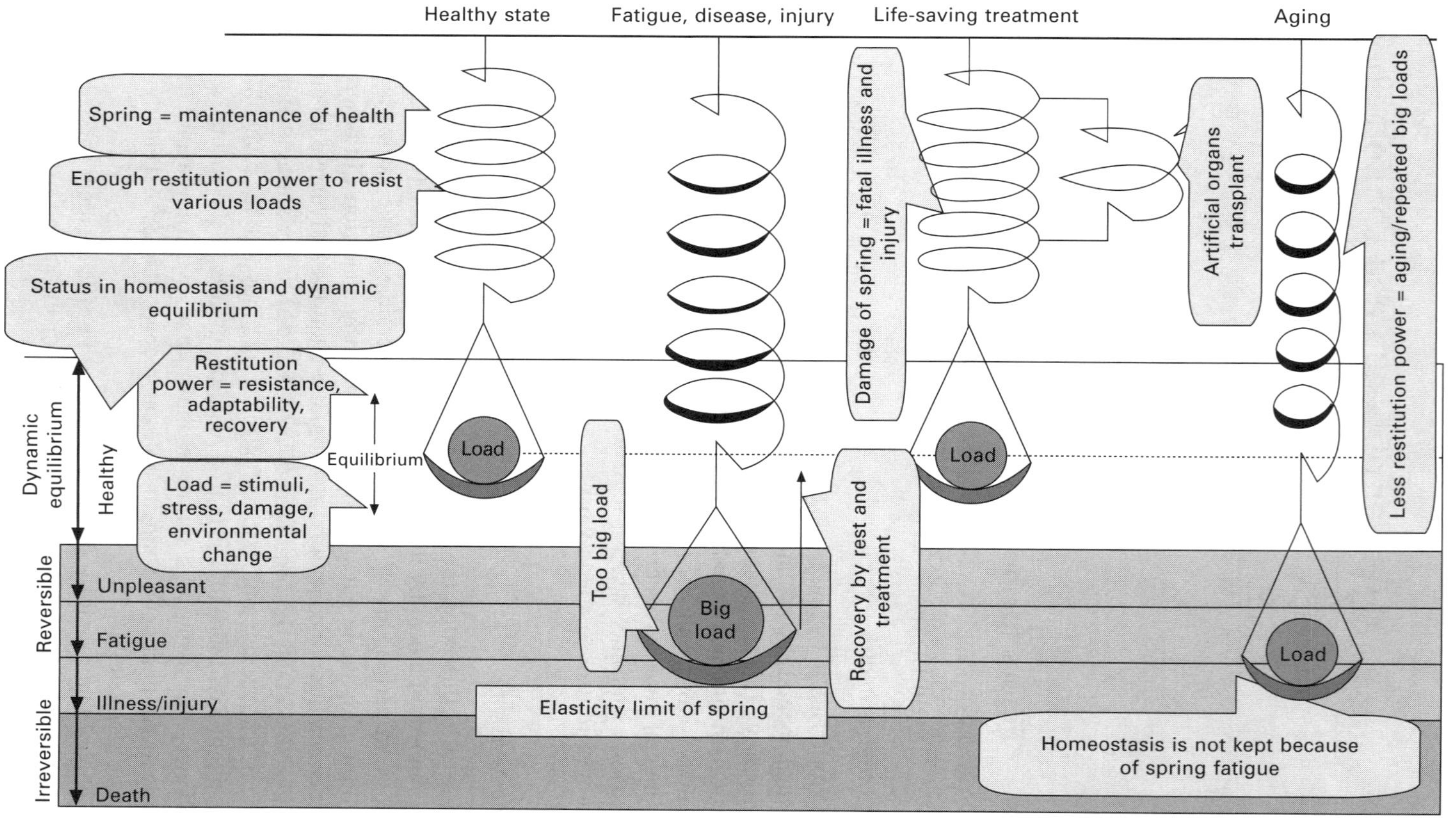


*6.1* Extension of life and healthy life.

repaired by life-saving treatment. An organ transplant law was approved for the first time in October 1997 in Japan. Since then, some patients could recover by internal organ transplant donated by a human donor.

*Repair by artificial organ*

Use of an artificial internal organ is another life-saving treatment which can restore the spring that does not function, and a dynamic balance can be obtained by organ transplant to recover health again. An example of an artificial organ is the artificial dialyzer using a 'hollow fiber'.

Japan is the world leader in fiber technology to make hollow fibers for the artificial kidney. Hemodialysis now saves the lives of more than 170000 patients every year in all Japan.

## 6.1.2 Healthy Japanese in the twenty-first century

*Healthy Nippon 21*

The Ministry of Health, Labor and Welfare in Japan recently announced the 'Healthy Nippon 21 Plan' (to extend from 2000 to 2010) aimed at improving general health in Japan. Health measures of the Ministry of Health, Labor and Welfare have evolved from the primary nation health measures (1978–1988), secondary nation health measures, the so-called 'active 80 health plan' (1988 to 1999) and now, as noted, the third plan. Until now, the measures have concentrated on nourishment and health promotion by increased exercise (Table 6.1).

'Healthy Nippon 21' aims to improve health and prevent disease. The plan was conceived because of the changing social background due to:

- smaller numbers of children and larger numbers of elderly people
- lifestyle-related diseases (see Fig. 6.2)

*Table 6.1* Change of measures to promote health

| | National measures | Tertiary prevention | Secondary prevention | Primary prevention |
|---|---|---|---|---|
| Healthy, life time | content | Treatment, life-saving, recovery and maintenance of functions | Early discovery and treatment | Promotion of health Prevention of disease |
| Change of measure | correspondence | | Active 80 health plan | Healthy Nippon 21 plan |
| | term | 1978–1988 | 1988–1999 | 2000–2010 |

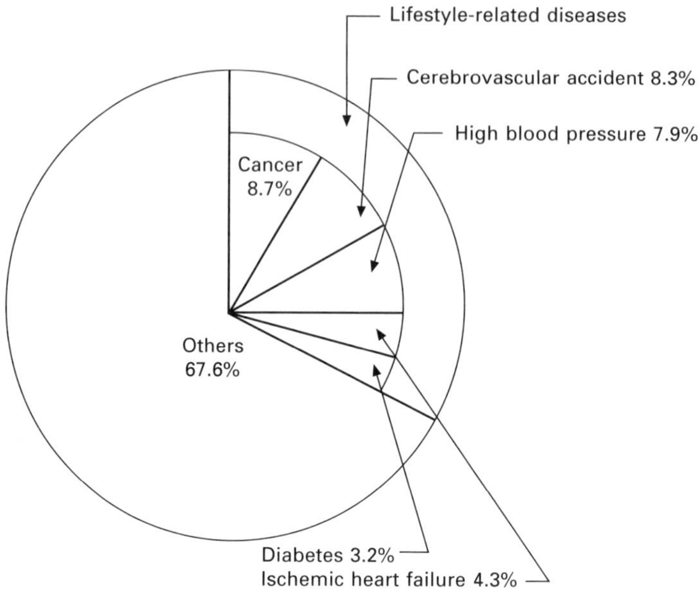


*6.2* General medical examination and treatment costs in Japan (1997).

- increase in private care of senior citizens (see Table 6.2)
- increasing medical costs.

In order to keep medical costs at a reasonable level and to maintain the quality of life, primary prevention is important. This plan was created by a study group of heads of medical care insurance. The 'Healthy Nippon 21 study meeting' consisted of 34 people of learning and experience in the field of diseases, habit diseases, medical insurance services, health insurance societies and local government.

### 6.1.3 The clothes which intercept stimuli from the natural world

*The environment surrounding health*

In Europe and the United States, the thesis that 'the thing which is bad for the environment is also bad for health' is commonly accepted. Here, the

*Table 6.2* Change of population composition in Japan (%)

| Age | 1950 | 1994 | 2025 |
|---|---|---|---|
| Older than 65 | 4.9 | 14.1 | 25.8 |
| 15–64 | 59.7 | 69.6 | 59.7 |
| 0–14 | 35.4 | 16.3 | 14.5 |

words environment and health are used together. The close relationship of environment, health and clothes needs study.

From the standpoint of protection of the body by clothing, external and internali environmental stimuli must be considered. In Japan, external stimuli include sweating in summer, pressure by layering of clothes in winter, changes of temperature and humidity caused by the Japanese four seasons, such as strong UV rays in May and abnormally hot temperatures in August. Furthermore, there are stimuli such as radioactivity, high temperature, ultra-low temperature, and extreme situation of high tension and high vacuum. In recent years the stimulus from the internal environment has attracted attention. The physical environment, such as illness, allergic trouble, and injuries, and mental environment, such as stress or neurosis, are examples of such stimuli.

### *Extreme environments*

Generally clothes are the barrier between the body and the outside natural world. In the natural world there are ground, sky, and sea. Normal everyday clothes need to be different from clothes for climbing Mount Everest, with a height of 8000 m, or those for climbing Mt. Fuji, with a height of 3000 m. In other words, the higher the mountain, the more advanced must be the clothing. Furthermore, space developments require completely different suits for the surface of the Moon and Mars than those worn on Earth.

There are special environments where a person cannot live even on Earth. In order to work under such conditions, special clothes are necessary. For example, clothing to protect from radiation, nuclear radiation and heat or to wear in clean rooms for manufacturing semi-conductors. Clothes used in such extreme environments put priority on functions.

The sea is an extreme environment. Ships have a duty to carry life-jackets to aid survival in the sea. Such clothing must float with constant buoyancy, and support life even at sub-zero temperatures.

### *Coexistence with artificial environments*

Clothes cannot intercept all hazards: temperature (heat), light (electromagnetic waves such as UV rays, infrared rays), air (wind), etc.

Life has been helped by the spread of mobile telephones, promotion of computerization, the development of information and communication technology. However, electromagnetic waves are emitted by such electronic equipment, which can change day by day. The effects on the health of the human body are not yet certain.

Turning off mobile phones in Tokyo Station began in April 2000 to protect people with cardiac pace-makers. Development of various fibers for protection, comfort, healthcare, medical care, and mental care is shown in Fig. 6.3.

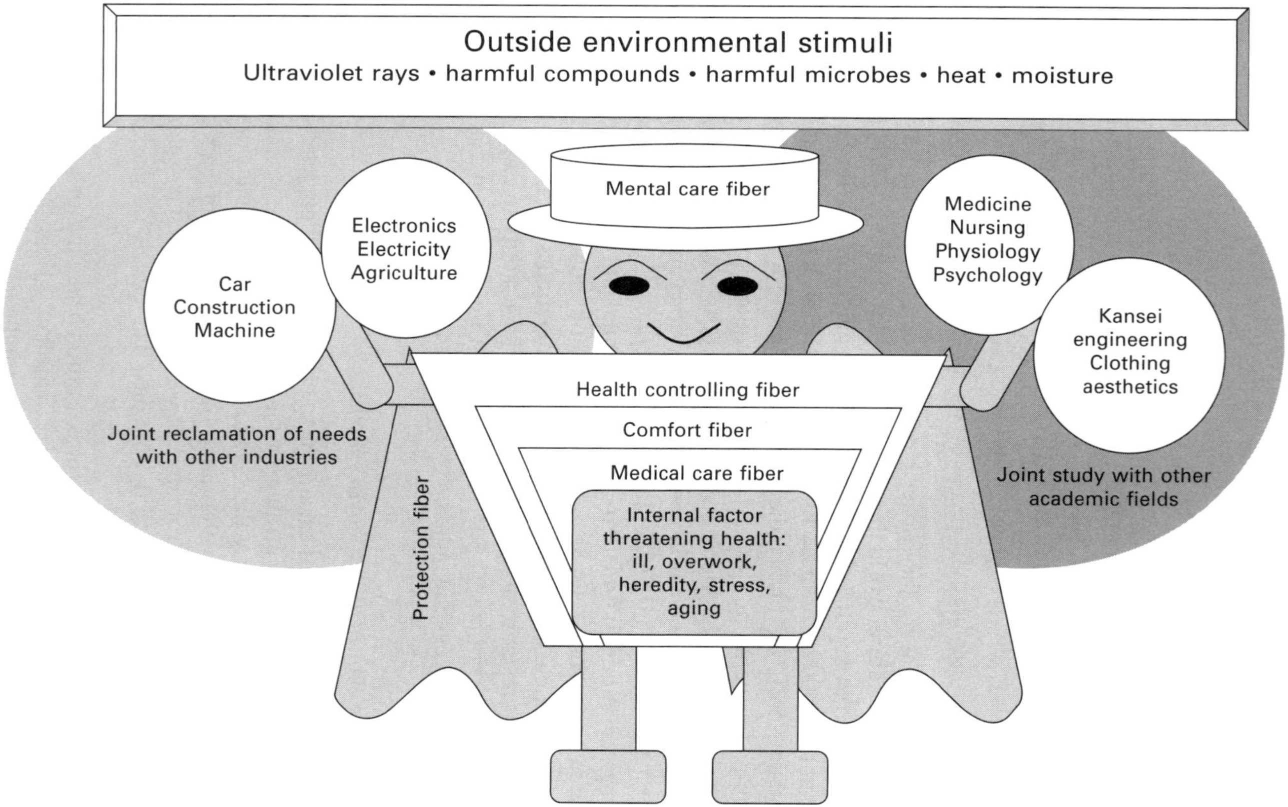


*6.3* Pursuit of fiber to keep physical and mental health.

## 6.2 Development of medical care materials to learn from 'smart fiber'

### 6.2.1 Learn from fiber of the living body

Fiber material design by copying fibers within the living body is very important in order to protect human health in an ever aging society. Varieties of artificial internal organs are used as shown in Fig. 6.4. The human body is a large user of fiber materials such as artificial kidney and artificial blood vessels. Of course, the human body is itself a fiber manufacturer and produces various kinds of fiber to protect our health. The communication between nanofibers in a cell (DNA) and nanofibers in clothes will be possible by the middle of the century.

A good example is the artificial kidney used for hemodialysis. About 170 000 patients use an artificial kidney and the number increases about 5000–8000 per year. Including potential patients, 400 000 people need the artificial kidney per year. The average age of the patient becomes younger year by year. The greatest factor is complication of diabetes, and there are many chronic diseases of nephritis.

The glomeruli in human kidney are a bulk material of special parallel capillaries. The artificial kidney imitates the glomeruli using hollow fiber.

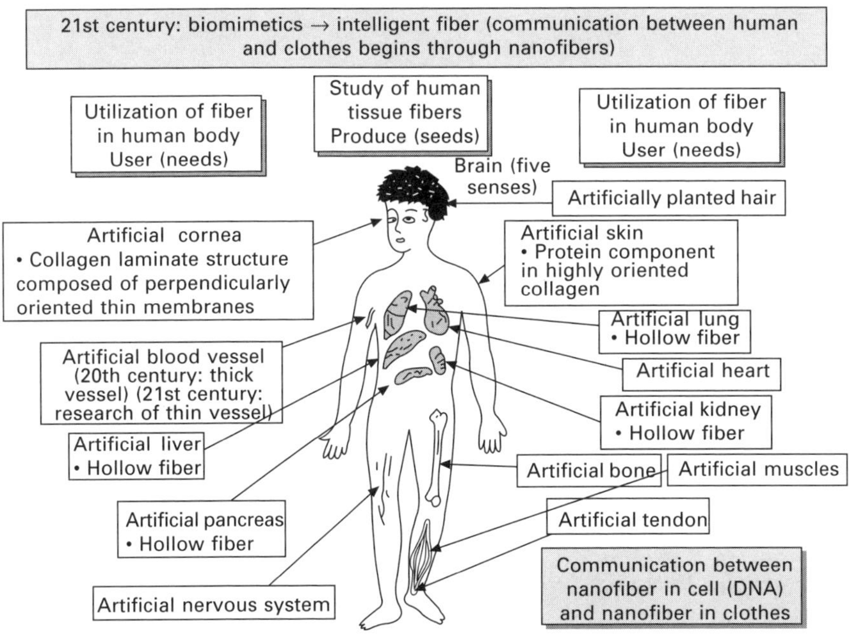


*6.4* Various artificial organs.

The original aim of treatment was only life extension. However, the emphasis now is not only the prolongation of human life but also rehabilitation. Since the artificial kidney is not perfect compared to the kidney itself, the development of new dialysis membranes with continuity of high performance remains a priority.

Hollow fiber is also used in an artificial lung and in artificial blood vessels. Furthermore, active study progresses toward production of artificial liver and an artificial pancreas. Study for an artificial skin, artificial muscle, and the artificial nervous system is also being carried out. Development of such smart fiber, which can learn from the functioning of the fiber of the living body is a great need for an aging society.

#### *Achilles' tendon*

The Achilles' tendon is an ultra super fiber composed of collagen fibers which have become a development objective of the next generation fibers. Muscle and bone are tied by collagen fibril at the Achilles' tendon. The components of the tendon close to the muscle and bone are different and show the necessary specific characteristics of each protein. Moreover, the tendon is a flexible and soft ultra superfiber and has the possibility of being used as an absorbent surgical suture which does not need to be removed. Other types of collagen fiber are used in the cornea, blood vessel, and liver.

#### *Fibrin*

Another smart fiber is fibrin. Water-soluble protein fiber of fibrinogen is usually dissolved in blood. When injured, bleeding occurs. Then blood coagulates naturally in about 1–2 minutes. This depends on the defense reaction which a body possesses when the fibrinogen changes to fibrin fiber network. When the wound recovers, the blood which hardens is degraded by an enzyme. Fibrinogen is an intelligent fiber which recognizes the environmental change, changes its physical properties and repairs itself. The role of the intelligent fiber is shown in Fig. 6.5.

Bio-fibers are soft and have no incongruity; they are completely different from metal embedded in the human body. Biomimetic fibers which are soft to humans are now being pursued to support an aging society.

### 6.2.2 Care products demand rises in aging society

#### *Care articles for senior citizens*

Development of care fiber for an ever aging society will become more and more important in the future. These are textiles for patients undergoing

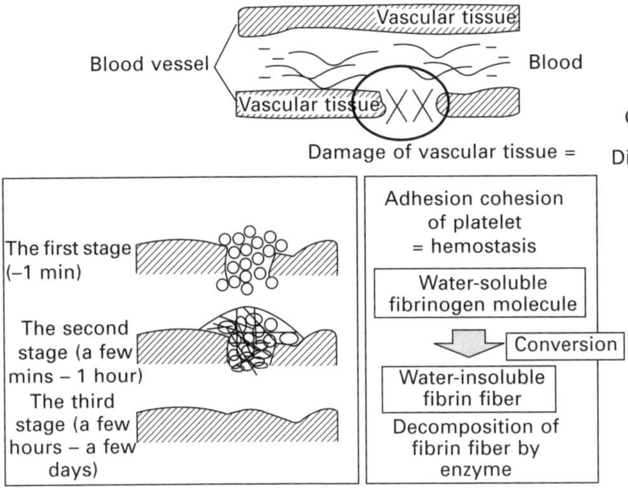


*6.5* Role of intelligent fiber (fibrin fiber).

medical care. Patients are different from people with normal and healthy bodies in that they are in hospitals or welfare facilities. Some might be immobile or incontinent and others, such as those suffering from dementia, can be very active. Their needs are very different. Accordingly it is important to collaborate with other fields of industry for the development of new fibers. This is, in effect, a new frontier for fiber development. The target can be very specific. For example, there are new fibers with the function of antimicrobial and deodorization to protect the target person whose immunity resistance becomes weak from infection. Furthermore, there are textiles for MRSA (methicillin resistant *Staphylococcus aureus*). Fiber manufacturers develop the textile and sell it in the market place.

Many companies enter the care fiber market using their own fiber materials. 'Full-scale expansion to health care products' by Toray, 'Well life plan' by Teijin, and 'the healthy cheering party series' (Toyobo) are good examples. These have grown to be big business in the United States. They will be big business in Japan too in the near future.

An example of an active clothes study interchange meeting the needs of handicapped people and senior citizens is shown in Fig. 6.6.

## 6.2.3 Clothes which are useful for health maintenance

### *Three elements to keep health*

Three elements to maintain health are nourishment, exercise, and rest. Moderate exercise is most suitable for a healthy life and stamina reinforcement. It is

*6.6* Clothes study meeting the needs of handicapped people and senior citizens.

utilized broadly without distinction of sex and age because of the ability to reduce stress and change feeling.

The first functions required for sports clothes are lightness and fitting well to the body. Furthermore, durability is important. With intense exercise, there are occasions when people fall. It is thus necessary to be friction and abrasion resistant and durable to repeated washing. In addition, it should be safe and comfortable, and absorb sweat immediately and evaporate to keep the body in a consistently dry state. In winter sports, clothes which keep the body warm are important.

Wearing clothes helps protect the body from various external stimuli and to give distinction to the wearer. This identifies the direction of new fiber material development.

### Sweat absorbent and easy dry material

Natural fiber has water absorbency, but polyester itself does not have water absorbency. By changing the textile structure, polyester can absorb sweat using a capillary action and allow the sweat to pass outside. MOISTY® (Toray), WELLKEY RMA® (Teijin), [eks]® (Toyobo), TECHNOSTAR® (Asahi Kasei), and AQUASTERUS® (Kanebo) are examples.

Moisture-permeable and waterproof material

This is the material which keeps the body comfortable in various kinds of bad weather. Mycroft R Rectus R® (Teijin), Goretex® (Goretex Japan) are examples.

Lightweight material

People feel more comfortable wearing lighter clothes. To get lighter clothes, the hollow part of fiber is increased, and as a result, the air in the hollow part contributes to lightness and heat-retention because of small heat conduction of air. Microart (Unitika), Aircube™ (Toyobo), LIGHTRON (Kanebo), Aerocapsule R (Teijin) are examples.

Low friction materials

To get a low friction material, various methods are available. It is better to have as flat a plane as possible, for example, making a dense plain weave with superfine thread, or making the surface water-repellent by modifying the surface with Teflon laminate. Another way is to make the surface uneven like a golf ball dimple to reduce resistance. Linplex (Descente) for skiwear and Aquabrade II (Mizuno) for swimming wear are examples.

Heat-retention and heat storage material

The passive heating material which does not allow heat to go out from the human body to the outside and positive material which takes in heat from the outside are typical of heat-retention and heat storage materials. The idea of the latter is to keep heat storage material in the hollow part of hollow fiber. Solar $\alpha$ (Unitika/Descente), Mobilethermo (Descente, Matsushita Denko), and Kerbinthermo (Toray) are examples.

The super strong material

An increasing tendency to use strong fiber, particularly for sports use, will continue in the future. T fiber in the shoes which Carl Lewis wears in a sports stadium is light and uses a strong polyethylene fiber, 'Dyneema' (Toyobo).

### 6.2.4 Development of a guard vest for jockeys in flat horse racing and steeplechases

Descente developed a guard vest for jockeys at the request of the Japan Racing Association (JRA). Since a fall from a horse in a steeplechase is more frequent than racing on the flat, the Japanese horse racing society requires a jockey to wear a guard vest. The purpose is two-fold. First, to reduce the shock to the jockey caused by the collision relaxation effect during a fall, and second, to reduce the damage to the jockey who might be stepped on by a horse after a fall. The vest is filled with polyethylene foam materials, so that it is stuffy when worn in summer, even though it has some holes for ventilation. A steeplechase is held only once a day, but flat races could take place several times a day. One jockey could ride several horses in a day. For this reason it is difficult to use the polyethylene foam vest for several flat races, and there was a request for a guard vest that includes heat control. The new vest must fulfill the following functions:

- not be stuffy and uncomfortable when worn
- more impact resistant than the existing available vests
- lightweight
- easier to move when worn than the existing vest
- price of the vest to be less than the existing vest.

To achieve these aims, an approach to introduce air into the vest was tried. The improvement of moisture permeability is possible by introducing many holes into the shock absorber, but then however, the strength of the material is reduced.

To solve the problem a honeycomb structure made of aramid was used by sandwiching the structure with foamed polyethylene of thickness 3 to 7 mm. The strength of the vest was then satisfactory. To improve the moisture permeability, apertures of 10 mm diameter were made by introducing 15mm apertures into the foam, with polyester mesh used for outer and lining materials. Elastic Velcro is used in the side adjuster to improve handling and fitness. The hips section is pushed into the pants usually. Accordingly mixed woven textile polyester/polyurethane was used for the side piece in order not to obstruct the movement of a jockey when riding a horse. The vest gross weight was 430 g, although the aim was initially 400 g. As shown in Fig. 6.7, the shock absorber is made of a rectangular block for the central part and equivalent hexagon for the shoulder and side parts not to disturb the movement of the jockey when using a whip. The honeycomb structure has more shock absorbency if the thickness of the structure is increased. The vest was provided to jockeys by the Japan Racing Association from June 1999.

## 6.3 Development trend of comfortable fiber for health

The trend of development of fiber for health is shown in Fig. 6.8. From the data, each manufacturer develops items which consumers demand. For example, the top five considerations required for clothes were:

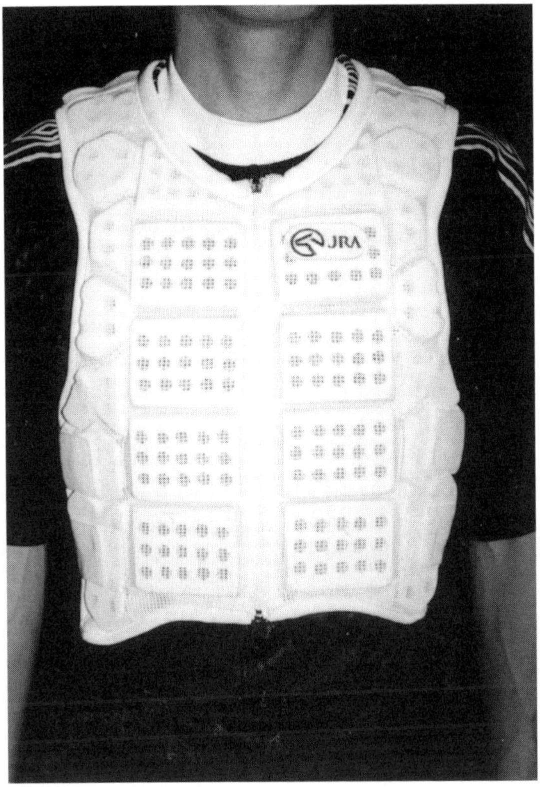


*6.7* Improved guard vest for jockey (Descente).

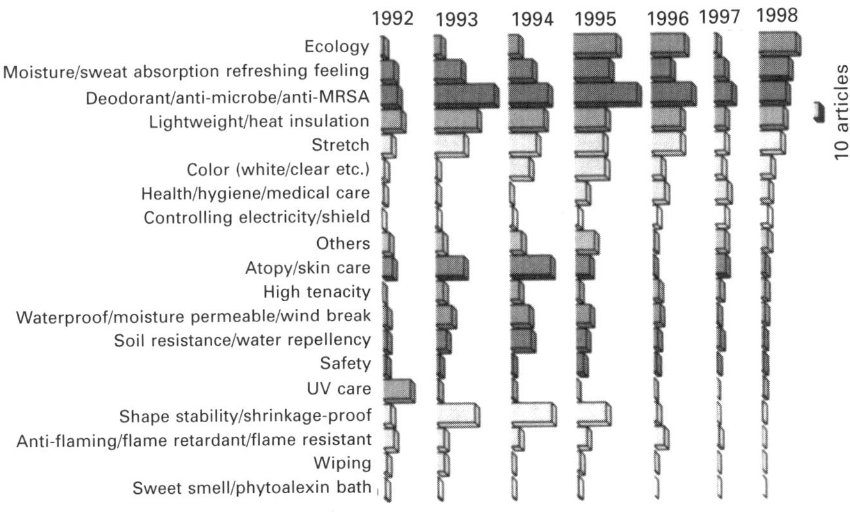


*6.8* Development trend of functionality fiber (number of articles).

(1) ecology
(2) moisture absorption/sweat absorption/refreshing coolness
(3) deodorant/antibacterial/anti-MRSA (anti-hospital infection)
(4) lightweight/heat insulation
(5) stretch.

## 6.4 Trend to seek for cleanliness and comfortableness

The high temperature and high humidity in Japan are suitable for bacteria and fungi to grow. Thus people who pursue cleanliness and comfort demand goods with deodorant and antibacterial functions. The demand is not only for textiles but also for many kinds of goods such as socks, curtains, underwear, washing machine, telephone, floor material, waste basket, lunch box, insole of shoes, toothbrushes, mechanical pencils, interior materials in cars, etc.

'Seiketsuhakusho of Toray', 'Toyobo's Hygiene Revolution of Toyobo', 'Series of Comfortable and Hygiene products of Unitika', 'Esthetique salon series of Kanebo and Clean Declaration of Kuraray' are examples of fibers designed for more healthy and comfortable life styles. A meeting to evaluate new functional needs of fiber products was convened by 177 manufacturers to unify the terms, evaluation method and standards of the evaluation. In this way consumer confusion is avoided, and the meeting is expected to contribute enhancement of antibacterial deodorization methods. There is no precedent for this kind of meeting in any other field of industry.

### 6.4.1 Evaluating new functions of fiber products

*Background of the establishment of the meeting*

In the high temperatures and humid climate of summer, microbes propagate very easily in the rainy season, particularly in June. Athlete's foot, bedsores and uncomfortable odor caused by the propagation of microbes when wearing clothes became the target in the development of new products to control these problems. The first product with sanitized finishing was marketed in 1955. After that many products with deodorant function were marketed. However, then there was no standard to show antibacterial and deodorant functions. Consumers complained about such conditions. In 1983, a 'textiles hygiene processing meeting' was held by fiber material manufacturers, processing companies, and clothing companies including 35 companies in industry, university, and officials under the directive of the MITI, with the aim of consumer protection and sound development of the fiber industry. They settled on a standard of evaluation method for antibacterial and deodorant processing in 1989, and gave certificates to products which had a higher level of processing than the standard. The mark was named the 'SEK mark'.

*Inauguration of meeting*

In April 1998, fiber products hygiene processing assembly changed its name to the Japanese Association for the Functional Evaluation of Textiles (JAFET) with the aim of international standardization. JAFET continued the authorization of 'antibacterial deodorization work' for more than 15 years, and started the certification mark of microbial control. The outbreak of pathogenic *E. coli* 'O157' in summer 1996 caused an antimicrobial boom not only in textiles, but also in kitchen utensils, toiletry (cosmetics) supply, and stationery (office supplies) manufacture.

*Definition of microbial control*

'Microbial control' means controlling microorganisms. 'Microbial control processing' means controlling specified microbes growing on fiber. The word 'antibacterial' is a wide-ranging term used within pasteurization, sterilization, antisepsis, disinfecting, preservation, mold proofing, and bacteriostasis. For this reason JAFET decided to use the term 'microbial control'. Furthermore, there was denotation to avoid consumer confusion by making a clear distinction between 'antibacterial goods' of different industries.

*What is microbial control?*

Microbial control means processing to restrain propagation of specific microorganisms on fiber such as *Staphylococcus aureus*, *Klebsiella pneumoniae*, a colon bacillus, *Pseudomonas*, methicillin-resistant *Staphylococcus Aureus* (MRSA). 'Propagation restraint' means to restrain propagation of microorganisms, not to make microorganisms extinct completely. If all microorganisms including skin normal bacteria are made extinct completely, it will cause other problems such as immune strength degradation and drug resistant fungus.

*The difference between microbial control and antimicrobial deodorization processing*

Antimicrobial deodorization processing is carried out for the purpose of 'deodorization', but 'microbial control processing' is classified into general and specific fields. The difference is shown in Table 6.3.

*The SEK mark*

SEK comes from the initials of Sen-i (fiber), Evaluation and Kinou (function). Granting the SEK mark is performed by a committee composed mainly of

*Table 6.3* Comparison of deodorization and microbial control processing (Report of new fiber evaluation meeting)

| Term | Deodorization processing | Microbial control processing<br>General use | <br>Specific use |
|---|---|---|---|
| Purpose | Suppression of microbe growth on fiber and prevention of bad smell caused by microbe | Suppression of microbe growth to improve environment for life, healthcare and medical care | |
| Definition | To suppress growth of microbe on fiber and to show deodorant effect | To suppress growth of microbe on fiber | |
| Target product | All textiles | Textiles used in general family | Textiles used in medical and related institution |
| Target microbe | □ *Staphylococcus aureus* | ■ *Staphylococcus aureus*<br>■ *Klebsiella pneumonia*<br>□ *Pseudomonas aeruginosa*<br>□ *Escherichia coli* | ■ *Staphylococcus aureus*<br>■ *Klebsiella pneumonia*<br>□ *Pseudomonas aeruginosa*<br>□ *Escherichia coli*<br>■ MRSA |
| SEK mark | Blue (DIC 66) (Deodorant effect) | Orange (DIC 156) (Improvement of life and care environment in general family) | Red (DIC121) (Improvement of medical environment in medical institution) |

■: Essential microbes to be examined
□: Optional microbes to be examined

experts. The SEK mark guarantees antimicrobial deodorant processing, microbial control for general purpose (orange SEK mark), and microbial control for specific purpose (red SEK mark).

### *Target products of SEK mark*

Microbial control for general purpose and antimicrobial deodorant processing target almost all application fields such as clothing, bedding, room design product and personal effects. On the other hand, microbial control for specific purpose is applied to white robes, nursing clothes, care clothes, underwear, pajamas, apron, socks, a sheet, cover, a blanket, curtain, masking, cap, a towel, a cloth diaper and dust cloths, for the needs of medical institutions.

### 6.4.2 Fiber with water and moisture absorbency

Sportswear, in particular, needs to absorb water and moisture after intense exercise. Synthetic fiber without such function is not suitable for sportswear because of stuffiness and stickiness. For this reason, development of functional synthetic fiber with water absorption/hygroscopic properties was a major objective over a long period.

One of the principles used to give a water absorption property is the use of capillarity. Two representative examples are in the fiber industry. One depends on the application of capillary movement, a phenomenon at liquid boundaries resulting in the rise or fall of liquids in narrow tubes or in a slit between two leaves. The level becomes concave or convex, and the contact angle becomes an acute or obtuse angle depending on whether the surface of the capillary tube gets wet or not. The level of the liquid balances with surface tension and external pressure. Generally, the phenomena that control the wetting of a surface can be classified as:

- dispersion wetting caused by liquid diffusing on the surface of solid
- infiltration wetting observed in a cotton dry cloth and towel absorbing water
- and the so-called adhesion wetting as raindrop sticks to the surface of a solid.

Wetting occurs when the contact angle is 0°, less than 90°, and less than 180° in dispersion, infiltration, and adhesion wetting, respectively. For example, spacing of liquid to infiltrate into the apertures of porous material and a sieve tube of bamboo becomes very large for the liquid to get the wall wet. The fiber manufacturer produces sportswear with capillary structure to let sweat go outside as soon as possible. Hollow fiber and fiber with the heteromorphology in cross-section of fiber are examples. Examples of water absorption fibers from several companies are shown in Fig. 6.9. The ability of polyester or nylon fiber to absorb water can be improved by making the fiber gap just like a capillary. Furthermore, the decrease of contact angle can absorb water highly by alkali treatment of polyester fiber.

#### *Porous water absorptive polyester fiber*

Teijin developed a porous water absorbing fiber 'WELLKEY' (Fig. 6.10) more than ten years ago. It was the first water absorbing fiber material and was awarded the Technical Prize of the Japan Fiber Science and Technology Society in 1990. The micropores of 0.01–0.03 μm in diameter on the surface run right through into the hollow part. The attached water on the side surface of fiber goes into the hollow part. The capillary phenomenon and rough surface make the fiber water absorbing. Textiles made of WELLKEY can

| Trade name (Manufacturer) | Cross-section | Properties |
|---|---|---|
| WELLKEY (Teijin) |  | Hollow fiber with many pores (the first sweat absorbent Tetron in the world), no sticky touch, comfortable to wear, no cold touch |
| AREOCAPSULE DRY (Teijin) |  | The cool touch spun yarn material, special polymer, dry touch, UV shielding |
| Technofine (Asahi Kasei) |  | Capillary phenomenon, absorb and spread sweat, dry touch |
| CEO • α (Toray) | 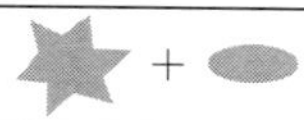 | Water absorption by capillary |
| SPACEMASTER (Kuraray) |  | Unique cross shape polyester, water absorption, quick drying, dry touch |
| Kilatt-P (Kanebo Gohsen) |  | Lightweight, moisture keeping, water absorbent, suitable for training wear and inner wear |
| COOLMAX (Du Pont) | 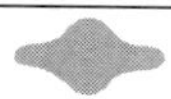 | UFO shaped (four channels) fiber, spread sweat quickly, dry and comfortable |
| Aegean (Mitsubishi Rayon) | 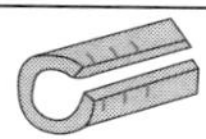 | Hollow fiber composed of two hydrophilic polymer, tear in longitudinal direction and have cut in lateral direction, water absorbency, quick dispersion and dry, not sticky |

Water absorption fits the following equation.
Water absorption rate (horizontal direction > $l^2 = (r\gamma \cos\theta/2\mu)t$
Introduction of capillary structure hollow fiber, different cross-section
Contact angle with water θ hydrophilic skin membrane

l: length of water absorption in horizontal direction (cm)
μ: viscosity of water (g/cm • sec)
r: radius of capillary (cm)
γ: surface tension of water (dyne/cm)
θ: contact angle of water and capillary surface wall (degree)
t: time

*6.9* Examples of water absorption fiber by various manufacturers (modified from T. Suzuki, 1999 The society of fiber science and technology Japan, basic lecture).

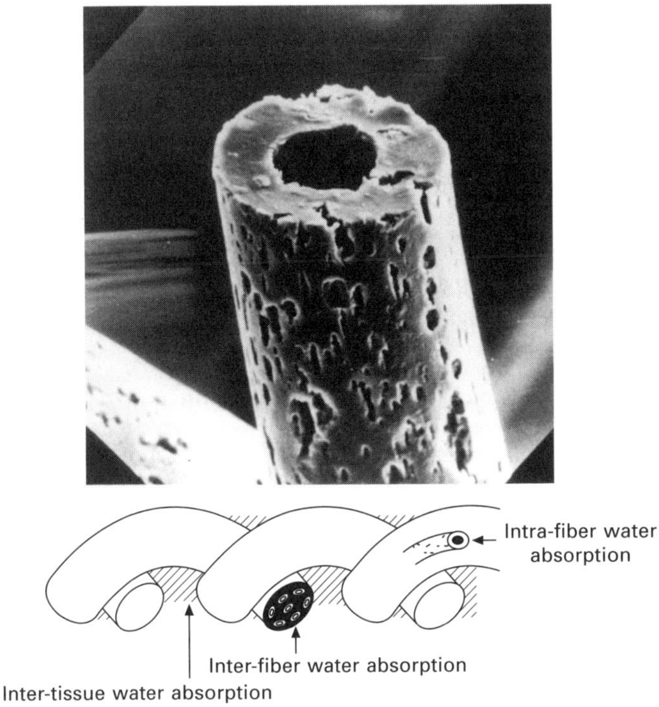


*6.10* Water absorbable polyester fiber 'WELLKEY' with many pores (Teijin).

absorb water and it is not sticky even after absorbing sweat. For this reason the uniform of the All Japan Women's Volleyball Team was made from WELLKEY.

### *Highly moisture absorbing and moisture releasing nylon*

Recently, highly moisture absorptive and highly moisture releasing nylon was developed independently by Toray and Unitika. Originally nylon itself had many good characteristics, but it was inferior in moisture absorbency. When nylon was used for clothes the lack of moisture absorbency caused stuffiness, stickiness and was uncomfortable. Both companies solved the problems and developed materials by completely different approaches. Toray has developed QUUP and used it for pantyhose and an undershirt wear. Unitika developed HYGRA and used it for sportswear and geotextiles.

### *Highly moisture absorptive and moisture releasing nylon for pantyhose*

Toray developed a new polymer material for nylon fiber, 'QUUP', with superior utility and highly moisture absorptive/releasing property, about double that of conventional yarn (see Fig. 6.11). It was achieved by a new polymer alloy technology development. A nitrogen-based special polymer was mixed with regular nylon molecules to give a fiber with high water absorptive and water releasing properties. The comfort of wearing QUUP used garments is very similar to that of cotton, and it has the added distinctive features of a soft and smooth touch and superior color intensity, while retaining the conventional functionality of nylon. QUUP is used for a wide range of applications, including pantyhose, underwear and sportswear. The price of the material is more expensive than ordinary nylon by about 20%. The evaluation result wearing pantyhose is shown in Fig. 6.12.

### Moisture absorptive/releasing synthetic fiber with skin-core structure

Unitika succeeded in making fiber from a highly water absorptive polymer, which can absorb water 35 times the polymer weight, and developed an epoch-making fiber HYGRA$^{TM}$. Conventional moisture absorptive fiber was made by modifying the surface of synthetic fiber to be hydrophilic or applying a hydrophilic polymer. Accordingly, no synthetic fiber could have superior moisture absorption to natural fiber. Many such water absorptive polymers were developed. However, they were applied only to disposable materials such as diapers and napkins because they do not have water releasing ability. Moreover, it was difficult to make such fiber by melt-spinning. Unitika

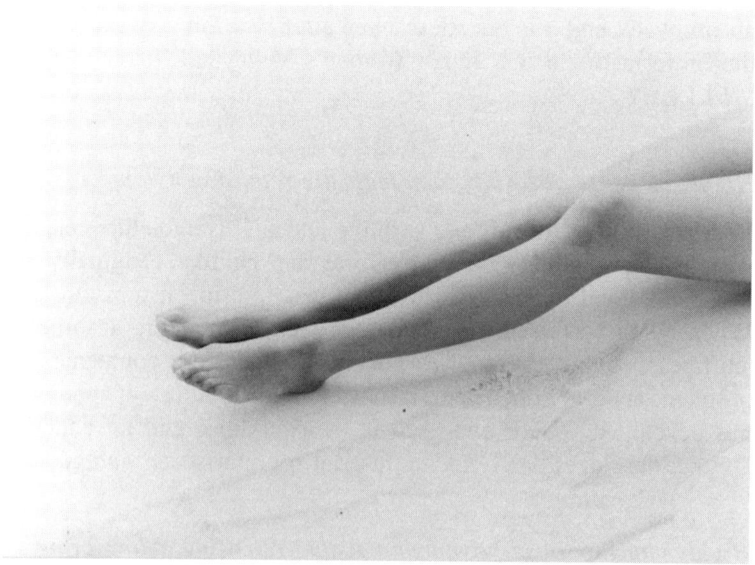

*6.11* Pantyhose made of highly moisture absorbent and moisture releasing nylon fiber (Toray).

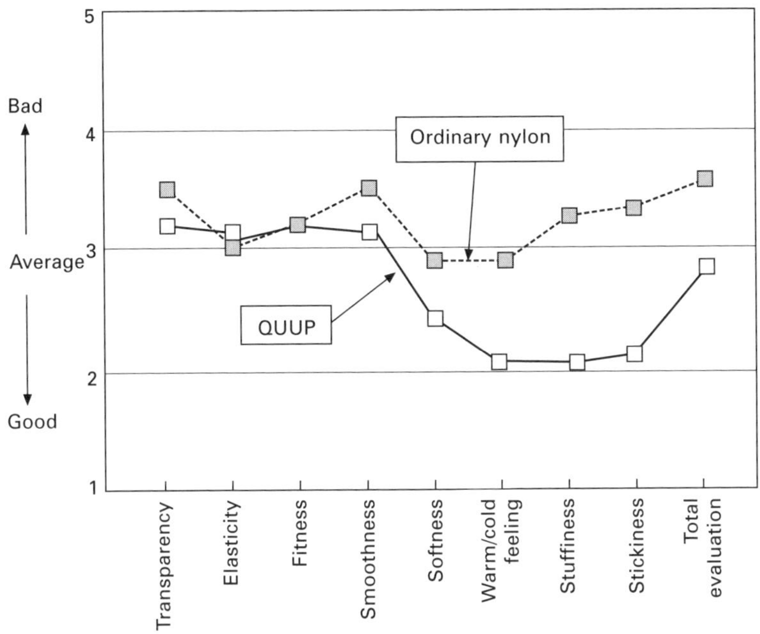


*6.12* Wearing evaluation of pantyhose (Toray).

developed a highly water absorptive/releasing polymer which can be fibrilized by melt-spinning. It has a skin-core composite structure. The core part has a special network structure and the skin part is nylon as shown in Fig. 6.13. Conventional water absorptive polymers retain water by an ionic bond between functional ionic groups such as sulfone groups and water. Even if it showed high water absorbency, it did not have water releasing ability and ability to be melt-spun. HYGRA does not absorb water by an ionic bond. Hydrophilic groups retain moisture. The network structure of polymer used for HYGRA is controlled. HYGRA has the ability to release water even after absorbing water fully. The skin-core structure of HYGRA consists of a nylon skin part and hydrophilic core part, so that the moisture/water on the surface of the fiber is permeated into and absorbed by the hydrophilic core part when the temperature increases. Conversely, when the temperature decreases, the nylon part shrinks and moisture absorbed in the core part is squeezed out into the nylon part. The ability of HYGRA to absorb water and release water is higher than with natural fiber and HYGRA is stronger, has more dimension stability and is more comfortable when worn compared to natural fiber (see Fig. 6.14). The combination of the core and skin polymer will give many kinds of products. For example, the combination of polyester will give the same level of electric controlling character with natural fiber at low moisture content. Unitika will apply HYGRA not only to clothes such as underwear,

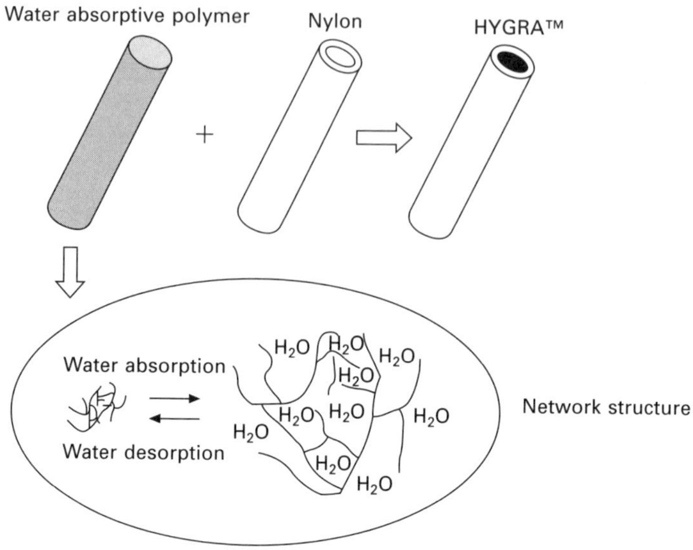


*6.13* Technical concept of HYGRA™ (Unitika).

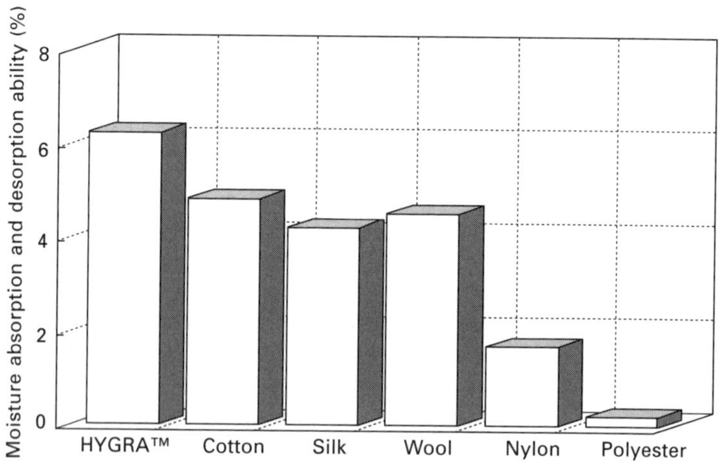


*6.14* Comparison of various materials on absorption and desorption of moisture (Unitika)

sportswear, and socks, but also to non-clothes fields such as life materials, civil engineering, construction, industrial materials, interior, and living materials.

### 6.4.3 Antimicrobial fibers

*The application demand for antibacterial products*

These have a very great range:

- Life materials: clothes, bedding, carpet, kitchen goods, kitchen mat, domestic electrical appliances, towel, toiletry (ceramics, mat), toy (stuffed animal)
- Environmental hygiene: antimicrobial sand in amusement park and sand box
- Food hygiene: kitchen utensils, wrapping material for food, tableware, clothes
- Labor hygiene: office supplies (stationery, telephone, facsimile, personal computer)
- Medical hygiene: bedding, white gown, surgical gown, curtains in hospital room
- Construction/public works: cement, building materials, woods with antimicrobial treatment
- Communication/transportation: coated electric wire, printed wiring, parts of car, hand straps
- Industry materials: clothes used in food and medical material factory, miscellaneous articles
- Agriculture/stock raising/fishery: wrapping material, container, working tools
- Sports: sportswear, accessories, tools for apparatus gymnastics
- Others: microphone for karaoke, cash card.

Antimicrobial goods are used not only in the field of food hygiene and medical hygiene, but also in everyday environmental hygiene fields including shirts and underwear, for which such treatment is not necessary if they are washed thoroughly. In addition, in the field of labor operation hygiene, as for the mechanical pencil, the ball-point pen, the telephone, antimicrobial treatment is applied. The microbial treatment for wood and leather is necessary to protect the material. Most of the treatment is done in one step.

*The term antibacterial*

'Sterilization' means that all microorganisms are sterilized. 'Disinfection' means decreasing some specific microorganisms. 'Repression' is used for textiles (see Fig. 6.15).

According to Professor N. Korai of the University of Tokushima, controlling microorganisms by antimicrobial reagents can be expressed by growth and growth inhibition curves, depending on the number of viable microorganisms and incubation time. There is an induction period, exponential phase, stationary

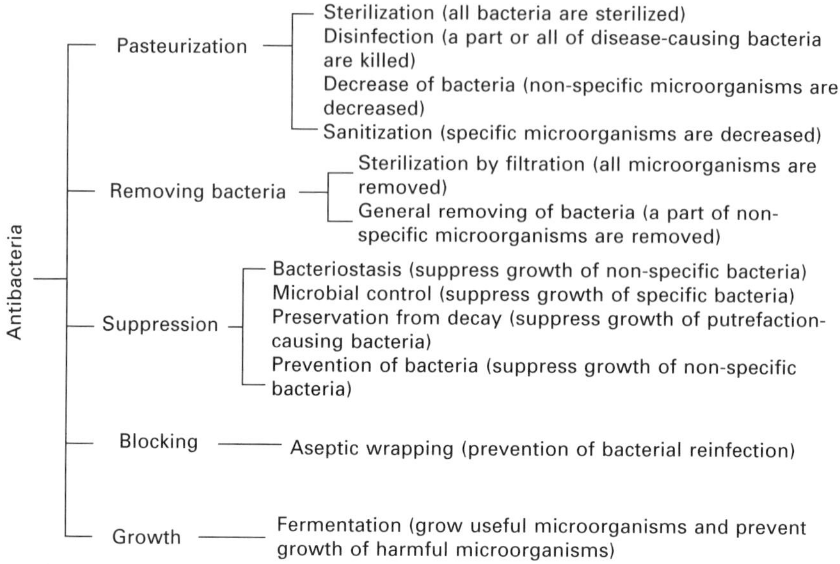


*6.15* Definition of antimicrobial-related terms (K. Urabe, Japan Chemical Fibers Association, Report of Investigation, No. 4, p. 37 (1999)).

phase, and death phase. Such growth curves and inhibition curves of microorganisms are shown in Fig. 6.16.

During the induction period, microorganisms take time to get used to the components of the medium, the pH, and temperature, and increase gradually. In the exponential phase, microorganisms grow very quickly. In the stationary phase, the rate of growth decreases because of the lack of nutrition in the medium, accumulated metabolites in medium, and change of medium pH, and the number of viable microorganisms becomes constant. In death phase, the number of viable microorganisms decreases with incubation without addition of antimicrobial reagent. If some chemical is added in the exponential phase, different effects appear depending on the kind and concentration of chemical added. Figure 6.16 shows the following:

① Without chemical addition.
② In the growth suppression region, the increase of microorganisms is less than the ordinary case depending on the kind and concentration of chemical.
③ Microorganisms do not increase at all in bacteria controlling region.
④ Microorganisms decrease by addition of chemical in the sterilization region.

Chemical usage in the sterilization region has appeared only recently. However, most of the chemicals belong to the propagation restraint region. As there are many products treated with antimicrobials, consumers should select what they need. The conditions necessary for antibacterial agent use in textiles are as follows.

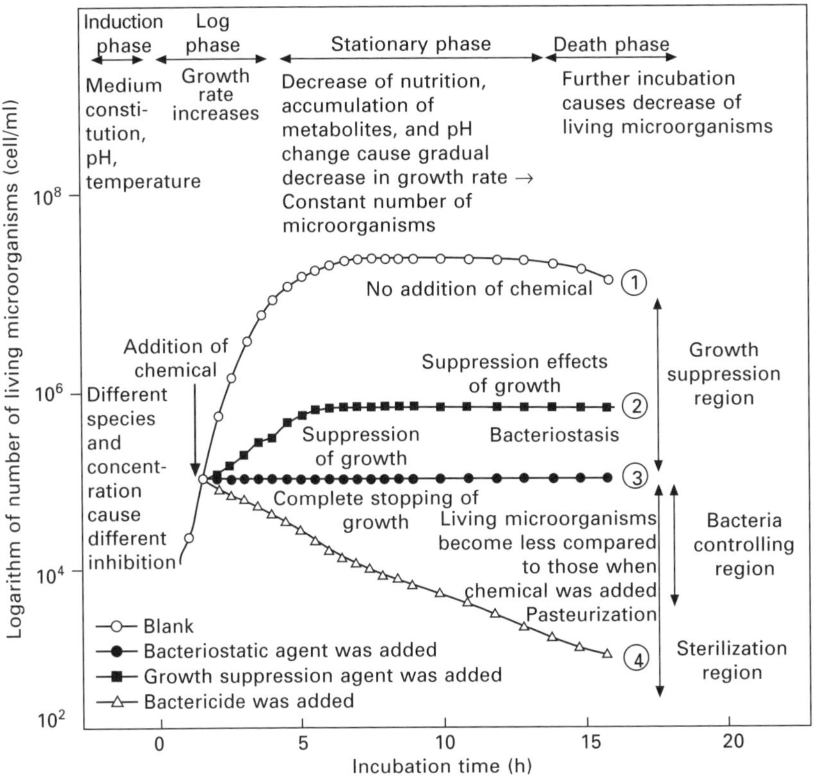


*6.16* Growth and inhibition curves of microorganisms (modified from H. Korai, Senshoku (Dyeing), No. 61, p. 1).

First to guarantee the effect. To pass the technical standard of SEK, which includes deodorant processing and bacterial control processing, a bacteriostasis activity value of more than 2.2 is necessary by a unification test in the case of antibacterial deodorization processing.

The second is durability, particularly during washing. For disposable goods, the number of washing times is zero, whereas ten or more washes are applied for most goods. In fact, the effect of antimicrobial reagents must be maintained after 50–100 washings. Then there is the application limit. For the general purpose of controlling microbes by SEK from 1998, *Staphylococcus aureus* and pneumonia bacillus must be tested for, but pseudomonas and a colonic bacillus are optional. For specific purposes, MRSA becomes a required microbe, in addition to microbes for general purposes. A specific purpose would be chiefly for use in the medical field.

In addition, the safety of a medicine is important. Acute toxicity, mutagenic property, and skin acrimony are specified, and the security in disposal and burning must be addressed. For example, there must be no dioxin generation.

There must be security also in decomposition of a medicine and the chemicals should not be endocrine disrupting chemicals and also must be friendly to the ecosystem.

### 6.4.4 Antimicrobial treatment against MRSA

Many kinds of antibiotics are not effective against MRSA. When such hospital infection occurs, local infection causes blood poisoning of a patient, and the situation may be life-threatening, and family and the medical care persons may also become infected. Measures against MRSA infection are taken very seriously by the Ministry of Health and Welfare. Periodical hospital sterilization and hand washing by soap and running water are executed at each hospital for infection protection.

MRSA infection transmits, not only within a special environment, but also via doctors and nurses. For this reason the fiber manufacturer has developed MRSA propagation restraint textiles for prevention of infection. A white gown using antimicrobial material is shown in Fig. 6.17.

*6.17* White robe made of microbe controlling material.

Tertiary ammonium and silver are the main chemicals for the treatment. 'Hinokichi' of Toho Textile uses a special microcapsule containing Cypress oil produced in Aomori Prefecture to treat the fiber. 'Chytopoly' of Fujibo uses chitosan from the shell of prawn and crab, mixed with rayon. In an investigation by the Ministry of Health and Welfare in 1992, MRSA was detected in about 90% of hospitals with more than 500 beds.

## 6.4.5 The causes and measures of unpleasant odors

*Good smell and bad smell*

Unpleasant smells from a person's sweat can be removed by taking a bath or shower. The microbe which lives in skin and clothes will multiply using sweat and soil as nutrition. The smelling component is isovaleric acid and comes from the decomposition of sweat and soil, and 'nonenal' which is peculiar to the old and middle aged. Possible causes of bad smells from a human body are shown in Fig. 6.18.

*Unpleasant odors in living spaces*

The Environmental Agency searched for the causes of 'unpleasant odor in living spaces' using 1200 private questionnaires. Top of the list is the smell of toilets, followed in order by decaying garbage, the smell of cooking, of drainage, musty odor (in winter dew condenses in a closet and mold grows), the smell of a shoe cupboard, from walls (due to formaldehyde and organic solvents used in the adhesives), cigarettes, a bad-smelling air-conditioner, and pet odor.

A mixed odor of volatile molecules comes from the living bodies due to protein, carbohydrate, and higher fatty acid being decomposed by microbes.

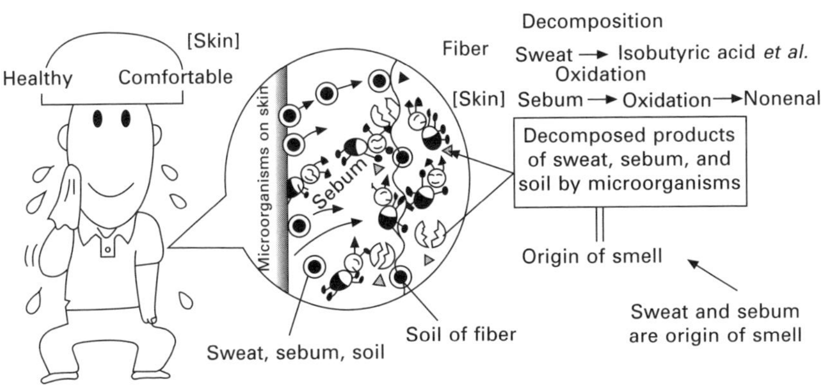


*6.18* Cause of bad smells from human body (modified from JAFET data).

These number between 300 and 400 types. Among them the strongest are the following four offensive odors:

(1) Hydrogen sulfide, as from rotted eggs.
(2) Methyl mercaptan, as from a rotted onion.
(3) Trimethylamine, as from rotting fish.
(4) Strong irritating smell of ammonia.

An example of the four offensive odors from humans is given in Table 6.4.

Additionally, research and development has been carried out into the effects of deodorant on sulfur, oxides of nitrogen, indole and skatole from excrement, a smell of perspiration, a smell of a cigarette. Such offensive odor not only gives discomfort, but can affect the nerves, irritate, prevent peace of mind, give headache, and obstruct job performance. Furthermore, if the stimulus is strong, it also causes stress in the nervous system, increased pulse rate, elevated blood pressure and a bad influence on various internal organs. The minimum amount of material producing a sense of a smell is called the threshold value. Threshold values of the four offensive human body odors are given in Table 6.5. Clearly only very small amounts are needed to cause very offensive smells.

### *The mechanism by which we respond to smell*

The chemical causing the smell volatilizes, enters into the nose, is absorbed in the upper part of the large chamber of the nasal cavity by the olfactory cells. Then a potential is generated in the membrane of the cell, a chemical stimulus causes an electrical impulse which is transferred to the nervous system, and the brain records the smell as unpleasant. If there is damage to the olfactory senses, it is debilitating and even dangerous since the taste of cooking can be affected and detection of gas leakage can be severely delayed.

*Table 6.4* Bad smells from human body

| Origin of bad smell | Main component of bad smell | Major cause |
|---|---|---|
| Bad breath | Methylmercaptan<br>Hydrogen sulfide | Bad breath of leftover fermentation by bacteria |
| Body odor | Trimethylamine<br>Acetic acid<br>Valeric acid<br>Caproic acid | Decomposition of sweat constituent by microbe |
| Excrement | Ammonia<br>Skatole<br>Indole<br>Hydrogen sulfide | Smell of feces and urine, and decomposition of feces by microbe |

*Table 6.5* Threshold of four main offensive smells (Professor Y. Shirai, Shinshu University)

| Chemical causing bad smell | | Bad smell | Threshold (mg/l) |
|---|---|---|---|
| Hydrogen sulfide | | Smell of rotted egg | $4.1 \times 10^{-4}$ |
| Methylmercaptan | | Smell of garlic | $7.0 \times 10^{-5}$ |
| Trimethylamine | | Smell of rotted fish | $2.7 \times 10^{-5}$ |
| Ammonia | | Strong irritating odor | 1.5 |
| Reference | Skatole | Smell of feces and urine | $5.6 \times 10^{-6}$ |
| | Acetaldehyde | Irritating odor | $1.5 \times 10^{-3}$ |
| | Musk | Fragrance | $5.0 \times 10^{-6}–10^{-10}$ |

*Approach to eliminate or reduce offensive odours*

Deodorization refers to removing odor in the atmosphere, and the word is used both for chemical and physical deodorization.

Using chemical deodorization the offensive odor is neutralized or immobilized by chemical reactions and is changed into material of low odor level. Fiber manufacturers immobilize such material on a fiber easily using blasting processing. However, the method is limited because chemical deodorization uses harmful chemical reactions. Photocatalysts such as titanium oxide have recently been developed which decompose offensive odors using UV light which initiates a redox reaction. The durability of the catalyst is theoretically semi-permanent. However, due to the energy of UV rays the fiber itself deteriorates.

By physical deodorization the chemical causing offensive odor is absorbed on to porous materials with a large surface area such as activated carbon, zeolites, and silica gel. These can absorb a large variety of chemicals. Such physical deodorizers also contain some chemical deodorizer.

Using biological deodorization, garbage can be decomposed in soil by microorganisms under aeration. Biomimetic deodorization was suggested by Professor Shirai of Shinshu University, utilizing an artificial enzyme which copies the way living organisms decomposes offensive odors. Deodorization can include masking of offensive odors with spice or oil of extracts of plant. Such deodorization can mask the odor immediately, but acute effect is high, and the smell itself is not removed.

## 6.4.6 The smell of cigarettes is very complicated

The smell of a cigarette is very complicated. From the cigarette, there is underflow smoke (very harmful) and mainstream smoke coming out from a point of a cigarette. When smoke comes out through a glass filter, some part of smoke like oil mist is trapped by the filter, and some like vapor is half trapped by the filter, and some like gas passes the filter. The components in the smoke have been studied and nicotine, acetaldehyde, acetic acid, catechol, hydrogen cyanide, phenol, ammonia, cresol, hydrogen sulfide, pyridine are included in high ratio. In reality, there are several tens of thousands of different kinds of chemicals in smoke from one cigarette. Among them, the biggest component is nicotine, which does not have a strong odor itself. It turns to pyridine or pyol and these smell.

There are deodorant fibers used with cigarettes. These include the strong deodorant material 'Cigernon' from Toray (Fig. 6.19), the highly durable deodorant polyester staple 'Freshcall$^{TM}$ II' by Teijin (Fig. 6.20). The deodorant fibers are shown in Figs 6.19, 6.20 and Fig. 6.21, respectively.

In reality, smells in everyday life do not come from a single source, but from many components mixed together.

### 6.4.7 Formaldehyde deodorant processing 'Deofor'

Indoor air pollution materials are classified by the World Health Organization into four categories: UV organic compounds, volatile organic compounds, non-volatile organic compounds, and particulate matter. Formaldehyde is a highly volatile organic compounds, and is used in building materials. No smell is sensed if the concentration of formaldehyde is less than 0.1 ppm.

As the concentration increases from 0.1 to 1 ppm, most people begin first to sense the smell, then become very aware, and later to feel a stimulus to nose, eye and throat and then discomfort. At concentrations greater than 1

*6.19* Curtains made of high cigarette deodorant ability, 'Cigernon' (Toray).

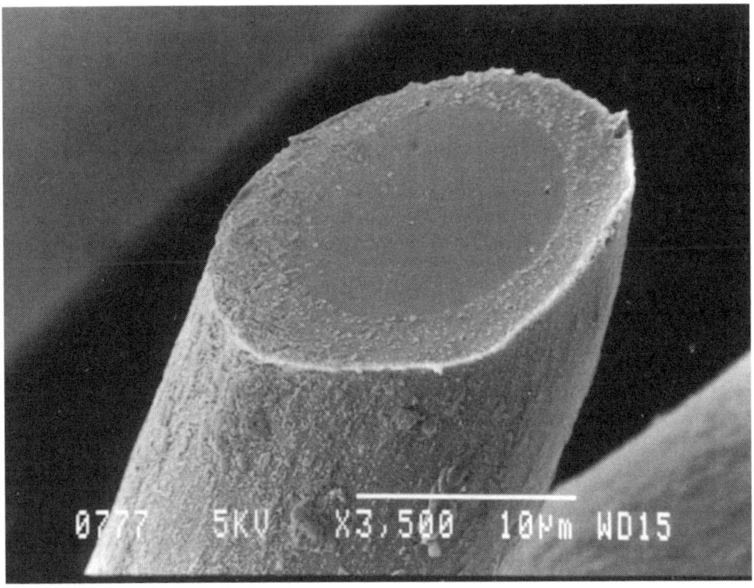


*6.20* Highly durable deodorant polyester staple 'Freshcall™ II' (Teijin).

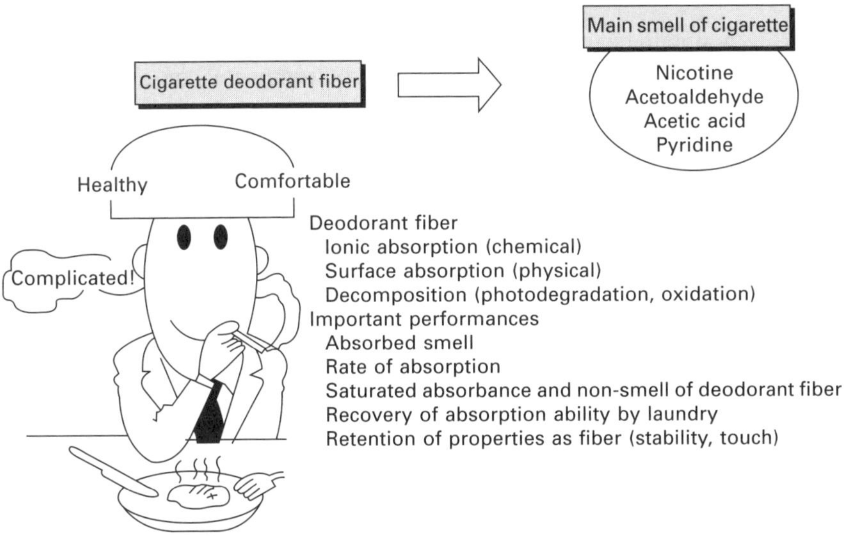


*6.21* Cigarette deodorant fiber.

ppm, tears are induced and a person may fall into dyspnea. At concentrations greater than 50 ppm, the experience could be fatal.

Daiwabo developed material to deodorize formaldehyde. An amine compound is coupled with formaldehyde by the Schiff's reaction which then deodorizes the formaldehyde. The deodorant effect of DEOFOR was evaluated by putting the test sample cloth (10 cm × 10 cm) in tedlar bag with formaldehyde. The concentration of formaldehyde was determined at various time intervals. Figure 6.22 shows the result when formaldehyde concentration was 4 ppm. The deodorant activity of a cloth having the DEOFOR processing improved compared to the untreated cloth. The curve of the cloth in the figure after washing three times showed same level of deodorant activity as the processed cloth before washing. Even after ten washes the result did not change.

Furthermore, the evaluation of DEOFOR treatment was done using 7 $m^2$ curtain in a room where formaldehyde gas was released from three pieces of plywood at 30°C and 60% humidity for one week. After three days, the concentration of formaldehyde using untreated cloth rose after only one exposure. The cloth with DEOFOR treatment kept up deodorant activity for one week. In addition, material with DEOFOR treatment has the activity to deodorize ammonia, acetic acid, and acetaldehyde as the main components of cigarette smoke.

## 6.5 Fiber to guard environment and health

### 6.5.1 Various kinds of approach to maintain health

Fibers to maintain health are classified into four categories: healthcare fiber, comfort fiber, stimulus relaxation fiber, and environmental conservation fiber. In the classification, the relation among environment, health and fiber is readily understood.

Utilizing healthcare fiber has three objectives as shown in Fig. 6.23. 'Making life comfortable' and 'relaxation of outside stimulation' help to maintain health. The effects aimed at are sweat absorption, high water absorption,

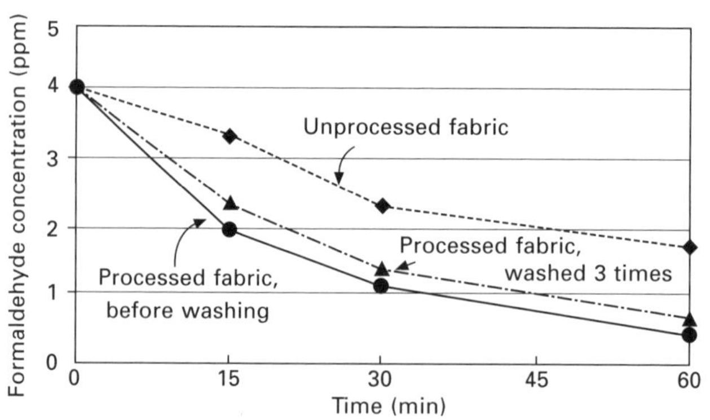


*6.22* Remained formaldehyde concentration (Daiwabo).

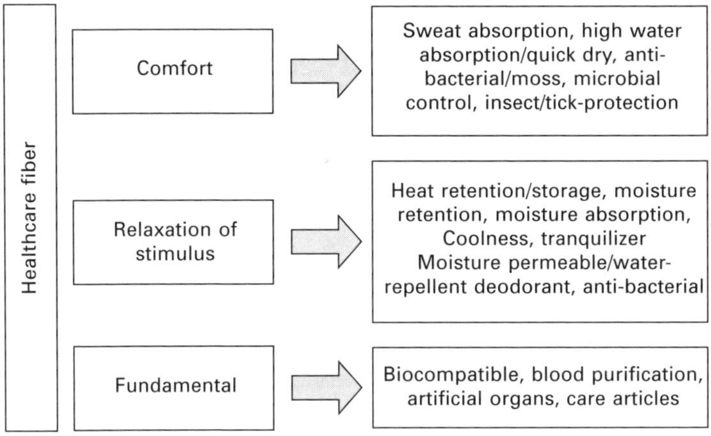


*6.23* Application of high functionality fiber to health.

fast-drying, anti-bacterial, mold-proofing, controlling bacterial growth, insect repellent, and acarid repellent, and the latter includes heat insulation, heat storage, moisture keeping, moisture absorption, refreshing coolness, moisture permeable, water repellent, deodorization, being antibacterial. Recently, fibers with complex antibacterial/deodorization functions have appeared. Functions required for fiber are shown in Fig. 6.24.

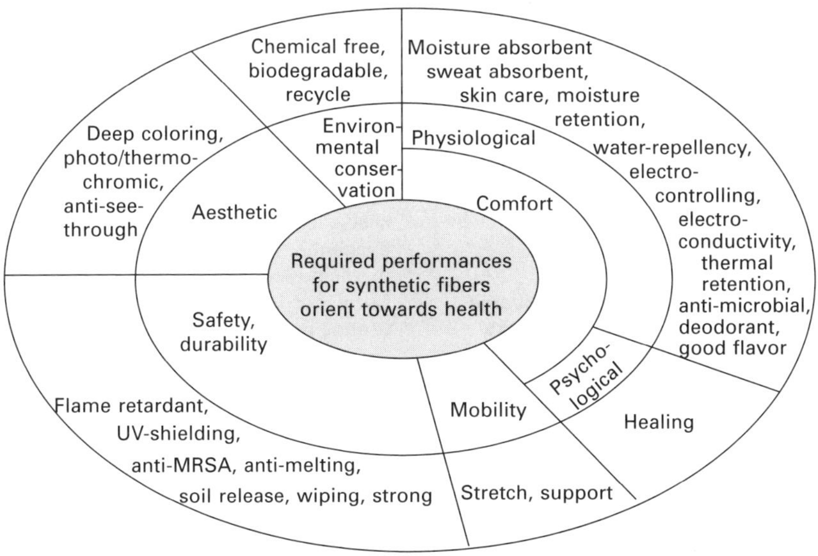


*6.24* Main functions required for recent health keeping fiber.

### 6.5.2 High needs of fiber for medical applications

Fibers used in the medical field and their applications are shown in Table 6.6. Surgical suture used in operations is classified into two categories: non-absorbed and absorbed suture by human body depending on the place of operation.

### 6.5.3 Comfort fiber

Application of high function fiber to become also a comfort fiber is shown in Fig. 6.25. Health can be kept by making people comfortable and able to relax to external stimuli. These two overlap within the definition of healthcare fiber. Furthermore, aims of comfort fiber include 'positive pursuit of comfort' and 'comfort by relaxing against uncomfortable external stimuli'. Such comfort fibers need to be lightweight and emphasize fitness and gentle feel.

### 6.5.4 Stimuli relaxation fiber

The application of high performance and high function fibers to stimuli relaxation fiber is shown in Fig. 6.26. Stimulus relaxation fiber, for example, are clothes worn in reactor accident, in fires and in outer space. They must be fireproof, extremely thermo-resistant, radioactive rays reflective, and of super high strength. In addition, they must be flame resistant, abrasion resistant, soundproof and vibration proof.

*Table 6.6* Medical fiber and its application

| Term | Material | Application |
|---|---|---|
| Short to middle | Hollow fiber | Dialysis, reverse osmosis, gas exchange, artificial organs, medicine industry |
| | Fiber absorbable by human body | Surgical suture, implanted artificial organs and blood vessel |
| | Anticoagulant fiber | Artificial blood vessel and organs |
| | Optical fiber | Autoscope |
| | Polymer/fiber releasing chemical gradually | Medicine |
| Middle to long | Anti-microbe fiber | Medical use |
| | ACM lightweight structural material | Bed, conveyance, artificial arm and leg |
| | Artificial muscle | Artificial arm, leg and heart |

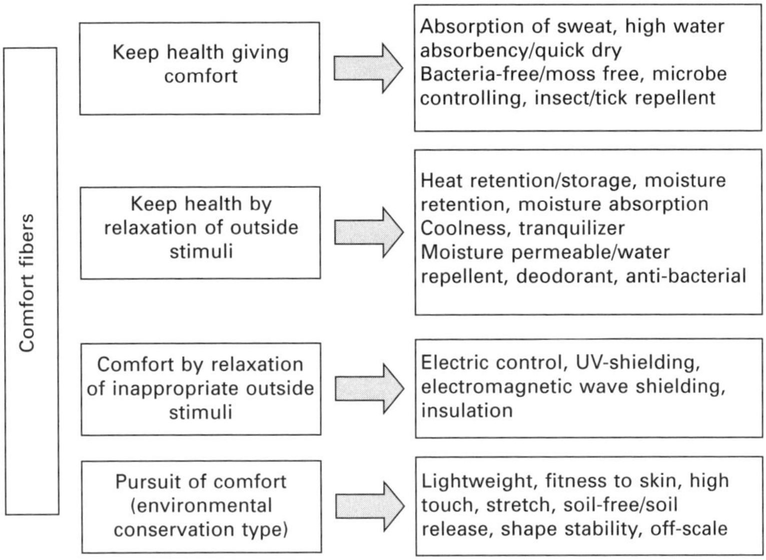


*6.25* Utilization of high functional fibers for comfort.

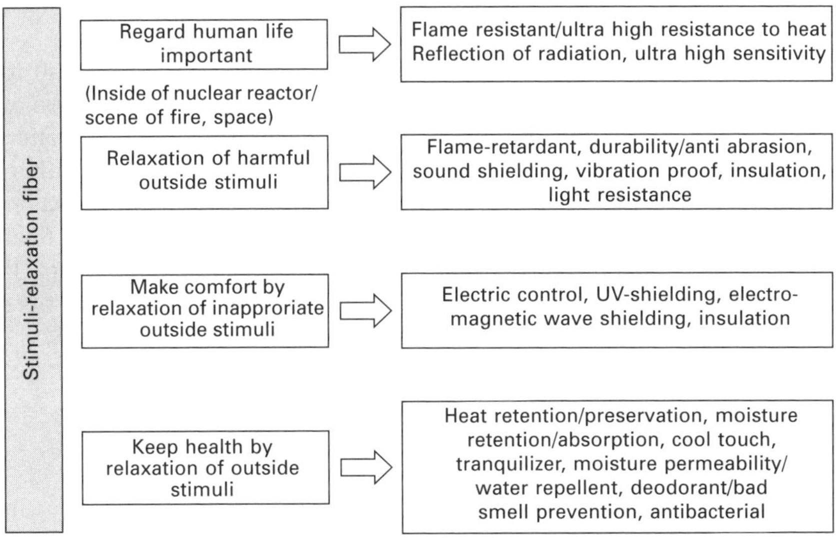


*6.26* Utilization of high performance/high functionality fibers for relaxation of stimuli.

## 6.5.5 Environmental conservation fiber

High performance fiber applied to environmental conservation is shown in Fig. 6.27 together with industrial fields of use. Protection from UV rays, radioactive rays, harmful matter, microorganisms, viruses, and moisture is

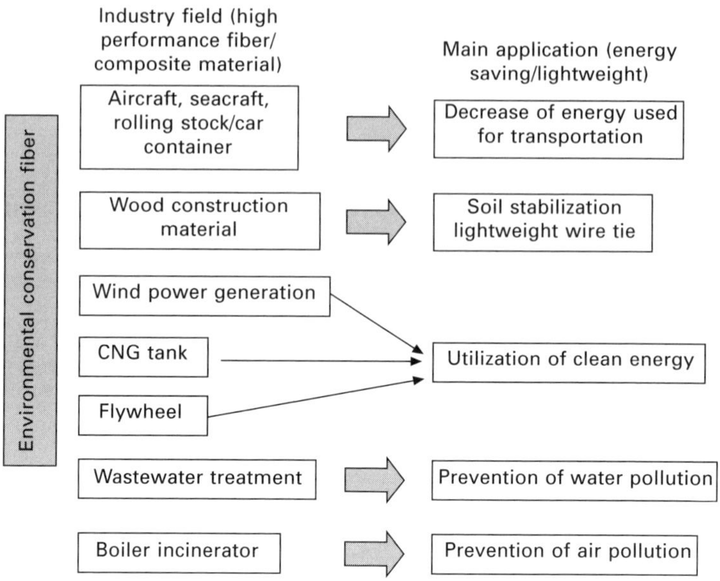


*6.27* Application of high performance fibers to environmental conservation.

necessary because these can be harmful to health. It is necessary also to consider internal stimuli such as heredity, aging, stress, and deterioration of human health. Fiber directly concerned with environmental conservation includes recycling fiber, resources saving fiber, biodegradable fiber, environmental conservation/formation/purification fiber, storing water/ waterproof fiber, and high oil absorption fiber. Stimulus relaxation fiber from outside is also included within environmental conservation fibers. If indirectly concerned fiber is included, healthcare fiber and, comfort fiber also belong to environmental conservation fiber, as shown in Fig. 6.28.

## 6.5.6 Pursuit of fiber to maintain physical and mental health

Figure 6.29 shows the relation of stimuli and fibers necessary to maintain physical and mental health.

### *Stimuli from internal environment*

The internal environment introduces stimuli composed from physical and mental environments. The physical environment includes illness, allergies, genetic disease, immune disorder, injury, after-effect of disease, functional disorder, and aging. Then there is the 'mental environment' which indudes

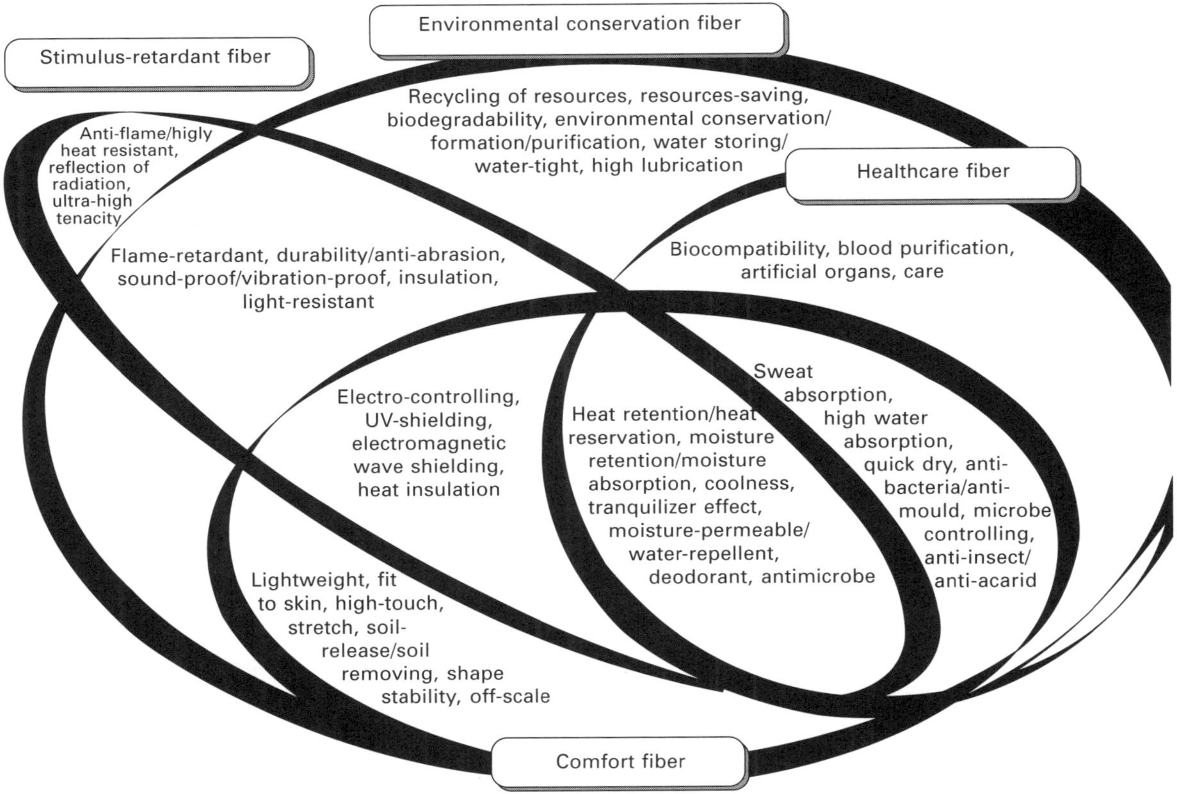


*6.28* Fiber to maintain environment and health.

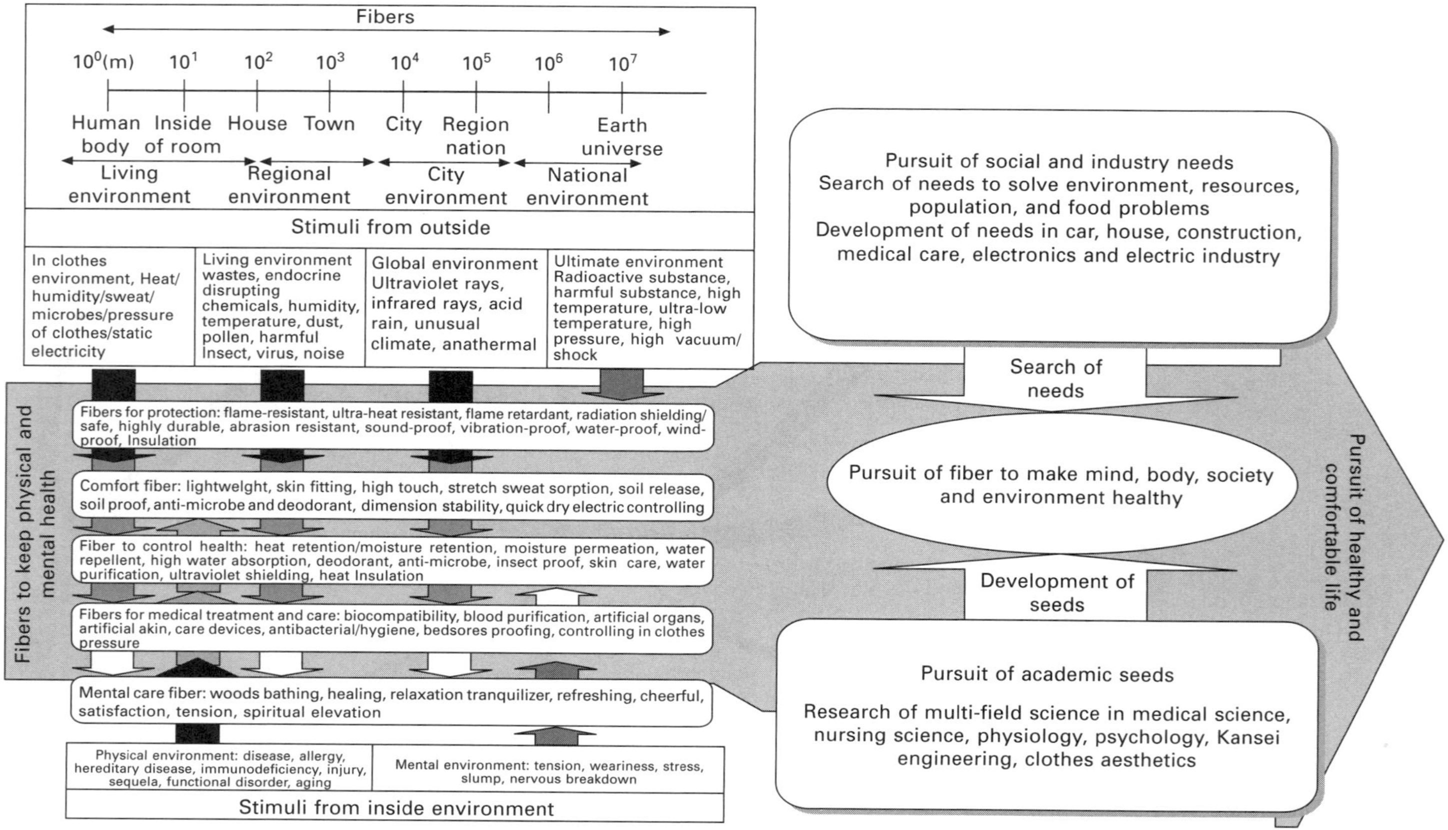


*6.29* Pursuit of healthy and comfortable life (T. Koyama, Shinshu University, modified).

tension, weariness, stress (stimulus added to the living body), slump (dull), and neurosis.

Fluctuations found in nature such as the murmur of a brook, or a breath of wind are called 1/f fluctuations. Honorary Professor of Tokyo Institute of Technology, T. Musha has studied 1/f fluctuation theory. Besides this aspect, there is geometry of a fractal of self similarity in natural form such as a cloud, a mountain or a river, and this is called fractal theory which has been studied by Professor Monday in the United States.

This kind of phenomenon and uniformity in nature gives a feeling of ease and comfort. This can be taken into textiles and each company develops and markets these products. This is a new way of thinking. An example of the healing materials applying 1/f fluctuations is shown in Fig. 6.30.

*External stimulus*

Various popular stimuli from the outside environment indude: 'clothes internal environment', 'living environment', 'global environment', 'ultimate environment'.

The 'clothes internal environment' includes heat, moisture, perspiration, various germs, pressure from clothes, and static electricity. The study of such factors on wearing clothes is the most advanced. In a company and a university, for example, Toyobo and Faculty of Domestic Science of Ochanomizu Women's University, they study these factors using mannequins. The result is reflected as 'climate in clothes' by manufacturer.

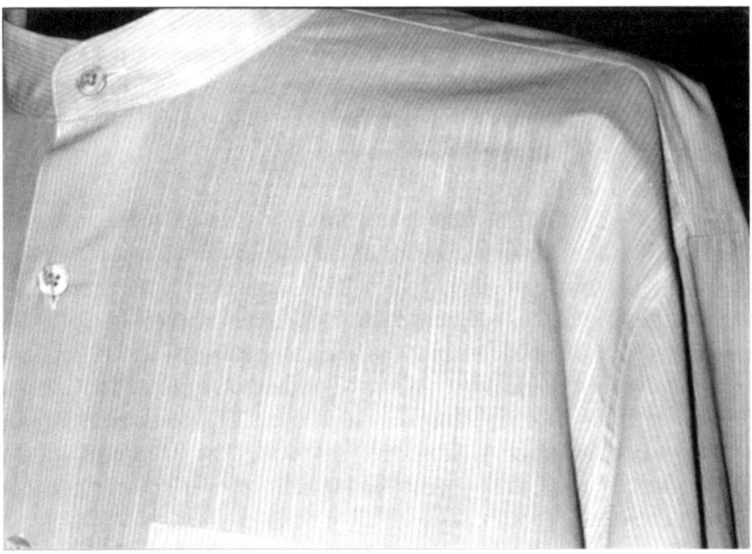

*6.30* Material using 1/f fluctuation (Nisshinbo).

'Living environment' includes the effects of waste or endocrine disrupting chemicals, moisture, air temperature, dust, pollen, a harmful insect, microorganism, viruses, and undesired sounds.

'Global environment' includes UV rays, infrared rays, acid rain, abnormal weather, and a thermal problem.

## 6.6 Technical concentration to achieve comfort

Following the Kyoto International conference on global warming, to prevent an increase of carbon dioxide concentration the government proposed that the temperature of air-conditioning should be maintained at 28°C and 20°C for cooling and heating the office, respectively. A person who has adjusted to the environment with traditional cool air-conditioning feels considerably hot and humid when the temperature is set to 28°C. Thus suitable clothing had to be developed to adjust to these conditions.

### 6.6.1 Nisshinbo and Teijin were at the forefront

Nisshinbo and Teijin jointly developed 'Ecosys 28°C' for a shirt in which a businessman feels comfortable even at 28°C in the office, and it was first marketed in November 1998. This required the development of (1) the material, (2) the texture of fabric and (3) processing in order to get a comfortable cool feeling. The polyester included in this material is different from ordinary polyester since it is made with hollow fiber with triangular cross-section. By combination of the hollow fiber and high-quality Egypt cotton, the fabric is light and does not stick to the skin. They examined the texture and fabric density, and adopted Lawn, Oxford and a Panama cloth in order to improve breathability. In addition, by using voile fiber, the fabric is cool and feels better than the ordinary fabric. In finishing, they adopted new shape stable finishing SSP to improve water absorbency.

The functions of Ecosys 28°C were examined in artificial meteorological conditions by Teijin. The volunteer wore long-sleeved shirts made with Ecosys 28, walking at 5 km/h for 10 min and the result was compared with that examined wearing an ordinary shirt. The sweat decreased by 20%, feeling of stuffiness and stickiness decreased by 5–7%, and moisture in clothes decreased more than 20% after 40 min motion. They also carried out different evaluations wearing the shirt made of Ecosys 28°C at temperatures of 28°C and humidity of 60% as in an office in summer. The subject first rests quietly in bed under windless condition for 10 min. He then walks for 10 min in wind and then rests quietly in bed under windless condition for 20 min. According to the organoleptic test, the 'Ecosys 28°C' shirt is comfortable because it is cool, not sticky and not stuffy when worn.

### 6.6.2 Development of 'Toyobo's Science in Comfort™ shirt'

Toyobo carried out research under the key word of 'comfort shirt' 20 years ago. They commercialized the product under the brand 'Toyobo's Science in Comfort'. 'Climate in clothes' became core, and they provided sportswear first, and then uniform, undershirt and shirt. Toyobo's comfort science 'comfort shirt' is shown in Fig. 6.31.

Toyobo gave the shirts the name 'Toyobo's Science in Comfort™'. The fiber 'Alsace™' uses cotton and polyester in three layers. It is made of hollow polyester, and 'Ramiluck™' (polyester and ramie), and is used as the material for shirts. Lawn, Dobby cloth and voile with rough texture are used for the fabrics of the shirts in order to give breathability. Besides that, treatments named 'Eternally™' 'Miracle Care™', and 'Deodoran™' are given for moisture absorbing/soil repellent, wash and wear, and odor-eliminating, respectively. In Toyobo, they combine material, texture and treatment to follow the needs of consumers.

### 6.6.3 New comfort business shirt material 'Azek™'

'Azek™', after 'Azekura', a storehouse built of logs, was first announced in 1999 as a joint development between Shikibo and Toray. Azek structure and function image drawings are shown in Fig. 6.32.

*6.31* Comfort shirt made of Toyobo's Science in Comfort (Toyobo).

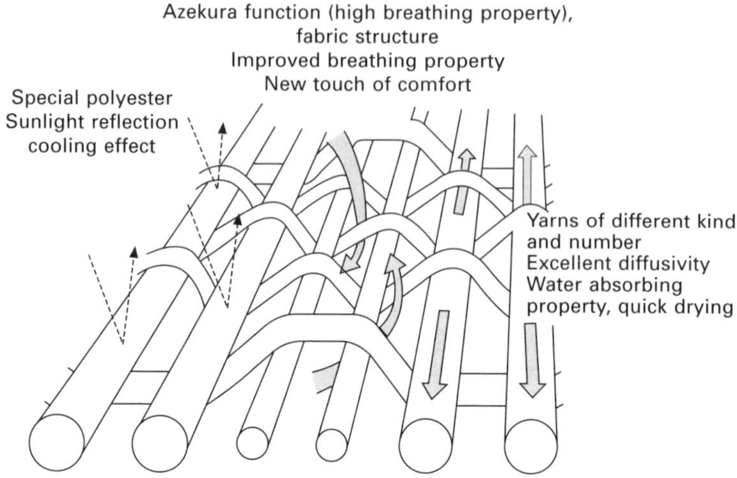


*6.32* Structure of Azek (Shikibo).

It was the golf shirt that led to the development of 'Azek$^{TM}$'. It was made of a high special ceramics/'Alfix$^{TM}$' composite material which can reflect sunlight, and has sold well using the catchphrase 'cool & dry' in 1999. Before then it was not thought that a business shirt could be made of 100% polyester, for previously they used natural fibers, such as cotton and ramie. Alfix was used to reflect heat by changing the thickness of the thread to realize the Azekura structure, square log architecture, within the fabric. As a result, the surface of the fabric becomes three-dimensional with wide space between thread to realize breathability. Moreover, it does not have the see-through nature which is disliked in men's shirts, nor the fluffy nature used for ladies' materials.

'Ecosys 28°C' and 'Toyobo's Science in Comfort' were developed combining existing material, texture and treatment, but 'Azek' was a new texture in fabrics.

### 6.6.4 A new comfort business shirt from various companies

Fuji Spinning developed 'BODY GUAR-°C' under the concept of being friendly to the earth. The project to develop the comfort shirt is shown in Fig. 6.33. They aimed to develop comfort and fashionable shirts, suitable for air conditioning at 28°C and 20°C, in summer and winter, respectively. The material used in summer, 'Escusa', is made of hollow fibers with triangular cross-section, increasing breathability by a factor of four, and lighter by 20%. 'Ecsandre' is made of high-quality French linen, and the antibacterial fiber 'Chitopoly' made of the natural polymers chitin and chitosan from crab

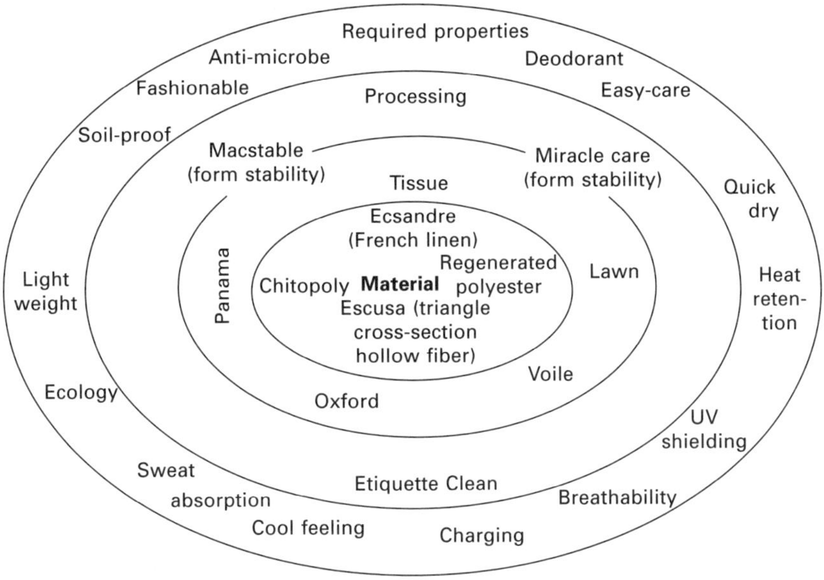


*6.33* Planning of comfortable shirt (Fuji Spinning).

and prawn shell. Lawn, voile, Panama, and Oxford are used as texture of the materials.

For winter shirts worn in a 20°C room, 'Incerared' is used. Here broadcloth, herringbone and Karsey are used as fabric texture, and functionalities such as shape stability finishing, moisture absorption/heat generation finishing and warm finishing are combined.

Kurabo developed, 'Feel More' and 'Charade' by mixed-spinning of hollow polyester and cotton, giving fine denier polyester covered with cotton in the 'Oasis project'.

Kanebo developed 'SPERANZA', 100% cotton, in the program 'Interface amenity material planning by Kanebo'.

Daiwabo developed polyester/cotton/rayon mixed-spinning material, 'Cooldry', so making rayon part anti-microbial.

## 6.6.5 Comfortable underwear

Gunze developed the underwear, 'YG28°C', for people to feel comfortable even in the 'hot' environment of air-conditioning at 28°C. It uses 'ACTICOT' made by blending 'Benberg' of Asahi Kasei, which is superior in moisture absorption, and 'Technofine', high water absorption polyester. It absorbs and diffuses sweat very quickly.

### 6.6.6 New warm shirts comfortable at temperatures of 20°C in autumn and winter

Three companies, Nisshinbo, Nippon Keori and Teijin, planned 'Ecosys 20°C' for autumn and winter together as a part of 'a triangle project' using distinct Tetron of Teijin, Egypt cotton of high quality by Nisshinbo, and spun wool of Nippon Keori.

Toyobo developed 'Toyobo comfort science 20°C'. The Toyobo 'comfort shirt' was 'warm, lightweight and tender'.

Shikibo developed the moisture absorbent and heat generative material, 'Thermo stock', for the use of a comfortable shirt to be worn in autumn and winter. When wearing the shirt, the temperature increases by 1 to 1.5°C and is made of modified cellulose, in which the hydrophilic groups of cellulose are bonded.

Kurabo developed 'warm process' using cotton for the core part and acrylic for the sheath part.

Fuji Spinning developed 'Incerared' and 'Escusa' made of cotton/polyester mixed with ceramics having an ultra-red radiation effect.

Kanebo used warm materials of cotton/wool 'RANA COTONNE', and hollow-type polyester/cotton 'Kilatt', together with insulating fiber 'ANGELUS FM'.

These shirts have contributed to a major boom for the industry and have been of great benefit and comfort to the user.

## 6.7 Bibliography

*General*

Harada R. and Sato H., *J. Text. Machine Soc. Jpn.*, **44**(3), 10 (1991).
Harada R. and Sato H., *J. Text. Machine Soc. Jpn.*, **46**(2), 91 (1993).
Urabe K., *Japan Chemical Fibers Association, Report of Investigation*, **21**(4), 37 (1994).
Suzuki T., *Japan Chemical Fibers Association, Report of Investigation,* **24**(10), 21 (1997).
Hayashi T., Hiyoshi T., Miyagi M. and Kawaguchi T., 'State and problems of medical care fiber', *Sen'i Gakkaishi,* **54**(10), 344–364 (1998).
*Proceedings of 13th Specialty Fiber Materials*, June 1999, Society of Fiber Machine, Japan.
Nojima Y., 30th Textile Finishing Seminar (at Fukui University), The Society of Fiber Science and Technology, Japan, p. 29 (2003).
Report on Healthy Noppon 21, Ministry of Health, Welfare and Labor, February, 2000.

*Artificial kidney*

Gejyo F., *et al., Biochem. Biophys. Res. Commun.*, **129**, 701 (1985).
Craddock P.R., *J. Clin. Invest.*, **59**, 879 (1977).
'Kidney Dialysis Equipment & Supplies Market' No. 173, Sept. 1991.

Humes H.D., *et al., Nature Biotech*, **17**, 451 (1999).
Asahi Medical Co. Ltd., Plasmaflo$^{TM}$ catalogue.

### *Artifical blood vessel*

Harrison P.W., *Textile Prog.*, **8**, 13 (1986).
AAMI, *American National Standard for Vascular Graft Prostheses* (1986).
Ookoshi T., *Artificial Organs*, **22**, 495 (1993).

# 7
# Polymer fibers for health and nutrition

## 7.1 The concept and effects of dietary fiber

A definition of dietary fiber is:

> '. . . the edible parts of plants or analogous carbohydrates that are resistant to digestion and absorption in the human small intestine with complete or partial fermentation in the large intestine. Dietary fiber include polysaccharides, oligosaccharides, lignin and associated plant substances. Dietary fibers promote beneficial physiological effects including laxation, and/or blood cholesterol attenuation and/or blood glucose attenuation'
>
> (American Association of Cereal Chemists, 2001)

Because there is a growing belief throughout the world that natural fiber foods are an integral part of a healthy lifestyle, food producers source an increasing proportion of their raw materials from nature itself. There is a growing demand from an increasingly health-conscious consumer for reduced fat and enhanced fiber foods of all types. If this can be achieved using materials that have low calorific value, further health benefits will result. Foods containing such ingredients will need to match the quality of the original product and without adverse dietary effects. This target cannot be achieved without the scientific use of thickeners, stabilizers and emulsifiers, particularly of the 'natural type'. This calls for fibers, which can interact with water to form new textures and perform specific functions, which calls for the use of 'hydrocolloids'. In 2003 the world market for such hydrocolloids of the fiber type was US$ 3.5 million and is set to grow significantly to meet the health aspirations of the consumer in the next millennium. It is the task of the food scientist to provide the hydrocolloids in the most appropriate form for inclusion in the food product. This requires an understanding of their structure and the way in which they act to produce the desired function in the food.

Dietary fiber was first described as the skeletal remains of plant cell walls, which are resistant to hydrolysis by the digestive enzymes of man. Since this excluded polysaccharide fibers in the diet, the definition was subsequently expanded to include all polysaccharides and lignin, which are not digested by the endogenous secretions of the human digestive tract. Dietary fiber thus mainly comprises non-starch polysaccharides, and indeed has been defined by Englyst and others as the 'polysaccharides which are resistant to the endogenous enzymes of man'. Industrialized countries now generally recognize the health-giving properties of increased consumption of fiber and reduced intakes of total and saturated fat. In this respect 'fiber' is used in a non-specific way, but is generally taken to mean structural components of cereals and vegetables. More recently the concept of 'soluble fiber' has emerged which assist plasma cholesterol reduction and large-bowel fermentation.

The physical and fiber properties of such soluble and insoluble fiber allow them to perform both in a physical role and also to ferment through colonic microflora to give short-chain fatty acids (SCFA), mainly acetate, propionate, and butyrate. These have a very beneficial effect on colon health through stimulating blood flow, enhancing electrolyte and fluid absorption, enhancing muscular activity and reducing cholesterol levels.

### 7.1.1 The physical effect

To be effective dietary fiber must be resistant to the enzymes of the human and animal gastrointestinal tract. If physically suitable it can work effectively as a result of its bulking action. In the stomach and small intestine the fiber can increase digesta mass, leading to faecal bulking, which readily explains the relief of constipation, which is one of fiber's best documented effects. It can increase stool mass and ease laxation very efficiently. This behaviour has considerable human and agricultural importance. The growth of the ruminant animal depends on the fermentable fiber content of the stockfeed.

Soluble as well as non-soluble fibers exert their actions in the upper gut through their physical properties. Those that form gels or viscous solutions can slow down the transit in the upper gut and delay glucose absorption, best explained in terms of 'viscous drag'. Thus the reduction in glycemic response by soluble fibers can be explained.

### 7.1.2 Fermentation product effects

Large bowel microorganisms attack the soluble fibers, in fermentation resembling that in the rumen of obligate herbivores such as sheep and cattle. The products too are similar: short-chain fatty acids (SCFA), gases (hydrogen, carbon dioxide and methane) and an increased bacterial mass. The principal SCFA are the same in humans as in ruminants, and the concentrations are similar too, particularly for omnivorous animals with a similar digestive physiology (for example, the pig). The increased bacterial cell mass, as noted, also has a positive effect on laxation. Faeces are approximately 25% water and 75% dry matter. The major components are undigested residuals plus bacteria and bacterial cell wall debris. These form a sponge-like, water-holding matrix which conditions faecal bulk and cell debris. The ability of different fibers to increase faecal bulk depends on a complex relationship between chemical and physical properties of the fiber and the bacterial population of the colon.

The production of SCFA and their beneficial effects in humans and ruminant species has been well established for a considerable time, but the effect was not thought to be relevant to the carnivorous dog and cat. Now this too has been demonstrated.

### 7.1.3 Health benefits

Whether by physical bulking action or through the production of SCFA, several health advantages are now established: Increasing fiber (20–30 g per day in humans) can eliminate constipation through increased faecal bulking and water-holding. The fermentation to produce SCFA can also assist, since propionate stimulates colonic muscular activity and encourages stool expulsion.

It was thought at one time that fiber lodged in the colon could lead to inflammation and herniation. This has now been disproved, and fiber can now relieve diverticular disease conditions, probably in the same way as it relieves constipation. Applying a solution of SCFA into the colon of ulcerative colitis patients or into the defunctioned portion of surgical patients has given rise to substantial remission in colitis. It could be that the condition arises due to a defect in the fermentation process in these patients or in the products.

SCFA stimulate water and electrolyte absorption by the mucosa and enhance their transport through improving colonic blood flow. Fiber fermentation also reduces the population of pathogenic bacteria such as *Clostridia* and can prevents diarrhoea due to bacterial toxins. Epidemiological studies have shown repeatedly that populations with high levels of fiber in their diet have reduced risk of colon cancer. Protection may be through the SCFA butyrate, which inhibits the growth of tumour cells in vitro.

When applied to the companion animal, the increased production of SCFA increases gut acidity marginally, which reduces the activity of putrefaction and pathogenic bacteria and so lowers toxin, thus reducing bad odors and bad smelling faeces. The low level of toxin production reduces the load on the liver and results in better coat and skin quality. Therefore, the aging animal can look better and produce less offensive faeces.

The behavior in the intestine can influence the immune system. Thus protection is possible against colonization by opportunistic bacteria, and the improved colonization of beneficial indigenous bacteria in the gut, which gives greater resistance to infectious bacteria.

### 7.1.4 Prebiotics

A balanced population of intestinal microflora needs to have a sufficient supply of substrates (most importantly carbohydrate fiber) to be able to grow. Part of this carbohydrate supply is bowel mucus and the rest consists of indigestible or only partially digestible carbohydrates or compounds derived from them. After consumption, these substances end up, wholly or partly, in the large bowel. There they are broken down further by bacteria into substances that are beneficial to humans, for example lactic, propionic, butyric and acetic acids. This results in the fall in the pH of the bowel contents while at the same time gases are formed (carbon dioxide, hydrogen, and methane).

Dietary fibers are important substrates (specifically their water-soluble fraction) including difficult to digest starch fraction (amylose), pectins, indigestible oligosaccharides (e.g. inulin) and substances like lactitol, other sugar alcohols, and the food hydrocolloids. The fall in pH caused by fermentation in the large bowel in turn leads to a reduction in the formation of secondary bile acids. These acids can damage bowel cells and this can lead to an increased risk of cancer. The pH influences the composition of the intestinal microflora and a reduction in the pH in combination with the volatile fatty acids that have been formed will protect against the settlement of pathogenic bacteria.

### 7.1.5 Probiotics and synbiotics

Probiotics are living organisms that, after being ingested by humans or animals, exert beneficial or health-enhancing effects by improving the characteristics of the intestinal flora. They are, for instance, used in fermented dairy products. The most important are lactic acid bacteria that (by resistance to gastric acid and bile) are able to survive passage through the stomach and small intestine. Several strains of *lactobacillus acidophilus, Lactobacillus casei* and certain bifidobacteria possess this characteristic. It has been found that in humans the duration of certain kinds of diarrhoea can be limited by the intake of *Lactobacillus casei* or specific bifido bacteria.

Thus the food fiber components that can exert the beneficial effects on the colon may be divided into three categories:

- nutrients (substrates) for the intestinal microflora, also called prebiotics
- living lactic acid bacteria with a beneficial effect on the intestinal microflora, the so-called probiotics
- synergistic combination of prebiotics and probiotics, also called synbiotics.

## 7.2 Hydrocolloid fibers

The food fibers belong to a class of chemical materials which can generally be termed 'hydrocolloids'. It is their ability to be presented to the body in various physical and functional forms that enables them to be so effective in a variety of food products. Their properties and main characteristics will be described. Table 7.1 gives the main hydrocolloid fibers and their food applications.

The term 'hydrocolloid' describes a group of water soluble naturally-occuring polymers found abundantly in nature. They have evolved to perform many different functions, e.g. act as structural agents and energy reserves in plants and animals, to facilitate cell recognition and adhesion processes, to provide lubrication in bone joints, to act as ion exchangers and blood anticoagulants, etc. Their key function in food products is to act as a source of fiber by controlling the texture and organoleptic properties mainly by enhancing the viscosity and gel characteristics. Even at concentrations of 1 wt% or less some hydrocolloids are capable of producing highly viscous solutions or forming gels with varying textures. Their thickening ability has led to their use as suspension and emulsion stabilizers where they function by retarding particle sedimentation and droplet creaming due to bulk viscosity effects. The hydrocolloids may also adsorb onto the surface of particles or droplets and inhibit aggregation by steric or electrostatic forces. Each hydrocolloid has its own unique functional characteristics, which is a consequence of its chemical structure, molecular size, and shape.

### 7.2.1 Thickening characteristics

The viscosity of hydrocolloid solutions shows a marked increase at a critical polymer concentration commonly referred to as C*. This concentration corresponds to the transition from the so-called 'dilute region', where the polymer molecules are free to move independently in solution without touching, to the 'semi-dilute region' where molecular crowding gives rise to the overlap of polymer coils and interpenetration occurs. Hydrocolloid solutions normally exhibit Newtonian behavior at concentrations well below C*, i.e. their viscosity is not dependent on the rate of shear. However, above C* non-Newtonian behavior is usually observed. A typical viscosity–shear rate profile for a polymer solution above C* shows three distinct regions:

1. a low shear Newtonian plateau
2. a shear thinning region
3. a high shear Newtonian plateau.

Microstructurally in the low shear region it is envisaged that the system is able to rearrange at a rate this is greater than the imposed deformation (i.e. the polymer molecules entangle at a greater rate than they disentangle). Above a critical shear rate, however, in the shear thinning region the rate of rearrangement is less than the imposed deformation and shear thinning results. The viscosity drops to a minimum plateau value at infinite shear rate (high

*Table 7.1* Source, function and main applications of hydrocolloids

| Hydrocolloid | Source | Function | Application areas |
|---|---|---|---|
| | **Botanical** | | |
| Carboxymethyl cellulose | Trees and cotton | Thickener | Dairy and desserts, ready-to-eat meals, bakery products, meat products, sauces and dressings. Reformed vegetables, fish cakes etc. |
| Methyl- and hydroxypropyl methyl-cellulose | | Thickener and gelling agent | |
| Modified starches | Corn, potato, etc. | Thickener and gelling agent | Ready-to-eat meals, dairy and desserts, meat products, soups, bakery products, sugar confectionery |
| Pectin | Citrus peel and apple pomace | Gelling agent | Jams, fruit preparations, sugar confectionery |
| Guar gum | Seed endosperm *(Cyamopsis tetragonoloba)* | Thickener | Dairy and desserts, bakery, petfoods, ready-to-eat meals, sauces and dressings |
| Locust bean gum | Seed endosperm *(Ceratonia siliqua)* | Thickener | Dairy and desserts |
| Tara gum | Seed end-osperm (*Cesalpinia spinosa)* | Thickener | As guar and locust bean gum but limited application at present |
| Gum arabic | Tree gum exudate *(Acacia senegal and seyal)* | Produces low viscosity solutions at gum concentrations Emulsifier | Sugar confectionery, beverages |
| Gum karaya | Tree gum exudate *(Stercula urens)* | Thickener | Dressings but usage limited |
| Gum tragacanth | Tree gum exudate (*Astragalus gummifer)* | Thickener | Dressings and sauces, icings but usage limited. |
| Konjac mannan | Tuber *Amorphophallus konjac)* | Thickener and gelling agent | Japanese noodles, jelly desserts |

*(Continued)*

*Table 7.1* (Continued)

| Hydrocolloid | Source Botanical | Function | Application areas |
|---|---|---|---|
| | **Algal** | | |
| Agar | **Red seaweeds** *(Gelidium, Gelidiella and Pterocladia)* | Gelling agent | Confectionery, dairy and desserts |
| Carrageenan *Kappa type* | *Euchema cottonii and Chondus crispus* | Gelling agent | Dairy and desserts, meat products, petfoods, sugar confectionery |
| *Iota type* | *E. spinosum* | Gelling agent | |
| *Lamba type* | *Chondus crispus* | Thickener | |
| Alginate | **Brown seaweeds** *(Laminaria hyperborea, macrocystis pyrifera)* | Gelling agent | Dairy and desserts, bakery products, petfoods, sugar confectionery |
| Propylene glycol alginate | | Emulsion and foam stabiliser | Salad dressings, beer |
| | **Microbial** | | |
| Xanthan gum | *(Xanthomonas campestris)* | Thickener | Dairy and desserts, ready-to-eat meals, sauces and dressings, petfoods |
| Gellan gum | *(Sphingo-monas elodea)* | Gelling agent | Sugar confectionery, dessert jellies, fruit preparations |
| | **Animal** | | |
| Gelatin | Cattle, pigs, fish | Gelling agent | Sugar confectionery, meat products, dairy and desserts |
| Milk proteins | Cattle | Gelling agent | Bakery products, dairy and desserts, confectionery |

shear region). A number of empirical mathematical models have been developed to describe the flow characteristics.

The most widely used model to describe the whole shear rate range is probably the Cross equation;

$$\eta = \eta_\infty + \frac{\eta_0 - \eta}{1 + (\tau\gamma)^m}$$

where, $\eta$ is the viscosity at infinite shear rate, $\eta_\infty$ is the infinite shear viscosity, $\eta_0$ is the zero shear viscosity, $\tau$ is a shear dependent constant denoting the onset of shear thinning, $\gamma$ is shear rate and $m$ is an exponent quantifying the degree of shear thinning. $m$ has a value of 0 for a Newtonian solution and increases to 1 with increased shear thinning.

The viscosity of hydrocolloid solutions is influenced significantly by the polymer hydrodynamic volume, which increases with radius of gyration (Rg). Rg increases with molecular mass, chain rigidity, and electrostatic charge density and is greater for linear compared to branched hydrocolloids. The main hydrocolloid thickeners used are listed in Table 7.2. The viscosity shear rate profiles for 1% solutions of some of these hydrocolloids are presented in Fig. 7.1. The most striking feature is the profile for xanthan gum. This hydrocolloid has a very high low shear viscosity, and hence it is good at suspending particles and oil droplets. In addition, however, it is also extremely shear thinning and, therefore, it readily flows on simple shaking. These characteristics have led to its widespread use in many food applications, notably, mayonnaise, dressings, and sauces.

## 7.2.2 Gelling characteristics

A number of hydrocolloids are able to form gels by physical association of their polymer chains through, for example, hydrogen bonding, hydrophobic association, cation mediated cross-linking, etc. and differ from synthetic polymer gels which normally consist of covalently cross-linked polymer chains. Certain helix forming hydrocolloids, for example, agarose, carrageenan,

*Table 7.2* Main hydrocolloid thickeners

| Hydrocolloid | Characteristics |
|---|---|
| Modified starches | Viscous solutions are formed depending on the type of modification. |
| Xanthan gum | Has an apparent yield stress and hence can prevent sedimentation and creaming, but solutions are highly shear thinning. Viscosity is not influenced by temperature, addition of salts or to pH changes. |
| Carboxymethyl cellulose | Viscous solutions are formed but the viscosity decreases on addition of salts and at low pH. |
| Methylcellulose and hydroxymethyl cellulose | Form viscous solutions which gel on heating. |
| Galactomannans (guar and locust bean gum) | Form highly viscous solutions which are not influenced by addition of salts or changes in pH. |

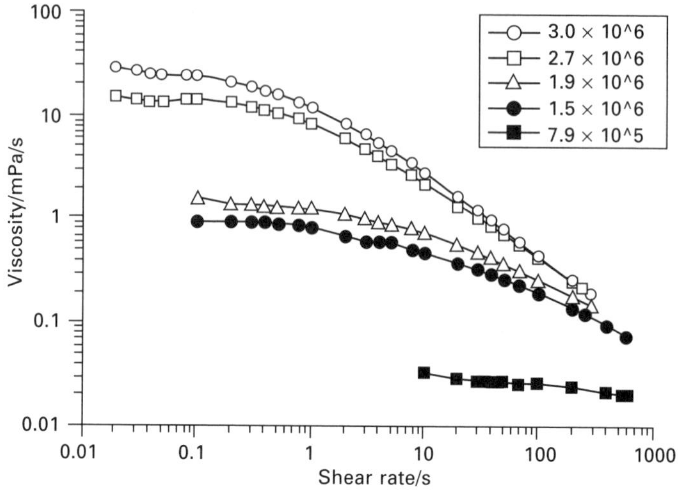


*7.1* Viscosity-shear rate profiles for 1% solutions of guar gum of varying molecular mass.

gellan gum and gelatin, form gels on cooling. These hydrocolloids adopt a disordered conformation at high temperatures but on cooling they undergo a conformational change and ordered helices are formed. The helices then aggregate to form a gel. The process is thermally reversible and hence the gels melt on heating. The melting temperature is often higher than the gelation temperature since melting only occurs after disaggregation of the helices. It is interesting to note here that mixing solutions of xanthan gum and locust bean gum (two non-gelling polysaccharides) results in the formation of very strong gels on cooling. This has been attributed to molecular association between ordered xanthan helices and 'bare' mannan regions along the locust bean gum chain. Non-thermoreversible gels can be formed by cross-linking chains with divalent cations. Alginate and LM pectin are typical examples. Some hydrocolloids, notably methyl- and hydroxypropylmethyl-cellulose form thermoreversible gels on heating. Chain association is believed to be due to hydrophobic bonding.

Gel formation only occurs above a critical minimum concentration, which is specific for each hydrocolloid. This concentration is not the same as the critical overlap concentration, C*, noted above. Agarose, for example, will form gels at concentrations as low as 0.2% while for acid-thinned starch a concentration of ~15% is required before gels are formed. The properties of individual hydrocolloid gels vary considerably in strength and elasticity due to differences in the number and nature of the junction zones and the degree of chain aggregation. The main hydrocolloid gelling agents and their characteristics are summarized in Table 7.3.

*Table 7.3* Main hydrocolloid gelling agents

| Hydrocolloid | Characteristics |
|---|---|
| Modified starch | Amylose containing starches will form thermally irreversible opaque gels. |
| Gelatin | Forms thermoreversible gels on cooling. Gels are elastic and melt at body temperature. |
| Agar | Forms thermoreversible turbid, brittle gels on cooling (~40°C). Gels melt only at high temperatures (~85°C). |
| Kappa carrageenan | Forms thermoreversible slightly turbid gels on cooling to 40–60°C, which is promoted by the presence of potassium ions. Melting occurs at 5–20°C above gelation temperature. Gels tend to be brittle and hence it is often used in combination with locust bean gum, which gives increased elasticity, improves clarity and reduces syneresis. |
| Iota carrageenan | Forms thermoreversible elastic gels on cooling to 40–60°C. Melting occurs 5–20°C above the gelation temperature. |
| Low methoxy pectin | Forms thermoreversible gels on cooling in the presence of calcium ions and sequesterant (e.g. citrate) at pH 3–4.5. |
| High methoxyl pectin | Gels are formed at high soluble solids content at pH < 3.5. Gels are not thermoreversible. |
| Gellan gum | Forms highly transparent gels on cooling in the presence of electrolyte. Low acyl gels are brittle and are often not thermally reversible. High acyl gels are elastic and thermoreversible. They set and melt at ~70–80°C. |
| Methyl and hydroxypropylmethyl cellulose | Form thermoreversible gels on heating. |
| Alginate | Gels are formed on addition of polyvalent ions (usually calcium). Homogeneous gels are formed by generating the calcium ions in situ. Gels do not melt on heating. |
| Xanthan gum | Forms highly elastic thermoreversible gels with locust bean gum and konjac mannan. |
| Konjac mannan | Forms non-thermoreversible elastic gels in the presence of alkali. |

## 7.3 The main hydrocolloids

### 7.3.1 Gum arabic

The main tree gum exudate is gum arabic, and it is also the most important commercially. Gum arabic occurs as a sticky liquid that oozes from the stems and branches of acacia trees (notably *Acacia senegal* and *Acacia seyal*) which grow across the Sahelian belt of Africa, principally Sudan. The gummosis process occurs when the tree is subjected to stress conditions such as heat, drought or wounding. The liquid dries in the sun to form glassy nodules, which are collected by hand Fig. 7.2.

This natural fiber remains the most versatile food additive now available because of its range of functionalities. Gum arabic is the acidic polysaccharide exudate currently derived from two acacia tree sources: *Acacia senegal* and *Acacia seyal.* The Gum arabic-yielding acacias grow in semi-arid areas and the vast majority of the product which enters international trade originates in the so-called gum belt of Sub-Saharan Africa. The belt occurs as a broad band from Mauritania, Senegal and Mali in the west, through Burkina Faso, Niger, northern parts of Nigeria, Chad to Sudan, Ethiopia and Somalia in the Horn of Africa. Modern gum arabic trade has been dominated by the Sudan. Thus production in Sudan over the years gives a good indicator of consumption world-wide. Towards the end of the 1960s, total gum arabic production in Sudan (hashab and talha) was in excess of 60 000 tons. Events, mainly drought,

*7.2* Collecting gum arabic from *Acacia senegal* trees.

locusts and political instability in the 1970s and 1980s led to fluctuation in both the supply and the price, and as a consequence led to changes in demand. The severe Sahelian drought of 1973–74 resulted in a world shortage and high prices, which in turn accelerated the search for substitutes such as gelatine, maltodextrins and modified starches. A low point of approximately 20 000 tons of Sudanese exports was reached in 1975, which recovered to around 40 000 tons during 1979. A further drought in 1982–84 saw levels of exports fall to below 20 000 tons in the mid-1980s and early 1990s. Sudan now faces an embargo on its products in the United States for its alleged terrorist supporting activities. In 2004 the civil war in Darfur further disrupted supplies and led to the price increasing dramatically.

Europe is the biggest regional market for gum arabic, and imports averaged 29 300 tons per year over the seven-year period 1989–95, with peaks of 32 100 tons in 1991 and 34 000 tons in 1994. France and the United Kingdom are the biggest markets, although both re-exported large proportions of their imports, which averaged 10 000 and 7900 tons per year respectively. France shows an upward trend over the seven years while the United Kingdom trend has been downward. Germany and Italy were the next biggest markets, averaging 4200 tons and 3700 tons, respectively. Outside Europe, the United States is the largest market for gum arabic. Imports averaged 10 000 tons in 1994. Japanese imports averaged 1900 tons over the seven-year period. The applications and use of gum arabic in the fiber health food market continues to increase and in 2001 income from this hydrocolloid was US$130 million. Detailed accurate current statistics are now readily available.

Of the other producers Nigeria is the next most important after Sudan, averaging exports of between 4000 and 7000 tons. Chad comes next, and has increased its production each year since 1990, reaching 5400 in 1995. The upward trend has continued, often due to gum originating in the Sudan finding its way out through Chad. Currently (2005) the best estimate of the overall annual usage of gum arabic is 40 000–50 000 tons.

The variability of supply over the past 20–30 years has led to dramatic fluctuations in price and in turn injected uncertainty into the user marker. When supplies almost dried up in the 1970s the price increased from about US$1500 to US$5000 per metric ton. It is impossible to evaluate the equivalence to today's prices since inflation in commodity prices has been uneven and less than in manufactured goods. The price stabilized in 1996 to around $5000 per ton, but then overnight the Sudanese dropped their price to $2500 per ton which led to consternation and extreme problems for the industry's processors, who were left holding stocks at the higher price. This cyclic nature of the supply and price has led to less utilization of this wonderful gum than there would otherwise have been. Large companies are uncertain whether they should have too many of their products dependent on this uncertain supply chain.

Gum arabic is a complex polysaccharide consisting of galactopyranose (~44%), arabino-pyranose and furanose (~25%), rhamnopyranose (14%), glucuropyranosyl uronic acid (15.5%) and 4-O methyl glucuropyranosyl uronic acid (1.5%). It also contains a small amount (~2%) of protein as an integral part of the structure. Analysis of the carbohydrate structure has shown that it consists of a core of β(1,3)-linked galactose units with extensive branching at the C6 position. The branches consist of galactose and arabinose and terminate with rhamnose and glucuronic acid. It has been shown that the gum consists of three broad molecular fractions, which differ principally in their size and protein contents. Most of the gum (~90%) contains very little protein and has a molecular mass of $\sim 2.5 \times 10^5$. A second fraction, ~10% of the total, contains ~10% protein and has a molecular mass of $1–2 \times 10^6$ and has been shown to have a 'wattle-blossom'-type structure where blocks of carbohydrate of molecular mass $\sim 2.5 \times 10^5$ are connected to a common polypeptide chain (Fig. 7.3). The third fraction, ~1% of the total, contains up to 50% protein and has a molecular mass of $\sim 2 \times 10^5$. The high degree of branching gives rise to a very compact molecular structure for all of the fractions and results in solutions of very low viscosity. The second fraction has been shown to be responsible for the gum's excellent ability to stabilize oil-in-water emulsions. A major use of gum arabic is in the confectionery industry. It is also used as an emulsifier for flavor oils for incorporation in soft drinks. Encapsulation of the flavor oil can be achieved by spray drying the emulsion to form a solid powder which can be added to dried soup and cake mixes.

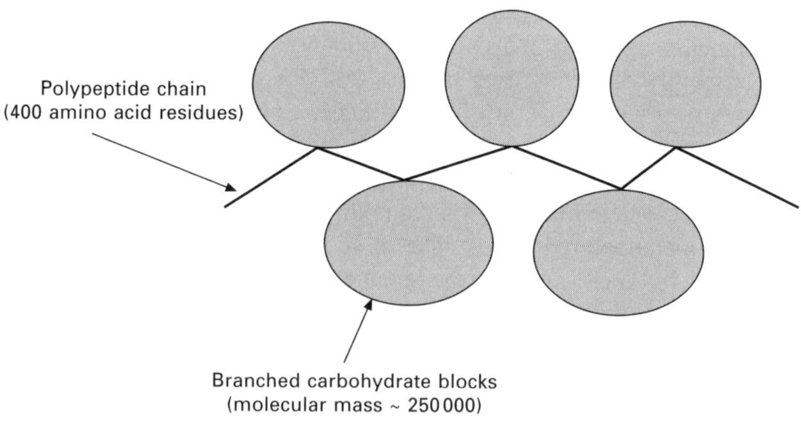


*Figure 7.3* Wattle blossom-type structure of the high molecular mass fraction of *Acacia senegal* gum.

### 7.3.2 Galactomannan seed gums

Locust bean (or carob), tara and guar gums are storage polysaccharides obtained from the endosperms of leguminous seeds of *Ceratonia siliqua*, *Caesalpinia spinosa* and *Cyamopsis tetragonoloba*, respectively. They consist of a linear main chain of β-(1,4)-linked mannopyranosyl units with galactopyranosyl units linked (1,6) to varying degrees (and have a molecular mass of the order of $10^6$) (Fig. 7.4). The mannose to galactose ratio, (M/G), is approximately 4.5:1, 3:1 and 2:1 for locust bean, tara, and guar gums, respectively. The galactose residues have been shown to be non-uniformly

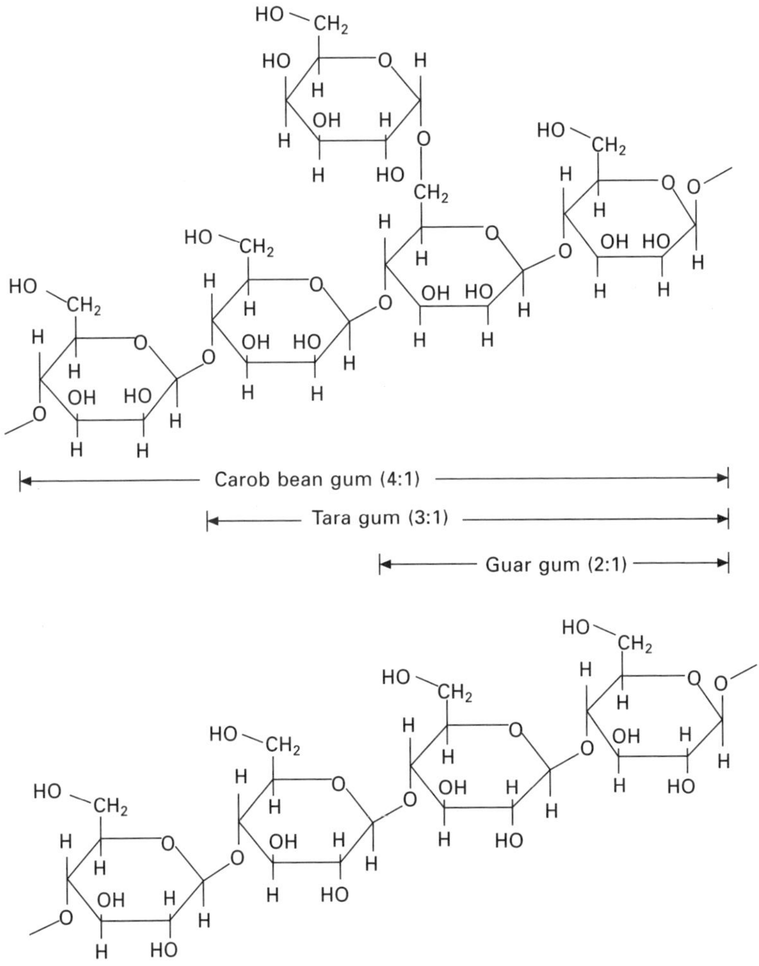


*7.4* Basic structural fragment of carob-, guar-, tara gum and cellulose.

distributed along the mannan chain. The presence of galactose tends to inhibit intermolecular association. Therefore, whereas guar gum is readily soluble in cold water, tara and locust bean gums have to be heated to high temperatures to achieve complete dissolution. Once dissolved all three yield highly viscous solutions. Locust bean gum will self-associate in solution and can form thermally irreversible gels on freezing. It is commonly used in combination with other polysaccharides, particularly kappa carrageenan, since it leads to the formation of stronger, more elastic gels, which have improved transparency and are less prone to undergo syneresis. Locust bean gum also forms strong thermoreversible gels with xanthan gum. For both mixtures it has been argued that the synergistic behavior is due to association of the ordered carrageenan and xanthan chains with mannose sequences along the backbone which are devoid of galactose residues. The model, therefore, explains why such behavior is not observed in mixtures containing guar gum.

Guar gum and locust bean gum are used as thickeners in a broad range of food products including dairy products, desserts, bakery products, pet foods, ready-to-eat meals, sauces, and dressings.

### 7.3.3 Konjac mannan

Konjac mannan (KM) is a glucomannan obtained from the tuber of the konjac plant notably the *Amorphophallus konjac* species which grows in Southeast Asia particularly Japan, China, and Indonesia. It is a high molecular mass ($>10^6$) polysaccharide consisting of linear chains of glucose and mannose units linked β-(1,4). Possible structures for the repeating unit are:

1. (G-G-M-M) M-M-G-M and G-G-M-G-M-M-M-M
2. (M-M-M-G-G)
3. G-G-M-M-G-M-M-M-M-G-G-M.

The main chain has branches, approximately every 10 residues, of up to 16 sugar units linked to the C3 position of the glucose and mannose. The mannose to glucose ratio is 1.6:1 and it is believed that there are no block sequences of glucose or mannose along the chain (Fig. 7.5). It has also been reported that KM has side chains and the branching position is considered to be the C3 position or C3 positions on both glucose and mannose. The degree of branching is estimated at approximately three for every 32 sugar units and elsewhere at one for every 80 sugar residues. The length of the branched chain was evaluated as 11 to 16 hexose residues.

Konjac mannan dissolves in water to form highly viscous solutions. It is acetylated (~1 acetyl group for every 19 sugar residues) and in the presence of alkali, deacetylation occurs and thermally irreversible gels are produced. Konjac mannan interacts with kappa carrageenan and xanthan gum in much

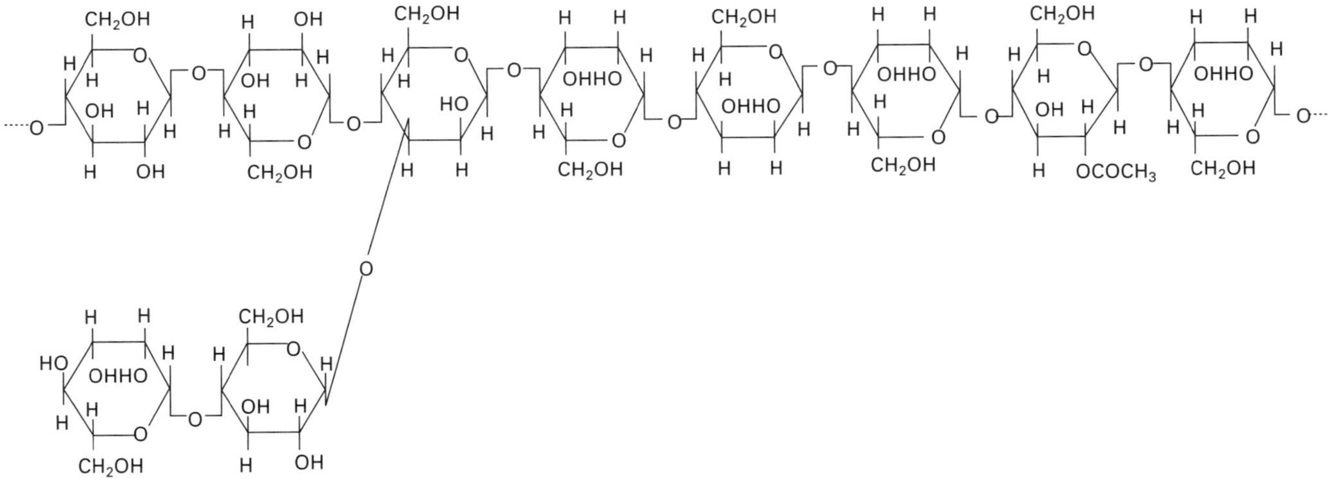

*7.5* Chemical structure of konjac mannan.

the same way as locust bean gum although the gels formed are considerably stronger.

Konjac flour has been used as an important food ingredient for more than a thousand years. With the addition of a mild alkali such as calcium hydroxide, konjac flour aqueous solution (ca. 3% of concentration) changes to a strong, elastic and irreversible gel. The alkali-treated konjac gel is quite popular in traditional Japanese food and is called Kon-nyaku in Japanese. Recently, synergistic gels prepared by mixing with other hydrocolloids are major products in the food industry as new types of healthy jellies.

Clinical studies indicate that konjac mannan solution has the ability to reduce serum cholesterol and serum triglyceride. Konjac mannan also has an influence on glucose tolerance and glucose absorption. However, the alkali-treated gel food does not have such effects.

The Food Chemical Codex lists the current uses of konjac flour in the United States as gelling agent, thickener, film former, emulsifier, and stabilizer. Konjac flour is also used as a binder in meat and poultry products. Konjac flour is suitable for thickening, gelling, texturing, and water binding. It may be used to provide fat replacement properties in fat-free and low-fat meat products. Applications and functional uses of konjac mannan are listed in Table 7.4.

### 7.3.4 Xanthan gum

Xanthan gum, discovered in the 1950s, was the second microbial polysaccharide after dextran to be commercially exploited and is now finding extensive application in the food industry. The gum is obtained from the genus *Xanthomonas*, notably *X. campestris* by aerobic fermentation. The xanthan molecules have a $\beta(1,4)$-linked glucopyranose backbone as in cellulose and in addition have a trisaccharide side-chain on every other glucose residue linked through the C3 position. The side-chain consists of two mannopyranosyl residues linked on either side to a glucuropyranosyl uronic acid group. The

*Table 7.4* Applications and functional uses of konjac mannan

| Application | Function |
|---|---|
| Confectionery | viscosity, texture improver, moisture enhancer |
| Jelly | gel strength, texture improver |
| Yogurt | fruit suspension, viscosity, gelation |
| Pudding | thickening, mouthfeel |
| Pasta | water holding capacity |
| Beverage | fiber content, mouthfeel |
| Meat | bulking, fat replacer, moisture enhancer |
| Edible film | water soluble, water insoluble |

inner mannose residue connected to the backbone may be acetylated while the terminal mannose residue may be pyruvated (Fig. 7.6).

The molecular mass of the xanthan molecules is very high (> $3 \times 10^6$) and the gum dissolves in water to yield highly viscous solutions. The xanthan molecules undergo a thermoreversible coil-helix transition in solution, which is shifted to higher temperatures by the addition of electrolyte. In the disordered coil form the side-chains are envisaged as protruding away from the backbone into solution, while in the ordered form the molecules form a stiff five-fold helical structure with the side-chains folded in and associated with the backbone. It is now generally recognized that the helix is double stranded. The stiffness of the xanthan chains gives rise to highly shear thinning rheological properties and unlike other polyelectrolytes the viscosity of xanthan solutions can actually increase rather than decrease on addition of electrolyte since the electrolyte will promote helix formation and association.

Xanthan gum is finding increasing use in a variety of applications including batter coatings, cake batters, frozen and chilled dairy products, sauces, and dressings. A comparison of the flow behavior of xanthan with other hydrocolloids is shown in Fig. 7.7. It is these excellent characteristics which make it the most versatile hydrocolloid currently used in food systems.

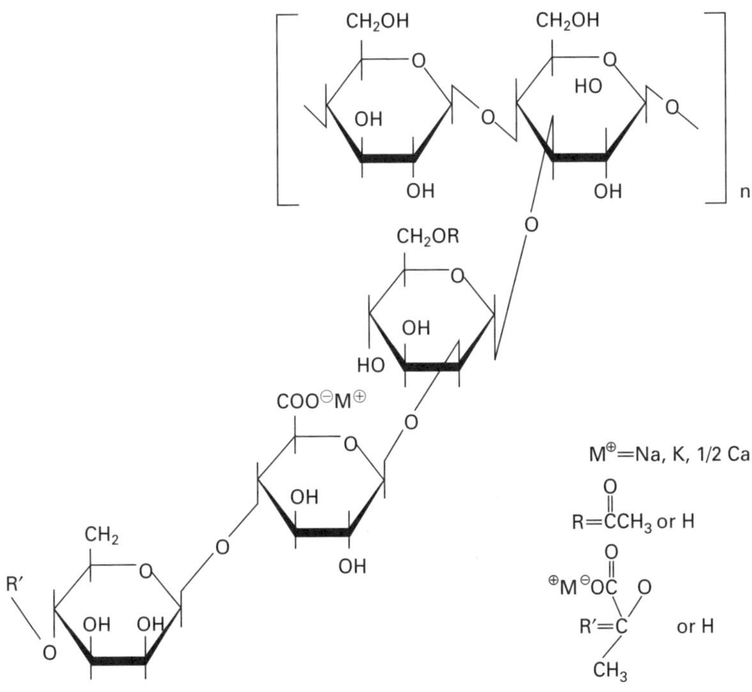


*7.6* Primary structure of xanthan gum

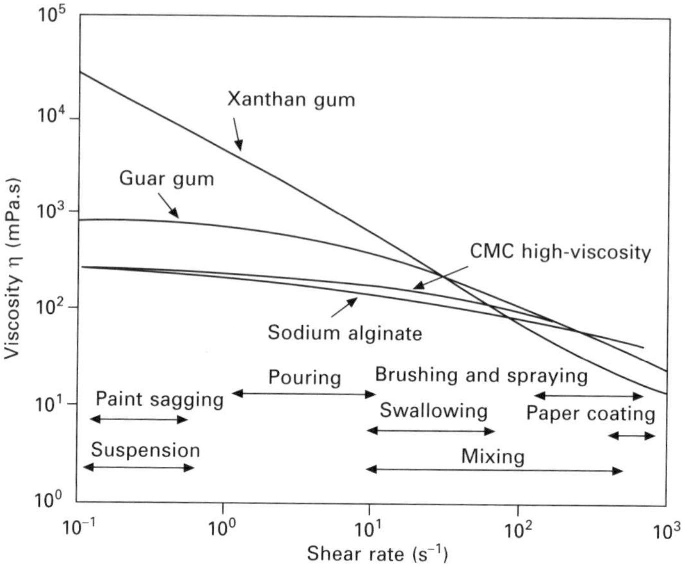


*7.7* Comparison of the flow behavior of xantham gum with other hydrocolloid solutions (0.5% concentration).

## 7.3.5 Gellan gum

Gellan gum has only relatively recently been commercialized. It is produced from *Sphingomonas elodia* by aerobic fermentation and consists of a linear tetrasaccharide repeat unit of β-D-(1,3)-glucopyranose-β-D(1,4)-glucuronopyranose-β-D-1,4)-glucopyranose-β (1,4)-rhamnopyranose. In the native form the β-(1,3) glucose residues contain glycerate and acetate moieties. X-ray fiber diffraction studies indicate that the gellan molecules form a three-fold double helical structure and in solution undergo a thermoreversible coil-helix transition. The transition shifts to higher temperatures in the presence of electrolyte. Once formed the helices tend to self associate leading to the formation of a transparent gel. Gels formed in the presence of monovalent ions are usually thermoreversible although the melting temperature is normally much greater than the gelation temperature, a consequence of the extensive aggregation. If gels are formed by the addition of divalent ions then they can be thermally irreversible. The native form produces soft elastic gels whereas the deacetylated material sold commercially forms hard brittle gels.

Gellan gum finds increased application at present in dessert jellies, sugar confectionery, dairy products, fruit preparations and as a suspending agent Fig. 7.8 illustrates the comparison of gellan gum with other common gelling agents.

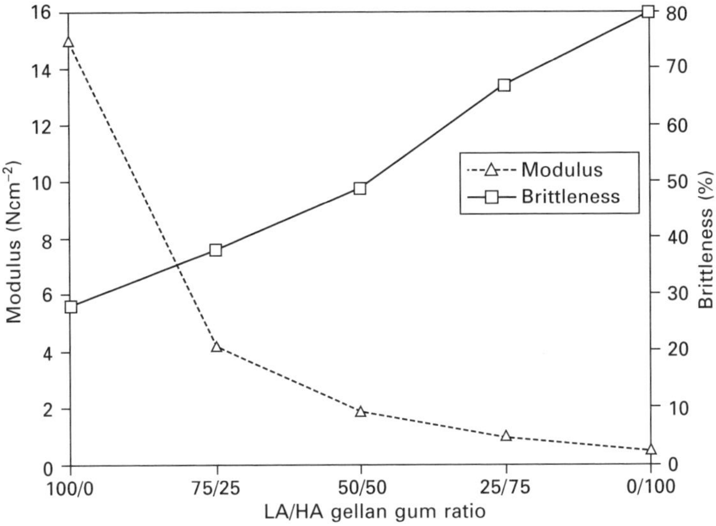


*7.8* Effect of high and low acyl gellan gum blend ratio on the modulus and brittleness of gels prepared at 0.5% total gum concentration.

## 7.3.6 Carrageenan

The carrageenans are a family of sulphated galactans obtained from red seaweeds (*Rhodophyceae*) where they have a key structural function. The traditional method of extraction is by treatment of the seaweed with hot alkali for 10–30 hours followed by precipitation with alcohol and then drying. The three major types are kappa, iota, and lambda carrageenan. Kappa is obtained from a species of seaweed called *Euchema cottonii* and occurs together with lambda carrageenan in *Chondrus crispus*. Iota carrageenan is obtained from *Euchema spinosum*. They differ essentially in their degree of sulphation.

Carrageenean is a high molecular eight linear polysaccharide comprising repeating galactose units and 3,6-anhydrogalactose (3,6AG), both sulphated and non-sulphated joined by alternating β-(1-3) and β-(1-4) glycosidic links (Fig. 7.9). Iota carrageenan differs only in that the latter residue is sulphated at the C2 position. Lambda carrageenan is further sulphated and consists of (1,4)-linked galactopyranose 2,6 disulphate and (1,3)-linked galactopyranose which are 70% substituted at the C2 position. The carrageenans are all soluble in water but whereas lambda forms viscous solutions, kappa and iota form thermoreversible gels. The gelation mechanism is similar to that described for gellan gum above, i.e. on cooling the molecules undergo a conformational

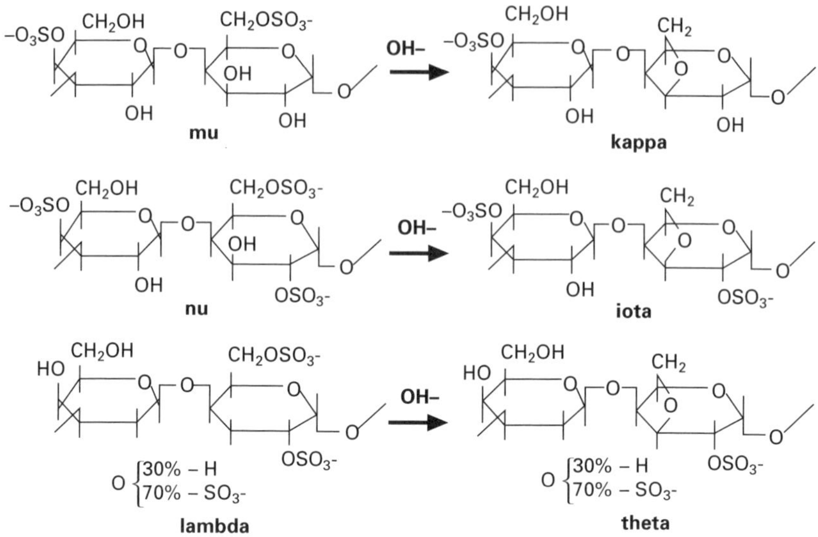


*7.9* Carrageenan structures (alkali conversion of mu > kappa, nu > iota and lambda > theta).

coil to helix transition and the helices self-associate giving rise to a three-dimensional gel structure. It is still not conclusively proven whether the helices are double or singly stranded. The temperature of gelation increases with increasing electrolyte concentration. It has been shown that potassium, rubidium, and caesium ions specifically bind to the helical structure of kappa carrageenan and hence promote helix formation and gelation at much lower concentrations than other electrolytes. As a consequence kappa carrageenan gels are much stronger in the presence of potassium chloride compared to, say, sodium chloride. This ion specificity is not observed for iota, which forms weaker more elastic gels compared to kappa. This is probably due to the fact that the increased charge on the iota carrageenan chains reduces the extent of helix self-association.

Carrageenan is used in dairy and dessert products such as puddings, milk shakes, ice cream and water dessert jellies. It is also used in meat products where it acts as a water binder.

Mention should also be made of a 'semi-refined carageenan' from the Philippines which has now gained acceptance after a controversial start. It is now an accepted regulatory ingredient for food use under the name 'Processed Eucheuma Seaweed'.

## 7.3.7 Agarose

Agarose (the major component of agar) is also obtained from red seaweeds notably *Gelidium and Gracilaria* species. It is a linear neutral polysaccharide and has a similar structure to the carrageenans consisting of alternating (1,3)-linked β-D galactopyranose and (1,4)-linked 3,6-anhydro-β-D-galactopyranose units. It dissolves in near boiling water and gels on cooling to ~ <40°C. The gelation mechanism is as described for the carrageenans but, since the agarose chains are non-ionic, extensive helix aggregation occurs resulting in the formation of very strong gels. The gels melt only on heating to 80–90°C.

Agarose applications include confectionery and baked goods such as icings and pie fillings.

## 7.3.8 Alginate

Alginate is obtained from brown seaweeds (*Phyophyceae*) and is a linear (1,4)-linked polyuronan consisting of β-D-mannuronic and β-L-guluronic acids of widely varying composition and sequence (Fig. 7.10). The acids are present as blocks of separate or mixed sequences along the chain depending on the seaweed source. *Macrocystis pyriferia* and *Ascophyllum nodosum* have a high mannuronic acid content (61% and 65% respectively), whereas *Laminaria hyperborea* has a high guluronic acid content (69%). Sodium

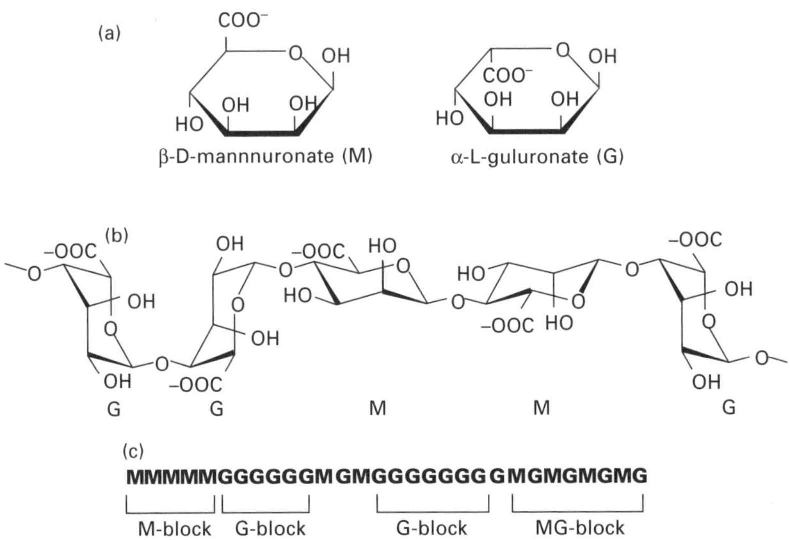


*7.10* Structural characteristics of alginates: (a) alginate monomers, (b) chain conformation, (c) block distribution

alginate dissolves readily in water to form viscous solutions and forms thermally irreversible gels in the presence of divalent cations (notably calcium). It is the guluronic acid residues that are responsible for gel formation and their proportion and distribution along the polyuronan chain have a major influence on the properties of the gels produced. Adjacent diaxially linked guluronic acid residues form a cavity which acts as a binding site for cations which interact with the carboxyl and hydroxyl groups. Intermolecular cross-linking of sequences results in the formation of junction zones and a three-dimensional gel network. This mechanism has been described as the egg-box model. If the divalent cations are added rapidly to sodium alginate in solution unhomogeneous gels are produced. In order to produce homogeneous gels a common practice is to generate the cross-linking ions slowly in situ. Soluble salts such as calcium sulphate are used sparingly and the release of ions is controlled by the presence of sequesterants and adjustment of pH.

Alginate is used in many food products including dairy products, desserts, fruit pie fillings, structured fruit, and sugar confectionery. Propylene glycol alginate is used to stabilize foam on beer and is particularly effective at stabilizing salad dressings.

### 7.3.9 Gelatin

Gelatin is denatured collagen, which is a protein and major constituent of the white fibrous connective tissue occurring in the hides, skins and bones of animals. It is obtained mainly from cattle and pigs although since the outbreak of BSE in the 1990s alternative sources, notably fish, have been investigated. The collagen is extracted from the raw material by subjecting it to either acid or alkaline treatment. Acid treatment involves immersing in cold dilute mineral acid (pH 1.5–3.0) for up to 30 hours while for alkaline treatment the raw material is steeped in saturated lime water (pH 12.0). The material is then washed with water leading to the Type A (acid treatment) and Type B (alkaline treatment) gelatins. The main amino acids are glycine, proline, alanine and 4-hydroxyproline for both types. Type A gelatin contains lower amounts of glutamic and aspartic acids and hence the isoelectric point for Type A is in the range 7–9.4 while for Type B it is in the range 4.8–5.5. In solution above ~40°C the gelatin molecules are in the form of random coils but on cooling the chains tend to order and form collagen-like triple helices which aggregate to form optically clear elastic gels.

Gelatin is widely used in confectionery, meats, dairy and dessert products.

### 7.3.10 Cellulosics

A variety of cellulosic fiber products can be produced from wood pulp. Since these are a good source of fiber and have a variety of functionalities in food, they seek to compete with the traditional natural hydrocolloids by offering security of supply, quality, and price. Novogel and Avicel are two examples. Both of these forms of cellulose fiber have uses in food. When properly dispersed in water systems, Avicel cellulose sets up a network with particles less than 0.2 micron in size. These particles are held together by hydrogen bonds which continue to form over a period of 24 hours, resulting in an increase in viscosity. Unlike soluble hydrocolloids the Avicel cellulose binds water to a much lesser extent. It is the physical network that affects the moisture in food systems. The applications are varied.

The gels made with Avicel cellulose readily break down with shear, but when the shear is removed the gel will reform (thixotropy). Temperature has very little effect on the functionality and viscosity of these dispersions and so they can be used during baking, processing, and microwave heating. Additionally the fiber is non-calorific and so can be used as a source of dietary fiber. Avicel cellulose can also be used to stabilize foams and emulsions. Novogel, a variant, differs from Avicel in that it capitalizes on the unique interaction between guar gum and microcrystalline cellulose. It is co-processing of cellulose gel and guar gum that imparts the special fat-like properties. These have a wide range of applications, as shown in Table 7.5.

Primacel is another cellulosic product, based on fermentation of cellulose, in combination with various co-agents and can be used as a highly efficient thickener, stabilizer, and film former. It can be used as a starch extender and in addition can be used to modify texture to give a more creamy mouthfeel in low-fat food systems.

Cellulose is produced by plants and it is a major structural part of the skeletons of plants. Fine crystal cellulose or fibrillated cellulose, obtained by pulverizing cellulose through chemical and physical treatments, is widely used as a food additive.

Cellulose can also be produced by microorganisms, such as *Acetobacter*. This cellulose of bacterial origin has been eaten for many years as a dessert food called nata de coco. The addition of fermentation-derived cellulose in a very small amount will give foods good dispersion and emulsion stability, and will also give foods short mouthfeel based on a good shape retainability. These functions are largely due to the three-dimensional network structure of cellulose fibers and are stable to physical and chemical treatments; they are also resistant to heat, acids, and salts. These characteristics enable fermentation-derived cellulose to be applied to foods for a wide range of purposes such as stabilization of thickening, dispersion, suspension, and emulsion, replacement of fat, prevention of protein aggregation, etc.

### 7.3.11 β-Glucans

The mixed linked (1→4), (1→3)-β-glucans (referred to as β-glucans) are linear polysaccharides composed of cello-oligomers separated by single (1→3)-

*Table 7.5* Examples of the applications of fermentation-derived cellulose

| Food | Effect |
|---|---|
| Cocoa drinks | Stabilizing dispersion of insoluble solid matters |
| Matcha drinks | Improving heat stability |
| Ca-strengthened drinks | Achieving good mouthfeel without raising viscosity much |
| Non-oil dressings | Stabilizing dispersion of insoluble solid matters<br>Achieving clear-cut mouthfeel without raising viscosity much<br>Achieving appropriate flowability<br>Improving stability at low pH<br>Achieving fatty mouthfeel |
| Sauce for roast meat | Stabilizing dispersion of insoluble solid matters<br>Decrease in threading<br>Achieving non-gluey good mouthfeel<br>Improving stability at high salt concentration<br>Improving stickiness |
| Low-fat mayonnaise | Improving creamy mouthfeel<br>Replacing fat<br>Achieving appropriate flowability<br>Achieving full body |
| Jellies with fruit flesh | Stabilizing dispersion of insoluble solid matters<br>Achieving heat stability<br>Achieving good meltability in the mouth |
| Retorted pudding | Preventing milk protein from aggregation<br>Achieving heat stability<br>Achieving good meltability in the mouth |
| Lacto-ice | Improving creamy mouthfeel<br>Improving shape retainablity |
| Soft-mix | Preventing whey from separating<br>Improving mouthfeel |

β-linkages. Generally those of high molecular weight form viscous solutions with the viscosity increasing with increasing molecular weight, while those of lower molecular weight show gel-like behavior, with some types of β-glucan able to form soft thermoreversible gels. Their major uses in foods to date have been as texturizing agents, especially as replacements for all or part of the fats in a range of dairy and bakery products, but there is also interest in including them in foods solely for their perceived health benefits.

β-Glucan and grains containing high levels of β-glucan have a useful physiological role as a dietary component and a significant part of research on β-glucans has focused on this aspect. For coronary heart disease, a leading cause of death in industrialized countries, high serum cholesterol levels and high levels of LDL (low density lipoproteins) cholesterol appear to be one of the risk factors. Consumption of foods rich in β-glucans significantly reduces serum cholesterol levels and LDL cholesterol. The FDA, recognizing the benefit of β-glucans in the diet, has allowed the health claims for foods that contain more than 0.75 g of oat derived β-glucan for each serving portion. Other health benefits of diet rich in β-glucans include moderating the glycemic response to the digestion of starchy foods and lowering serum lipids levels. β-Glucans are also a source of soluble fiber because humans produce no β-glucan degrading enzymes and they are therefore not hydrolyzed in the small intestines. They are degraded by microbial fermentation in the large intestines. The fermentation produces beneficial short-chain fatty acids which appear to have a role in guarding against colo-rectal cancer.

The health benefits of β-glucan has stimulated interest in including β-glucan in food products. Normally, cereal grains do not contain high enough levels of β-glucan to provide the most desirable intake of soluble fiber. So there has been interest in obtaining enriched sources of β-glucan either through supplementation with extracted β-glucan or using cereal flours or brans enriched in β-glucan. For this type of use any functionality that the β-glucan may impart to the food as a hydrocolloid is minor and the major reason for including the flour or bran is because of the perceived health benefits.

β-Glucan has been extracted commercially, mainly from oats, but the extraction process is expensive so that the addition of the β-glucan extract to foods has been uneconomic. Since the early 1990s there have been several key developments that should see the extensive use of significant amounts of β-glucan in processed foods, not only on account of their previously perceived health benefits, but also because the β-glucan itself imparts significant functionality to the food, that is, it is a useful food hydrocolloid in its own right. And the major functionality that they have in processed foods, that is the replacement of fats, is a considerable health benefit, since the average western diet is overloaded with calories and too little fiber.

## 7.4 Dietary fiber – in health and disease

The Vahouny Symposia which have been held since 1981 have made an enormous contribution to our knowledge of the action of dietary fiber in foods. Reference will be made here to information from the 7th International Symposium published in the *Proceedings of the Nutritional Society* (2003, Volume 62, No. 1), from which further reading can be obtained.

### 7.4.1 Methods of measuring dietary fiber

Since no single scientific definition of dietary fiber is universally recognized, the methods of measurement are similarly in a state of flux. It is clear that the importance of non-digestible oligosaccharides (NDO) as dietary fiber components, based on their physiological behavior, has been recognized.

There are both soluble and non-soluble dietary fiber, such as cellulosics and the soluble hydrocolloids which have been described above. To these should also be added the fructo-oligosaccharides and inulin, which can be extracted from chicory and Jerusalem artichoke on an industrial scale. High levels are also found in onion and leaf tissue of many grasses.

Two types of dietary fiber methodology are required: enzymic gravimetric for food labelling and control purposes and enzymic chemical for research purposes. These methods must yield similar dietary fiber results. The combination of Association of Official Analytical Chemists (AOAC) methods (2000) 985.29 and 991.43 (total dietary fiber) and the Uppsala enzymic chemical method (AOAM 2000 method 994.13) form a basis for fulfilling these requirements.

### 7.4.2 Colonic diverticula

Martin Eastwood has made a fundamental contribution to this subject through his work. Diverticulosis is a condition that is associated with aging. The older the person, the more likely they are to have diverticulosis. Diverticula are a herniation through the wall of the sigmoid colon and are likely to be a consequence of intracolonic pressure consequent to a low dietary fiber intake.The findings of Dr Eastwood show that a high-fiber diet from birth and preferably a maternal high-fiber diet lessens the risk of diverticulosis with age.

### 7.4.3 Anti-cancer effects of butyrate

Butyrate is an abundant material in colonic contents. As with the other short-chain fatty acids (acetate and propionate) butyrate is an end-product of bacterial fermentation of carbohydrates which have escaped digestion in the small bowel. Epidemiological evidence suggests that a high intake of dietary fiber protects against colo-rectal cancer. The mechanism underlying this protection are thought to be mediated by the short-chain fatty acid butyrate.

## 7.5 The appropriate molecular features to achieve Effective colon performance

*Plantago ovata* or P. *espaghua* (Plantaginaceae) is native to the Canary Islands and the Mediterranean regions of Southern Europe and is also indigenous to the Indo-Pak subcontinent. Its seed husk (ispaghula) is a rich source of effective polysaccharide (dietary) fiber for promoting the healthy functioning of the colon. A factor in its physiological fiber behavior is its high viscosity and gel-like character in water. This in turn is related to the molecular parameters associated with this extremely high molecular weight polysaccharide. This is now the basic material of commercial products for treating constipation and for promoting health in the colon. One of the main products is FYBOGEL (Reckitt Benchiser). It was, therefore, important to study the molecular features which are responsible for the effective action of this dietary fiber in the colon. We first studied the molecular parameters associated with the polysaccharide. Its behavior in water is unusual. Immediately it is suspended in water fine whiskers grow out of the solid within one minute. This behavior continues and the fibres grow with time until the solid has been converted, first into a soft and then into a stiff gel. It is this interaction with water which controls the rheological behavior of this polysaccharide which in turn is partly responsible for its effectiveness in increasing stool bulk and decreases stool transit time. There are, however, *in vivo* changes and modifications, which occur after exposure to the fermentation activity of the colonic microflora. To mimic this behavior controlled degradation studies were undertaken which studied the effects on rheological performance. The objective is to follow the breakdown of the tertiary matrix via the gel into the water soluble state and evaluate how this progression might relate to its functionality within the colon and lead to the various physiological benefits which have been reported.

The investigation showed that the ispaghula husk matrix is capable of interacting with water to yield a viscoelastic system which is capable of being preserved throughout its transit through the colon. This arises because the solid matrix can interact with water to give a gel fraction and a completely soluble fraction capable of rapid fermentation to yield short-chain fatty acids. In turn, these physical and chemical states can initiate a sequence of physical and physiological processes during transit through the colon, which enable a product such as Fybogel to be an effective aid to color health.

## 7.6 Bibliography


Al-Assaf S.A., Phillips G.O., Williams P.A., Takigami S., Dettmar P. and Havler M., 'Molecular Weight, Tertiary Structure, Water Binding and Colon Behaviour of *Ispaghula husk* fibre', *Proceedings Nutrition Soc.*, **62**, 211–216 (2003).

Aspinall G.O., *The Polysaccharides, Volume 1*. Academic Press (1982).

Aspinall G.O., *The Polysaccharides, Volume 2*. Academic Press (1983).

Aspinall G.O., *The Polysaccharides, Volume 3*. Academic Press (1985).

British Nutrition Foundation, 'Complex Carbohydrates in Foods', *The Report of the British Nutrition Foundation's Task Force*. London, Chapman and Hall (1990).

Harris P., *Food Gels*. Elsevier Science Publishers (1990).

Imeson A., *Thickeners and Gelling Agents for Food*. Blackie Academic and Professional (1992).

Lapasin R. and Price L., *Rheology of Industrial Polysaccharides; Theory and Application*. Blackie Academic and Professional (1995).

MacFarlane J., *Human Colonic Bacteria*. CRC Press, London (1995).
Phillips G.O. and Williams P.A., *Handbook of Hydrocolloids*. Woodhead Publishing Ltd (2000).
Phillips G.O., *Food Additives and Contaminants*, **15**, 251–264 (1998).
Ross Murphy S.B., *Physical Techniques for the Study of Food Biopolymers* Blackie Academic and Professional (1994).
Royall T., *Am. J. Gastroenterology*, **85**, 1307–1316 (1990).
Stephen A.M., *Food Polysaccharides and their Applications*. Marcel Dekker (1995).
Sunvold S., *J. Anim. Sci*, **73**, 1110, 2329, 3639 (1995).
Topping D., *Nutrition Rep. Int*, **32**, 809–814 (1985).
Trowell H.C., *Am. J. Clinical. Nutrition*, **29**, 417 (1976).
Whistler R.L. and BeMiller J.N., *Industrial Gums: polysaccharides and their derivatives,* third edition. Academic Press (1993).
Williams P.A. and Phillips G.O., *Gums and Stabilisers for the Food Industry 11*. Royal Society of Chemistry Special Publication No. 278 (2002) (and other issues in
Williams P.A. and Phillips G.O., *Gums I. Properties of Individual Gums; II. Food Uses Encyclopedia of Food Sciences and Nutrition 2003* (a revision of the Encyclopedia; First published in 1990 by Elsevier Science; Benjamin Caballero, Luiz Trugo and Paul Fingl (eds), 2003, 2992–3008.

# 8
# Fibers in medical healthcare

Fibers have always occupied a fundamental position in medical healthcare, by providing textiles, in the form of fibers, mono- and multi-filament yarns, woven, knitted, nonwoven and composite materials. The uses are many and varied, from the non-implantable materials such as bandages, dressings, etc., implantable materials such as sutures, vascular prostheses, hygiene products such as bedding, clothing, operating room garments, wipes, etc., to the more specialized items such as the artificial kidney. Most are disposable, but an increasing proportion of the products are reusable. In keeping with the objectives of this volume, some of the developing and possible future uses will be outlined.

## 8.1 Nonwoven

The official definition of nonwoven supported by the European Disposable and Nonwoven Association is:

> Nonwoven is manufactured sheet of directionally or randomly oriented fibers bonded by friction and/or cohesion and/or adhesion forces, excluding paper and products which are woven, knitted, tufted, stitch bonded incorporating binding yarns of filaments or felted by wet milling whether additionally needled or not. The fibers may be staple of continuous filament or be formed 'in situ'.

Traditional uses are for the many textile products used in operating rooms and hospital wards generally. Less obvious uses are:

*Filtration*

- Liquid, for example, blood, body fluids and water
- Air inwards and operating theatres
- Anaesthetic gases
- Odor removal, dressings, ostomy
- Anti-allergic bedding

*Clothing*
- Disposable gowns for patients
- Re-usable components of uniforms
- Components of shoes/footwear

*Building*
- Insulation for sound, heat
- Flame retardant materials
- Components of furniture
- Carpets

Their use is characterized by their ability to meet the huge variety of needs. Nonwoven technology allows for continuous production with minimal intermediate stages, whereas traditional textiles may require several distinct discontinuation batch processes, for example spinning, winding, beaming, sizing, before knitting or weaving.

The process for nonwoven is simple, productive, versatile, economic and innovative. It requires only the formation of the fibrous web and then binding together of the fibers. Their versatility will surely keep the nonwoven medical materials at the forefront well into the new millennium.

## 8.2 Alginate fibers

Alginate nonwoven dressings have many advantages over traditional dressings, particularly for moist wound dressings. Alginate is a hydrocolloid present in brown seaweed. The structure, shown in Fig. 8.1, is a polyelectrolyte made up of two building units: mannuronic acid (M) and guluronic acid (G). The counter ions are generally sodium or calcium fibers made from alginate either consisting of a high proportion of G or M units. In the high G alginate the calcium ions are firmly bound and consequently the fiber is slower to gel because of the slow rate of ion-exchange. High M alginate, on the other hand, gels more quickly because of the relative lower strength of binding of the calcium ions. To accommodate different dressing requirements, both high G and high M alginates are commercially available.

Alginates are indicated for all types of exuding wounds, although doubts have been expressed about their use on infected wounds. Their construction and availability in ribbons, ropes, and strips makes them suitable for many types of cavity wound. In the dental environment alginates are indicated as haemostat dressings. Most will require a secondary dressing to retain them in the wound. The England & Wales Drug Tariff (TSO) indicates they may be used on heavy exudate wounds, but it should be noted that the fluid handling capacity of some brands is lower than the foam dressings in terms of direct size comparison. However, their fast fluid uptake enables the alginates to cope with fast flowing wounds, at the cost of frequent dressing changes,

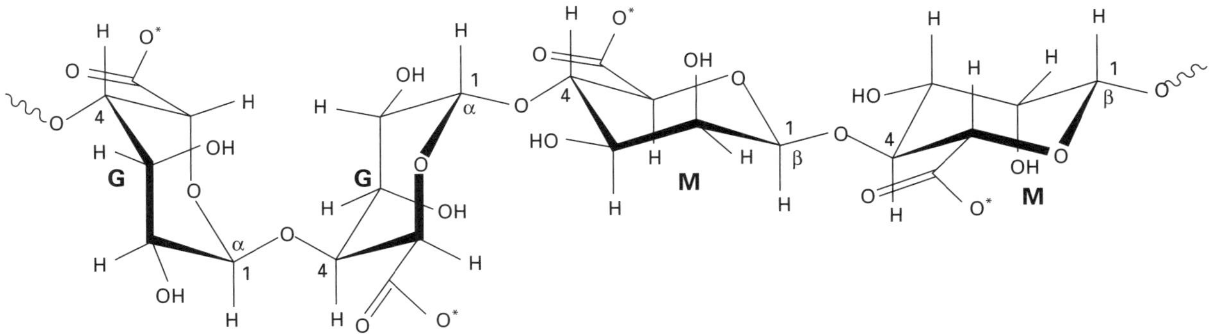

*8.1* The structure and functional groups in alginate in the acid form.

where foams are sometimes overwhelmed. In this situation the use of a foam dressing as a secondary dressing can prove to be an effective, if expensive, solution. Data from Kings College Hospital London (Edmonds and Foster) suggests that foam dressings may be more appropriate for diabetic foot ulcers.

Alginate dressings are constructed from interwoven strands of calcium alginate mixed, in some brands, with varying proportions of sodium alginate. Normally they are presented as a flat sheet of a few mm thickness, although ribbons or ropes are available. Some brands may have a backing fixed to the alginate mass for structural strength.

On contact with wound exudate the fibres turn to a gel trapping moisture at the wound surface to create a moist environment. This also makes them easy to remove at dressing changes. This mode of action makes them unsuitable for dry or necrotic wounds, even when soaked with saline. Some concern has been expressed that under some conditions fibres that do not fully gel may act as a foreign body in the wound. This may increase the length of the inflammatory stage. Table 8.1 show the various commercial type of polysaccharide dressings available.

When gel is formed the fiber structure disintegrates and the dressings lose their strength. In many such situations a used alginate dressing cannot be removed in one piece. Thus reinforced wound dressings refer to the type of alginate dressings that contain a web of continuous non-gelling materials, which when wetted will act as a firm structure to the dressings to facilitate a complete and clean removal.

## 8.3 Superabsorbent fibers

The use of superabsorbent materials in fiber form have now become a commercial reality. A collaboration between Courtalds Fibres and Allied Colloids has produced a cross-linked copolymer of acrylic acid. It is marketed under the name OASIS. Their properties are given in Table 8.2.

*Table 8.1* Various brands of commercial polysaccharide dressing

| Brand | Type | Manufacturer |
|---|---|---|
| Algisite M | | S+N |
| Algosteril | | Biersdorff |
| Aquacel | Cellulose fiber | Convatec |
| Comfeel SeaSorb | | Coloplast |
| Kaltostat | 20% sodium alginate | Convatec |
| Kaltogel | | Convatec |
| Melgisorb | Calcium/sodium mix | Molnlycke |
| Sorbsan | | Maersk |
| Sorbsan Plus | | Maersk |
| Tegagen | | 3M |

*Table 8.2* Properties of cross-linked acrylic acid co-polymers OASIS

| Property | Saline (g/g) | Water (g/g) |
|---|---|---|
| Free swell absorbancy | 40 | 80 |
| Retention (0.5 psi) | 30 | 60 |
| Absorbancy under load (0.25 psi) | 23 | 45 |

The advantages that fibers offer compared to powders is due to their physical form, or dimensions rather than their chemical nature. They will absorb many times their own weight, even when under pressure (Table 8.2). This is due to the small diameter of the fibers which is about 30μm, which gives a very high surface area for contact with the liquid. Also the fiber surface is not smooth. It has a crenulated structure with longitudinal grooves. These are beneficial in transporting moisture to the surface. The lubricant has also been selected to enhance this wetting effect and results in a very high rate of moisture absorption. Typically the fiber will absorb 95% of its ultimate capacity in 15 seconds. It is possible to process OASIS in blends with other fibers on most of the conventional nonwoven processing routes to produce fabrics suitable for medical products such as disposable incontinence products, wipes and absorbent pads, drapes, ostomy bags, and in wound care.

## 8.4 Wound healing and polysaccharide fibers

The process by which tissue repair takes place is termed wound healing and is comprised of a continuous sequence of inflammation and repair, in which epithelial, endothelial, inflammatory cells, platelets and fibroblasts briefly come together outside their normal domains, interact to restore a semblance of their usual discipline and having done so resume their normal function.

The process of wound repair differs little from one kind of tissue to another and is generally independent of the form of injury. Although the different elements of the wound healing process occur in a continuous, integrated manner, it is convenient to divide the overall process into three overlapping phases and several natural components for descriptive purposes.

Wound care management is an extremely complex medical operation and no single dressing can provide for all eventualities. The successful wound dressing must satisfy several criteria:

1. Seal the wound and prevent introduction of external stresses and loss of energy
2. Remove excess exudates and toxic components
3. Maintain a high humidity at the wound-dressing interface

4. Provide thermal insulation
5. Act as a barrier to microorganisms
6. Be free from particulates and toxic wound contaminants
7. Be removable without causing trauma at dressing changes.

Polysaccharide hydrocolloid systems have been devised to fulfill these criteria but they are seldom produced from a single polysaccharide. Commercial considerations protect the details of the various formulations.

Chitin (poly-1,4,2-acetamido-2-deoxy-β-D-glucose) is the second most abundant natural polymer, existing widely in the cell walls of fungi and crustacean shells. Chitin fibers has been suggested as a material which can accelerate wound healing, but due to its chemical and physical nature is very difficult to dissolve. New solvent systems have now been developed which allow fibers to be produced from chitin, but these have yet to make an impact on the market.

Another useful material is carboxymethyl cellulose (Fig. 8.2) which can be produced chemically from cellulose, the most abundant carbohydrate in nature. It has the merit that it can be produced in a variety of molecular weights and can then be integrated with other materials such as alginate to produce synergistic behavior in terms of physical properties.

### 8.4.1 Natural systems

Many natural systems containing polysaccharides have been suggested. One of these is the aloe vera leaf, which has polysaccharide in the parenchymal cells which is used by the plant for energy. When placed over a wound, the wound remains moist and does not dry up as dry wounds do. Epidermal and fibroblast growth factors come from the mucilage arising from the leaf and stimulate the fibroblast directly for growth and repair. The cells as a result migrate within the wound in a proper manner to increase wound healing. The occlusive nature of mucilage increases wound healing from a mechanical and endocrine viewpoint.

To follow up this observation, there has been further research and there is now a US patent covering this area, which claims the following:

> The present invention provides a rapid and efficient method for the preparation and isolation of biologically active polysaccharides from Aloe. The present invention includes the activated mixture of polysaccharides (referred to herein as 'Inmuno-10'), produced by the methods of the invention. The invention also includes the use of the polysaccharides as immunostimulating, immunomodulating and wound healing agents. The resulting immunomodulatory complex has a higher activity and is more stable than bulk carbohydrates isolated using prior art alcohol precipitation schemes.

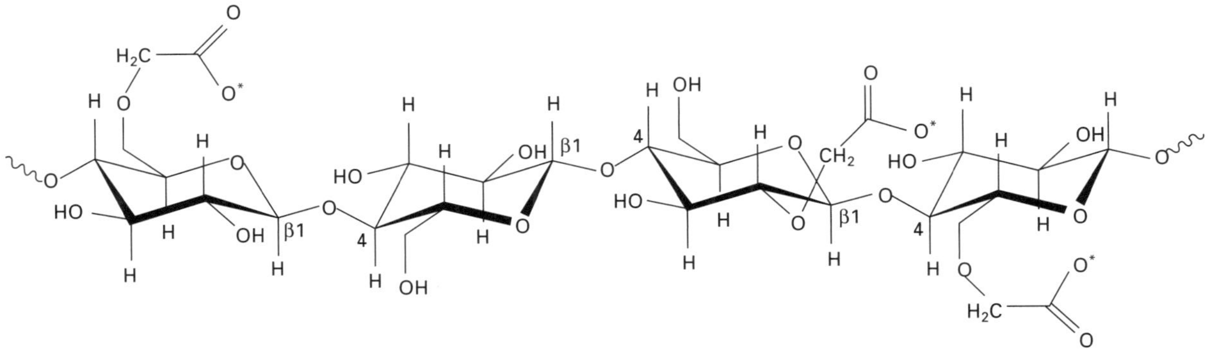

*8.2* The structure of carboxymethyl cellulose.

Chemically the polysaccharide consists of:

- primarily (>95%) of polysaccharides derived from aloe having an average molecular weight of 70–80 kDa with a range between 50–200 kDa;
- D-galactose (approx. 5% or less), D-glucose (approx. 5% or less) and D-mannose (approximately 90%);
- monosaccharides having primarily β-1,4 linkages and
- highly acetylated, having approximately 1 acetyl group per monosaccharide, with the acetyl group on the 2, 3 or 6 position of the monosaccharide unit.

Another is CM101, an anti-pathoangiogenic polysaccharide derived from group B streptococcus, has been shown to inhibit inflammatory angiogenesis and accelerate wound healing in a mouse model and minimize scarring/gliosis following spinal cord injury.

### 8.4.2 Cellulosic membranes

*Artificial kidney – Haemodialysis*

Over the past 20 years cellulosic membranes have improved considerably, due to the ability to form:

- thinner membranes
- controlled pore size
- improvement of surface properties.

These are now the basis of the production of a range of artificial kidneys for the treatment of chronic renal failure, and for this purpose the membranes are made from cuprammonium solution and saponified cellulose triacetate. The world market for artificial haemodialysis is expanding at a rate of 5% per year and in 1988 the world consumption was $3 \times 10^7$ units. Of these 66% were cuproammonium, and 15% cellulose acetate membranes, with the remainder using synthetic polymers. The number of patients receiving monthly haemodialysis in Japan at the end of 1990 exceeded 100 000 and 77% used the cuprammomium membranes. By the mid-1980s new cuproammonium-regenerated cellulose membranes with controlled pores sizes (4–10 nm) were developed by Professor K. Kamide of the Asahi Chemical Industry, Japan.

Nowadays most cellulose membranes are of the hollow-fiber type and fall into two categories:

- conventional hollow fibers (AM-SD series)
- biocompatible artificial kidney with standard and middle flux range (AD-Bio series)

*Virus removal filters for human blood*

The porous cellulose membrane (BMM$^{TM}$) with mean pore diameter ranging from 10 to 100 nm and having a sharp pore radius distribution enables the exclusive removal of disease viruses, such as acquired immune deficiency syndrome (AIDS) virus, human immunodeficiency virus (HIV), and hepatitis C virus. The composition of the filtrate of human plasma, separated through BMM$^{TM}$ with mean pore size greater than 20 nm is very similar to that of the original plasma.

Plasma separation membranes can separate red cells (6–9.5 μm in diameter), white cells (6–20 μm in diameter) and platelets (2 μm in diameter) from plasma, mainly composed of proteins like albumin and γ-globulin.

*Conclusion*

Breakthroughs in cellulose membrane technology have arisen because the membrane formation mechanism is better understood and pore characteristics can now be controlled. The thermodynamics of membrane formation based on particle-growth concept and lattice theory is better understood with respect to the solvent-cast method (that is the phase separation method). The pore size distribution can be calculated numerically by the lattice theory. Breaking strength of membranes is explained by considering development of bonding between the secondary particles and the total number of contact between the nearest secondary particles per unit surface of the membrane. The three-dimensional multilayered structure derived from the concept is the most suitable model for progressing the pore characteristics of the membrane.

New cellulose haemodialysis membranes having a larger sieving coefficient for $\beta_2$–microglobulin can avoid chronic haemodialysis associated syndromes such as carpal tunnel syndrome. Furthermore, cellulose membranes have already been improved to suppress the activation of complement using the new techniques to decrease free OH groups on the surface. The progress of the cellulose membrane industry in the future looks extremely promising.

## 8.5 Hyaluronan – a new medical fiber

Once a scientific novelty, this connective tissue polysaccharide is now central to the understanding and treatment of many intractable diseases. Two major international conferences were organised in Wales (2003) and the the United States (2003). The books arising from these landmark meetings have shown how wide the medical interest is in this material, as detailed in the reference section to this chapter. Previously regarded as the matrix which acted as a shock absorber in the body, it has now been found to bind into cells via specific receptor sites, and to instruct the cell how to behave. It is, therefore, at the centre of normal and disease-related cell proliferation.

Among the reported new developments are its uses and potential in the following areas:

- breast cancer
- as a switch to control cell-cancer behavior
- as an anti-cancer material
- in corneal wound healing
- cartilage maintenance
- repair of lung injury
- healing of chronic wounds
- engineering of new tissues
- treatment of diabetic foot ulcers
- kidney diseases in relation to diabetes
- in the control of eye surgery

Hyaluronan (HA) is a linear polysaccharide, with a fiber structure, which consists of repeating disaccharide units of D-glucuronic and N acetyl-D-glucosamine residues (Fig. 8.3). The residues are both β-linked in the polymer, D-glucuronic acid being linked at carbon 1 and 4, the glucosamine residue at positions 1 and 3.

Hyaluronan occurs naturally in vitreous humor synovial fluid and umbilical cord and in many animal tissues in smaller concentrations. It has been reported that the molecular weight of naturally occuring hyaluronan varies within the range $10^5$ to $10^7$. Hyaluronan can be produced from biological sources such as bovine vitreous, umbilical cord and rooster comb. The highest molecular weight of hyaluronan produced from animal sources commercially available is about $5 \times 10^6$ and marketed under the trade name of Healon GV. Hyaluronan can also be produced from certain strains of *Streptococcus* bacteria, but the molecular weight of bacterial hyaluronan is lower, despite an easier production process of fermentation, extraction, and purification from the broth. The highest molecular weight produced by this method is about $2.5 \times 10^6$.

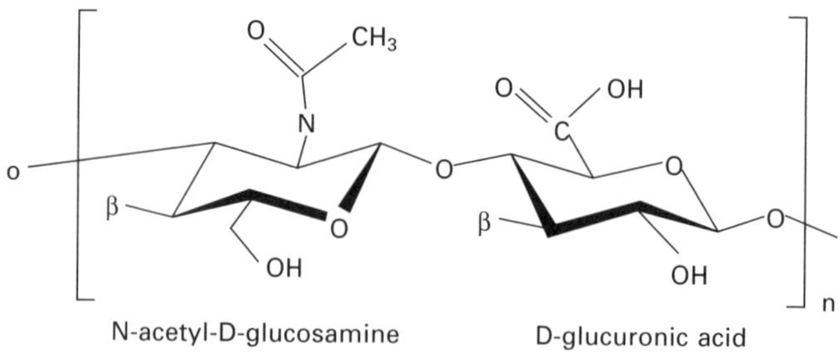


*8.3* The component sugars and repeating disaccharide unit of hyaluronan fiber.

Balazs and co-workers have developed a family of cross-linked hyaluronan derivatives called hylans (HY) which can be produced either in water soluble form (hylan A) or as viscoelastic gels (hylan B). In the first procedure, formaldehyde is used at neutral pH to produce a permanent bond between the C-OH group of the polysaccharide and the amino group of a protein with relatively small molecular size and specific affinity to the hyaluronan chain. The protein forms a bridge between two polysaccharide chains. Under appropriate conditions, the cross-linking process will yield a molecular network consisting of permanent association of two to eight HA molecules. The average molecular weight of HY molecules is $6–24 \times 10^6$ with a protein content of 0.4–0.8% of the total polysaccharide weight. The second cross-linking process utilizes vinylsulfone. The HY molecules obtained from this technique are insoluble but are produced in the form of a viscoelastic gel. The two cross-linking procedures retain the biocompatibility and physical functionality of the unmodified hyaluronan, but physicochemical parameters such as molecular weight, molecular size and rheological properties, of the polymer solution or suspension, on hydration are substantially affected. Thus HY can be used in applications for which high molecular weights are needed such as viscosupplementation for the treatment of osteoarthritis of the knee joint.

The term *matrix engineering* was first used by Dr Endre A. Balazs (1971) to describe the use of natural and chemically modified biopolymers, derived from hyaluronan, to control, direct and augment tissue regeneration processes. Such biomaterials, therefore, can be regarded as a specialized type of non-inflammatory, biocompatible tissue graft, capable of a wide range of applications for the augmentation, protection and repair of human tissues. The base material, the glycosaminoglycan hyaluronan, is present in virtually every tissue in the human body and if the natural repeating structure and conformation are preserved and inflammatory fractions removed, the resulting biomaterial is not recognized as immunologically or otherwise foreign. This technically difficult task, achieved by Balazs, enabled the routine production of a highly purified, non-inflammatory hyaluronan fraction, known by its acronym, NIF-NaHA.

The first clinical studies with NIF-NaHA were initiated in 1968 using racehorses with traumatic osteoarthritis. In equine arthritis, the elastoviscosity of the synovial fluid in the joints decreased, and by using NIF-NaHA with an average molecular weight of $2–3 \times 10^6$, the pathological synovial fluid of low elastoviscosity was replaced with a fluid that had significantly higher elastoviscosity than the normal healthy synovial fluid. For this supplementation of the synovial fluid, the term *viscosupplementation* was coined. Since that time, it has become known that the medical benefits of a viscosupplement depend on its rheological properties, rather than on the chemical nature of the viscosupplement. The efficacy of the viscosupplement depends on the elastoviscosity of the fluid or, at equal polymer concentrations, on the average molecular weight.

The first human clinical use of NIF-NaHA was reported by Balazs and coworkers in 1972 using this elastoviscous fluid as a viscosurgical tool during corneal transplantation and as a viscosupplement for the vitreus after retinal detachment surgery. The other clinical use of NIF-NaHA, at that time, was for the treatment of the painful osteoarthritic joints in humans.

In the early 1980s, after the introduction of intraocular lenses after cataract surgery became a widely used medical procedure, Healon® and viscosurgery found a new application. In a very short time, viscosurgery with Healon became routine in ophthalmic surgery, and provided elastoviscous protection for the corneal epithelium and the iris against mechanical damage due to instruments and implants. In viscosurgical procedures, the elastoviscous fluid also makes and maintains space for surgical manipulation, prevents tissue adhesion, controls bleeding by its barrier effect, and helps in the removal of tissue debris, such as pieces of cataractous lens. Today, in addition to Healon, at least ten different preparations are available worldwide for ophthalmic viscosurgery, but because of their lower molecular weight, none have the same high elastoviscous properties of Healon Viscosurgery. The use of 'viscoelastics' in ophthalmic surgery has been accepted internationally, and the International Standards Organization now provides standards for 'Ophthalmic Viscosurgical Devices for use in Anterior Segment'.

The first 'viscoelastic' used in medicine was marketed worldwide in the late 1970s and early 1980s by Pharmacia AB (Uppsala, Sweden, now Pharmacia & Upjohn Company). It was a 1% solution of NIF-NaHA with an average molecular weight of approximately 3 million and was used in ophthalmic viscosurgery (Healon®) and for the treatment of arthritis in horses and other animals (Hylartil®).

It was not until 1987 that the first hyaluronan (NIF-NaHA) viscosupplementation product became available to human patients in two countries for treatment of osteoarthritis. Two companies, Seikagaku in Japan, and Fidia in Italy, produced low molecular weight NIF-NaHA (average MW $0.5–0.8 \times 10^6$) and introduced it in their respective countries. The lower molecular weight and the lower elastoviscosity meant that these products required more injections (5–10 injections weekly) than the more elastoviscous NIF-NaHA tested decades earlier (2–3 weekly injections).

The need for a more elastoviscous hyaluronan product to produce a viscosity as close as possible to that of the synovial fluid of a healthy young adult human, and for strong, soluble and insoluble viscoelastic gels and solids, led to the development by Balazs and his coworkers at Biomatrix, Inc. (USA) of a new family of hyaluronan derivatives named hylans. Hylan is a generic term used to refer to a class of hyaluronan derivatives produced by cross-linking of the polysaccharide chains via hydroxyl groups of the hyaluronan, leaving the carboxylate, the acetamido, and the reducing end groups unaltered. The retention of the carboxylate group is particularly important because it is this group that confers the polyanionic character to the molecule, which is critical to its physicochemical and biological properties. Hylan polymers are available in physical forms which range from highly elastoviscous solutions to viscoelastic gels and solids. This allows the physical properties and residence time of hyaluronan to be controlled without loss of biocompatibility.

The hylan polymers used in medical practice today are called hylan A and hylan B. Hylan A is produced *in situ* by treating hyaluronan-rich tissue sources with aldehydes before extraction. The aldehyde activates the hydroxyl groups of the hyaluronan, which interact with a small amount of specific protein, forming a covalent bond (protein content is $\leq 0.5\%$ of polysaccharide) and resulting in soluble hyaluronan polymers with enhanced molecular weight. Hylan B is synthesized by treating hyaluronan or hylan A with divinyl sulphone under mild alkaline conditions. The reaction conditions can be varied to produce materials with properties which range from soft deformable elastic gels to solids of various shapes with long residence times in tissues.

Based on these two new biopolymers, a range of products has been developed and marketed worldwide. Synvisc® (hylan G-F 20) is a highly elastoviscous synovial fluid supplement with elastic and viscous properties similar to those of the synovial fluid of young, healthy humans used for the treatment of osteoarthritis. Hylaform® (hylan B) is a solid hydrated gel implanted into soft tissues for viscoaugmentation of facial wrinkles and depressed scars. Hylashield® (hylan A) is a dilute elastoviscous solution applied to the surface of the eye for comfort and protection from noxious environmental conditions (viscoprotection).

Tendon adhesion following injury or after surgical repair is a significant clinical problem. Ever-expanding procedures associated with tissue banking include the replacement of the cruciate ligament with human patellar allografts, spinal fusion, and revision hip surgery (using bone allografts) where this problem is frequently encountered. Dry sheets and high concentration NIF-NaHA solutions have been tested in various animal models of tendon regeneration and in primate tendon surgery since the early 1970s. Hylan B gels were used more recently as a viscoseparation device to reduce postsurgical tendon adhesion and scar formation.

Hylans have been used in percutaneous embolization to produce blood coagulation in vessels feeding arterial venous malformations in the brain or in facial tissue, and to alleviate arterial bleeding in the lung and other organs of the body. In this process, hylan B gel is used as a vehicle to deliver hemostatic agents with x-ray-opaque material to the target tissue, and blocking blood flow and causing the permanent blockage of the vessels by connective tissue formation.

Hylan A fluid and hylan B solids provide today the most biocompatible and most versatile intercellular matrix system to be used as a molecular framework alone or with other matrix molecules (collagen, elastin, proteoglycans) for cell populations to form artificial organs. The clinical use of medical devices made from hylans is well documented.

Matrix engineering introduced the concept of using elastoviscous fluids and viscoelastic solids as therapeutic agents in medicine. The various uses have been described as: *viscosurgery* for protection and manipulation of tissues during surgery, *viscoseparation* after surgery to prevent adhesion formation, *viscosupplementation* to replace or supplement dysfunctional tissues or tissue fluids, *viscoaugmentation* to add viscoelastic molecular matrices to augment and build up tissues, *viscoprotection* for coating of the tissue surfaces in order to protect them from environmental damage, and *viscoregulation* to regulate implants, tissue regeneration and new tissue and organ development with viscoelastic molecular matrices.

Hylan implants can be regarded as a new class of allografts because they are made from the molecules which exist in the human body and therefore fulfill non-definition of homeographs that originate from genetically not identical individuals of the same species. Matrix engineering can be redefined as the allogra use of molecular matrices made of building blocks of the body to be populated with homologous cells before or after grafting in order to replace tissues or regulate and stimulate their regeneration.

Hylans and their co-polymers, together with other glycosaminoglycans and a variety of molecular matrices and pharmacologically active agents, can be used as scaffolds, carriers and matrices of cells for implantation. Their unique biocompatibility and rheological properties combined with the greater resistance to free radical degradation and high water-binding properties of these new solid biopolymers, makes them an ideal material for tissue allografts for implantation and for control of tissue regeneration.

The interest in hyaluronan systems has been intense in respect to wound healing. Hydrogel dressings, based on hyalronan and another glycosaminoglycan from connective tissue (chondroitin sulphate), have been developed by University of Utah medical researchers. If the hydrogels work as well in people as they have in mice, millions of diabetics, elderly, burn victims and surgical patients may benefit from faster-healing diabetic ulcers, skin grafts, surgical incisions and other wounds. The researchers reported their findings in the journal *Biomaterials*.

The hydrogels look like a piece of clear, thin plastic when dry, but expand six times in volume and become pliable when wet. Trials with mice show hydrogel wound dressings accelerate healing in the epithelium – the outer layer of skin – in young, healthy mice by up to 33%, with complete healing of deep wounds in five to seven days. If the success in young, healthy mice is indicative, the researchers say the hydrogels would help older, less healthy mice, or people, even more profoundly.

The researchers hope to gain US Food and Drug Administration approval to begin human trials of the hydrogels. Conventional dressings and bandages serve mainly to keep moisture in and germs out as a wound heals. But hydrogel dressings slowly break down and reintegrate into the wound and extracellular matrix that surrounds human cells. According to Glen Prestwich, who led the project 'The hydrogel actually becomes integrated into the wound. It's a scaffolding that enhances healing.'

Hyaluronan and chondroitin sulphate were also mixed with a reactive version of a waxy polymer called polyethylene glycol (PEG). Within seconds of being mixed, the substances cross-linked – a chemical process akin to weaving cloth – and within minutes they became gels.

## 8.6 Other fibrous scaffolds for tissue engineering

The tissue engineering process starts with a scaffold and a supply of cells. The cells could be the patient's own cells, a donor's cell or taken from a tissue bank. These are seeded into a proposed scaffold. The cells themselves attach to the scaffold and are cultured within a mini bio-reactor. Tissue culture media provide nutrients for the cells and remove waste products. The cells increase in number and lay down the new extracellular matrix to form neo-tissue. If bio-resorbable fibers are used, then these will start to degrade. At the end of culturing the tissue engineered may require preservation and storage before used as an implant.

In addition to the polysaccharides the most important synthetic group which have application as bio-resorbable scaffolds are poly(glycolic acid) and poly(lactic acid). These have an established history as sutures or surgical devices. One such tissue engineered and now on the commercial market is Dermograft™, a joint venture between Advanced Tissue Services (USA) and Smith and Nephew, used for the treatment of diabetic foot ulcers. These wounds are difficult to heal and can often lead to serious complications. The scaffold is produced from multifilament yarn, a 90–10 co-polymer of poly(glycolic acid) and poly(lactic acid). Specially shaped scaffolds are produced for articular cartilage and meniscal cartilage.

## 8.7 Collagen: medical applications

Collagen is a major structural protein, forming molecular cables that strengthen the tendons and vast, resilient sheets that support the skin and internal organs. Bones and teeth are made by adding mineral crystals to collagen. Collagen provides structure to our bodies, protecting and supporting the softer tissues and connecting them with the skeleton. It is composed of three chains, wound together in a tight triple helix, each chain being over 1400 amino acids long. A repeated sequence of three amino acids forms this sturdy hydroxyproline structure. Every third amino acid is glycine that fits perfectly inside the helix. The special amino acid sequence makes the tight collagen triple helix particularly stable. Every third amino acid is a glycine, and many of the remaining amino acids are proline or hydroxyproline.

The collagen molecule is a triple helix assembled from three individual protein chains. The triple helix is further assembled into larger structures known as fibers. The collagen fibers play an important role in binding platelets under conditions of blood flow. These 'type I' collagen molecules associate side-by-side, like fibers in a rope, to form tough fibrils. These fibrils criss-cross the space between nearly every one of our cells. These form a basement membrane (collagen-2), which forms a tough surface that supports the skin and many organs. A different collagen ('type IV') forms the structural basis of this membrane.

Collagen has found widespread medical uses. Urology, dermatology, orthopaedics, vascular and general surgery utilize collagen in various forms ranging from injectable solutions to sponge-like materials. In addition, collagens extracted from animal species, primarily bovine, are used in the preparation of a wide variety of commercial products including:

- biological dressings
- tissue culture applications
- dermal injectables.

Collagen is an ideal biomaterial for the development of medical and other commercial products because it is highly biocompatible, is readily available at high purity, and can be manufactured in such diverse forms as pastes, gels, films, sponges, and felt-like sheets using a variety of process methods. In cosmetic treatments it is able to:

- smooth facial lines and wrinkles
- add definition to lip line borders
- smooth smile lines
- improve 'marionette' lines
- decrease frown lines
- improve vertical lip lines
- fill shallow acne scars

The role of collagen in blood clotting is complex and multi-factorial. Platelet-collagen interactions have received considerable attention because collagen is considered to be the most thrombogenic constituent of the vessel wall. After injury, platelets exposed to collagen in the sub-endothelial layer adhere rapidly to the exposed collagen fibrils. Platelet binding to collagen can occur through a direct platelet-collagen interaction or can be mediated via von Willebrand factor forming a bridge between collagen and platelets. Platelets bind to collagen, aggregate, adsorb, and concentrate clotting factors.

Platelet-bound fibrinogen is converted to fibrin which forms a cross-linked network. The fibrin network which forms reinforces the otherwise friable platelet plug. The bound, activated platelets are completely degranulated, releasing ADP, thromboxane, and other secretory products which facilitate clotting. Since the discovery of the role of collagen in blood clotting, collagen obtained from animal skin and tendon has been processed into loose fibrillar forms and felt-like sheets or collagen fleece and used to stop bleeding in an increasing number of procedures including spleen repair, laparoscopy, oral surgery, and general surgery.

Collagen also plays an essential role in the wound healing process. Acting as a tissue scaffold, it is used as a carrier vehicle for cells in tissue engineered products for dermal wound repair and as a carrier vehicle for growth factors in bone repair. Collagen fibers are one of the best scaffolds for cell migration and proliferation. Collagen interacts with fibronectin and other adhesion proteins to promote cellular in-growth which speeds up wound healing. Type I collagen has been shown to attract fibroblasts in cell culture and appears to cause directed migration of cells. Drugs, cells, and growth factors use collagen as both the delivery vehicle and structural support for tissue development and in-growth.

## 8.8 Medical textiles

A Centre of Excellence for Medical Textiles is located in the Bolton Institute, in the United Kingdom, led by Professor Subhash Anand. Regular update conferences are organized in Bolton and as noted in the reference list the 1996 and 1999 Symposia have been published. Another excellent meeting was held in 2003 (MEDTEX 03) and the Proceedings will be incorporated into a new book due to be published in 2005. Some direct reference will be made here to a few contributions made at that meeting with details of the presenters of the papers for reference.

### 8.8.1 Textile medical sensors (Lieva van Langenhove and Carla Hertleer, Ghent University)

Textile materials cover a large surface area of the body. Consequently, they are an excellent measuring tool. Biosignals that are mentioned in literature are:

- temperature
- biopotentials: cardiogram, myography
- acoustic: heart, lungs, digestion, joints
- ultrasound: blood flow
- motion: respiration
- pressure: blood.

A significant breakthrough can only be achieved, however, when the sensors and all related components are entirely converted into 100% textile materials. This is a big challenge because, apart from technical considerations, concepts, materials, structures, and treatments must be focusing on the suitability for use in or as a textile material. This includes criteria like flexibility, water (laundry) resistance, durability against deformation, radiation, etc. As for real devices, ultimately most signals are being transformed into electrical ones. Electroconductive materials are consequently of utmost importance with respect to intelligent textiles.

One area of success is the Intellitex suit for measuring heart and respiration rate. Instead of using metal plates, the Intellitex suit uses a conductive textile as an electrode. To measure the heart rate and even an ECG, the Textrodes were developed. The Textrodes have a knitted structure and are made of stainless steel fibres (by Bekintex). They do not require any electrogel. This enables long-term monitoring but has an negative impact on the contact with the skin. For children, attractive design makes them want to wear the suit, and they can be monitored without disturbing them.

The Textrodes make direct contact with the skin. Test results have shown that the electrode's textile structure is an important parameter. When changing the structure, a different contact surface with the skin is obtained. Finer structures with more protruding fibres for instance will adapt more easily to the heterogeneous skin surface, which results in a more intense contact between the electrode and the skin. In turn, this results in a lower impedance of the skin electrode system. So a compromise has to be found between the sense of comfort and the intensity of the contact with the skin. A knitted structure has the advantage of being stretchable. Elasticity is a required property for close fitting of the suit around the thorax.

The Intellitex suit combines heart and respiration rate measurements in one garment. The respiration sensor is a knitted belt called 'Respibelt'. It is also made of a stainless steel yarn. The basic concept could also be used as a strain sensor, for instance to control tension applied in pressure bandages

### 8.8.2 Textiles in burns treatment (J. Edwards, Wythenshaw Hospital, Manchester)

The care of burns patients has made steady progress. Until the 1960s, even moderate burn injuries were usually fatal. The introduction of fluid resuscitation and the establishment of burns units have had a major impact on mortality. Burns patients have subsequently benefited from many developments, including the introduction of systemic and topical antimicrobial agents, progress in intensive care and nutritional support, changes in surgical philosophy, advances in wound care and methods of achieving skin cover, and the concentration of treatment of patients with serious burns in specialist care. Alongside these improvements, the use of textiles in the patient's journey from injury to recovery has been crucial. Textile materials act as support surfaces, dressings, splinting, skin substitutes, pressure garments and silicone gels needed to enable a burns patient to travel the road from injury to recovery

The problem of hypothermia in burns has long been recognized. Plasticized Polyvinyl Chloride (PVC) is advocated by the Emergency Management of Severe Burns (EMSB) as a means of preventing hypothermia, reducing pain, and preventing desiccation and infection. The film is very thin and permeable to water vapor, oxygen, and carbon dioxide, easy to apply, and allows for visual inspection of the wound.

On admission, severely burned patients need to be nursed on specialized beds. These can either be air-fluidized beds or low air loss beds. Mattress coverings have developed greatly since the mid 1980s when regulatory changes were made to the flame retardancy requirements. These new coverings are water/moisture vapor permeable and have the ability to transmit water vapor molecules through itself, whilst at the same time remaining a complete barrier to liquid water. There are two main types of materials: microporous materials, and hydrophilic materials. Microporous materials are membranes made from special polymers that have tiny holes in them, e.g. Gore-Tex. Gore-Tex, is vapor permeable, it has pores 700 times larger than water molecules, which let water vapor pass through. This helps to eliminate moisture, friction, shear, infection, contamination, and heat. As burns have copious exudate during the first 24–48 hours, this function is important in a bed. Hydrophilic materials attract water into them and transmit the moisture through the coating by a chemical mechanism. They have no holes in them and are a complete barrier to liquid water. All polyurethanes are hydrophilic to some degree or other, and some can be formed into coatings that are tough, as well as being water vapor permeable. Polyurethane-coated fabrics provide greater patient comfort, are generally resistant to most cleaning agents, and complement carefully engineered support mechanisms.

Burns, after being dressed can be treated with a covering called Exu-Dry. Exu-Dry is a one-piece, multilayer, highly absorbent, non-adherent wound dressing. It incorporates a non-adherent wound contact layer, and an antishear layer, which helps to reduce friction. The absorbency of the product comes from the inner layer, which is highly absorbent. The outer layer is permeable and non-occlusive, allowing the wound to breathe. These dressings overcome many of the problems of traditional dressings of layered paraffin gauze, gauze and gamgee, which was time consuming and had problems of strike through. They also have a marked tendency to adhere to the surface of drying wounds. This is due in part to the exudate sticking the dressing to the wound as it dries and also if left in place long enough, the in-growth of capillary loops within the granulation tissue into the dressing. This in effect incorporates the dressing into the new tissue, which will inevitably be damaged when the dressing is removed. Exu-Dry dressings come in body shapes such as arms, legs and chests, which enable even inexperienced nurses to dress extensive burns efficiently.

A useful alternative is a product called Telfa Clear, a non-adherent contact layer made from a Mylar perforated polyester (polyethylene terephthalate) film. The specially designed film is composed of hundreds of minute perforations that act as a selective membrane. The size of these holes allows the passage of wound fluid into the secondary dressing, but blocks the entrance of larger epithelial buds. Telfa Clear comes in large sizes and is used as a primary contact layer for the application of Flamazine, enabling large awkward areas to be dressed that are not perhaps covered by the use of Exu-Dry.

Regardless of the dressing type used, most dressings for major burns will require bandages to secure them. Bandages perform a number of functions including retention, support and compression. In burns patients the main functions are retention of the underlying dressings, and support to prevent oedema formation and provide joint support. The type of bandages used traditionally are crepe, which are made of a cotton fabric of plain weave with a characteristic appearance made from the crepe twisted cotton yarns (e.g. Elastocrepe, Elvic). More recently white knitted bandages have been used. These are made from a white knitted conformable fabric containing 93% viscose, 4% nylon and 3% elastomeric yarn (K Lite). This means that these bandages can give more support than traditional crepe bandages.

Having assessed and dressed the burns, the next most important area of care is splinting. Splinting is used for a number of reasons; to increase function, to prevent deformity, correct deformity, protect healing structures, restrict movement and allow tissue growth or remodeling. A number of splinting materials are used, but the majority are made from polycaprolactones (Polyform, Aquaplast). These are low temperature thermoplastic materials, which provide greater conformability and ease of splint fabrication.

### 8.8.3 Textile finishing for the production of new generation medical textiles (N. D. Oltargevskaya and G. E. Krichevsky, Educational Textile Institute, Russia)

A recent trend in textile chemistry has been the development of the theoretical technological and manufacturing principles for prolonged action textile-based materials for medical purposes. 'Koletex®' is an example of a prolonged action medical bandage, based on the technology used in printing and textiles.

The efficiency of medical bandages is determined by the choice of textile material type, type of a polymer – the medicine carrier (thickener), and medical properties depend on the chosen medical product introduced into a textile material together with the thickening. Printing technology allows the obtaining on a base of bi-porous textile material a 'double depot', from which medical product can be released controllably and transported into a human organism (into a wound, through skin and into a malignant tumour as well) and performs a medical effect.

'Koletex'® bandages are widely used in Russia in medical practice: 20 types are in product in various areas of medicine for different purposes (surgery, neuralgia, stomatology, oncology, dermatology, gynecology, etc.).

### 8.8.4 Wound care dressings from chitin (K. Van de Velde, L. Szosland and I. Krucińska Department of Textiles, Ghent University and Department of Textile Metrology, Technical University of Łódź, Poland).

Chitin, a ubiquitous biopolymer found in the exoskeleton of insects and marine invertebrates, shows the extraordinary capability to promote the ordered healing of tissues and is well suited to use in wound dressings. However, chitin is insoluble in common organic solvents, therefore direct industrial applications of chitin are very difficult. Recently the synthesis of dibutyrylchitin (DBC), an ester of chitin, was developed. DBC is easily soluble in common organic solvents and has film- and fiber-forming properties. This invention opens the way for production of a wide assortment of novel functional biomaterials made from DBC and pure chitin regenerated (CR) from DBC, which would promote the wound healing process and can find other medical applications.

### 8.8.5 A spider silk supportive matrix used for cartilage regeneration (Kris Gellynck, Peter Verdonk, Fredrik Almqvist, Els Van Nimmen, Domir De Bakker, Lieva van Langenhove, Johan Mertens, Gust Verbruggen, Paul Kiekens, Ghent University)

Injured cartilage often decreases quality of life. Repair can be effected if the chondrocytes of the cartilage can be grown and implanted. The chondrocytes need an implanted support to bridge and recover the wound with extra-cellular matrix products forming fresh cartilage. Advances in cell biology and biomaterial research have lead to new possibilities in tissue engineering. Transplanted scaffolds, holding a 3D cell culture, should copy the cartilage characteristics. Strength and flexibility are important, but even more important is an adequate porosity, so the chondrocytes can migrate through the matrix, but are not able to float around.

Looking for regeneration and not a repair, the scaffold material should disappear while real cartilage is healing the wound. In this way the material and its hydrolysis products have to be biocompatible and harmless. In the case of synthetic polymers, the hydrolysis products are frequently toxic, but spider silk is a promising fibre for many applications. Completely made out of protein a possible biocompatibility has already been proven. The harmless amino acid hydrolysis products make the silk a good candidate for creating a bioresorbable textile scaffold.

The chondrocyte cells adhere quite well on the spider cocoon silk threads. Cocoons can be obtained each autumn in large numbers from the *Araneus diadematus* garden spider. The mechanical properties of the silk are more appropriate than polymeric gels, like hyaluronan, collagen, alginate, which proved to be successful in 3D immobilization and maintaining the differentiated phenotype of chondrocytes. The phenotypical products collagen II and aggrecan were also detected around the cells growing on the spider cocoon silk. A silk 3D textile could possibly be applied in combination with a polymer gel, probably alginate, in order to achieve some biomechanical stability. While biodegradation is occuring, the silk textile is overgrown with real cartilage and eventually the wound will recover without any definitive synthetic implants.

## 8.9 Bibliography

Anand S. (ed.), *Medical Textiles.* Woodhead Publishing Ltd, Cambridge (1996, 1999).

Gilbert R.D. (ed.), *Cellulosic Polymers.* Hanser Publishers, New York (1994).

Phillips G.O., Kennedy T.F. and Williams P.A. (eds), *Cellulosics: Materials for Selective Separations and other Technologies.* Ellis Horwood (1993).

Kennedy J.F., Phillips G.O. and Williams P.A. (eds), *Hyaluronan: Volume 1: Chemical, Biochemical and Biological Aspects; Volume 2: Biomedical, Medical and Clinical Aspects* (Guest Editor: Vince C. Hascall). Woodhead Publishing Ltd, Cambridge UK (2002).

Phillips G.O. and Williams P. (eds), *Handbook of Hydrocolloids.* Woodhead Publishing Ltd, Cambridge (2002).

Balazs E.A., Al-Assaf S. and Phillips G.O., 'A New Biomaterial Tissue Allograft: Hylan, a Hyaluronan Derivative', in, *Advances in Tissue Banking* (ed. G.O. Phillips), 3 (1999) 357–397.

Al-Assaf S., Meadows J., Phillips G.O., Williams P.A. and Parsons B.J., 'The effect of hydroxyl radicals on the rheological performance of hylan and hyaluronan', *Inter. J. Biol. Macromol.*, **27**, 337–348 (2000).

Gunning A.P., Morris V.J., Al-Asaaf S. and Phillips G.O., 'Atomic force microscopic studies of hylan and hyaluronan', *Carbohydrate Polymers*, **30**, 1–8 (1996).

# 9
# Developments in nanofibers for the new millennium

## 9.1 Background

While the history of fibers extends far back in time, developments over the past 100 years have been particularly significant. The Industrial Revolution opened with cotton spinning in Britain in the eighteenth century and brought about a new manufacturing process. Modern chemistry emerged in the nineteenth century and developed in the twentieth century with the growth of the petroleum industry, which led to the emergence of polymer science and three important synthetic fibers.[1] These multipurpose fibers were polyamide fibers (Nylon 66, Nylon 6), poly(ethylene terephthalate) (PET) fibers and acrylic fibers, with spandex fiber appearing in the first half of twentieth century. These fiber materials became the basis of the clothing that consumers wear every day.

The aim of the development of synthetic fibers was imitation and replacement of natural fibers. Developments in engineering from the 1960s followed two directions. One was innovation in processes, improvement in quality and investment of functions in production of the multipurpose fibers. In the 1980s the high performance fibers were developed – a group of carbon fibers and aramid fibers, known as superfibers. The other led to high functional fibers and high sense fibers.[2] In 2004, additional keywords such as environment care, using biodegradable fiber and recyclable fibers, and health, comfort and safety have been added in material development. Life needs have led to multifunctional fibers and attention has moved, over the past ten years, from ecosystem imitation technology (biomimetics) to environment response fibers and intelligent fibers. Nanofiber technology will now be developed for practical needs. Nanofibers of carbon nanotube and GI-type POF have already been produced on an industrial scale. Toray have succeeded in spinning nylon nanofiber where molecular arrangement of polymer chains is controlled in nano-level.[3] Teijin reported a light interference colored fiber, MORPHOTEX® during the 1st International Congress on Nanofiber Science Technology.[4] Professor Y. Atomi *et al.* (The University of Tokyo) have proposed clothes made with nanofiber which activates the body during the 2004 Expo/ Fibers for New Era.[5] This new technology will be applied to multipurpose polymers of polyesters and polypropylene, and polylactic acid. New technology for nanofibers will surely appear continuously in future. These are expected to become the new main trends in fiber science in the twenty-first century.

## 9.1.1 From biomimetics to super biomimetic

It is not going too far to say that the history of synthetic fibers is that of biomimetics. Many synthetic fibers have evolved out of a desire to produce novel fiber such as silk. Today, mimetic technology develops and produces synthetic fibers excellent in performance and function characteristics, mimicking more elaborate bio-function as well as structures and functions of organizations in nature to enrich our life. We must not forget that we have succeeded in mimicking nature. For example, synthetic fiber mimics the excellent fiber structure of cotton, wool, and silk and in some features improves on nature. Figure 9.1 shows the outlook for developments in fiber technology in the twenty-first century.

From the view of fiber science and technology, the organized structure found in natural fibrous materials (cotton, wool, and silk) ('fiber system') are made up from a hierarchical structure of micro-fiber, or to use the keyword of today, the nanofiber, as the basic unit.

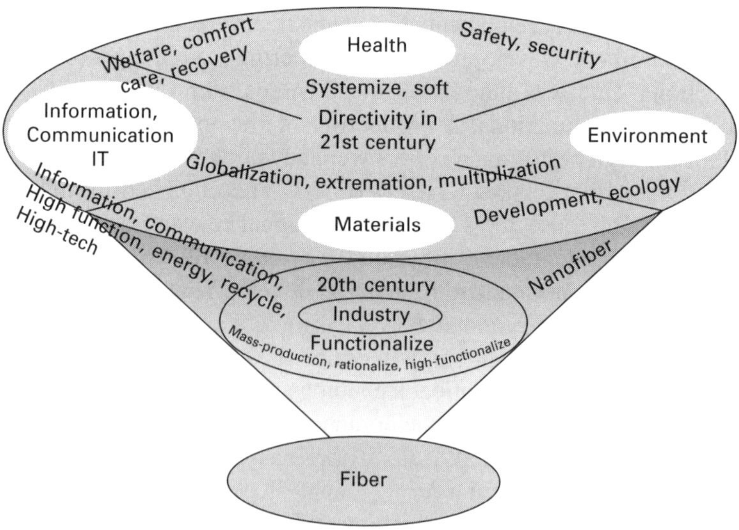


*9.1* Directions for development of new frontier fibers in the twenty-first century.

Today, fiber-related fields develop widely in relation to the living body (bio), sugar chains, organs, environment, IT (information). There are a vast number of structural models in nature: the cobweb, iridescence, supra-structures in the skin of tunicates, the flexibility of bamboo, leaves of plants and the structure and functions of bio-organs. Since developments of new functionalized fibers are based on sophisticated biomimetics, the developments lead to second-dimensional functions. Then on to supra-organized fibers or fibers amalgamated in third-dimensional functions to fourth-dimensional functionalized intelligent fibers. These can control environmental changes, and utilize various 'new system' in fiber science and its related fields.[6] These could lead to the creation of ultra-fiber mimicking biosystems and having functions and structures better than those of the original biosystem (super-biomimetics fibers). Here the 'nanofiber' becomes important. Figure 9.2 shows the possible development of fiber technology in the twenty-first century by integrating different fields.

## 9.2 Nanotechnology, materials and nanofiber[7]

The Japanese government has decided to promote several major projects for improving the economy by developing the following four basic technologies:

(1) life sciences
(2) information and communication
(3) environment and
(4) nanotechnology and materials.

The nanotechnology and materials field can be the basis for extensive science and technology innovation. An example of assimilation technology development with nanofiber is shown in Fig. 9.3.

### 9.2.1 Generation of innovative technology using nanofibers

A feature of any useful fiber should be characterization and control of its structure at nano-level. Diameters of existing useful fibers are those over 1 μm. With a constant volume, the surface area can increase a few thousand times by decreasing the fiber diameter to an order of nanometers. 'Nanofiber' with its diameter in the other of nanometers would have a huge surface area. Considering the fiber surface as a two-dimensional area film, new characteristics will be found when the structure of the surface and depth direction (radius direction of the fiber) is well controlled. The surface functions can thus increase over 100 times and multi-functions can be set into the system.

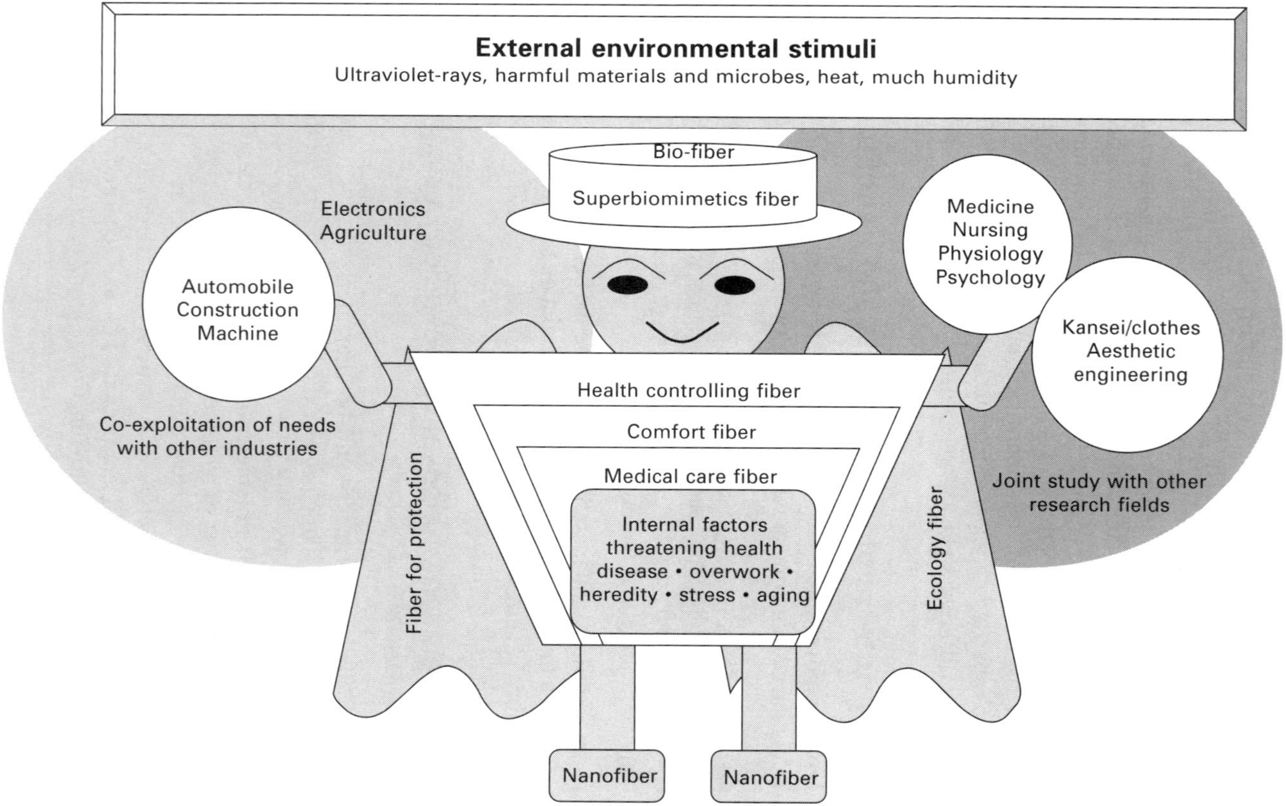


*9.2* Development of fiber technology in the twenty-first century with integration of different fields.

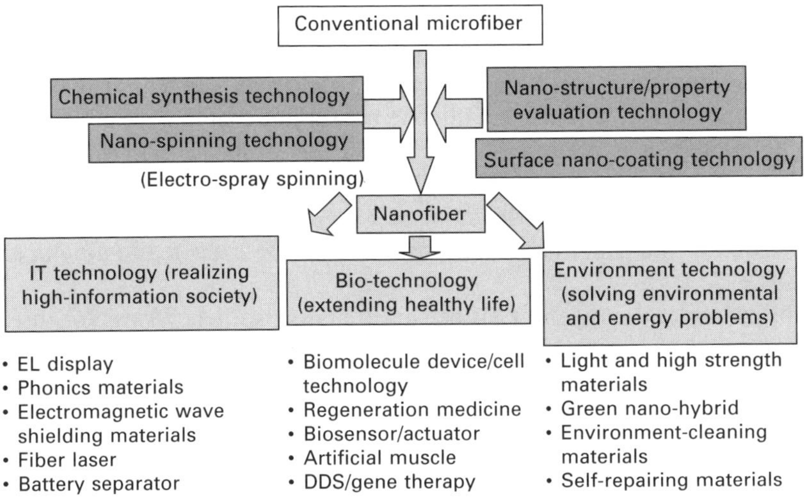


*9.3* Proposal of research project on nanofiber technology development, aiming to create new industries.

### 9.2.2 Defining characteristics of nanofiber

Although there are various definitions for nanofiber, we select the following.

*Nano-size fiber with nanometer order in its dimensions*

Nanofiber

In fiber diameters, the range from 0.1 nanometer to 1 nanometer is of angstrom size, that from 1 nanometer to 10 nanometer is nano size, that from 100 nanometer to 1000 nanometer is sub-micron size, and that from 1000 nanometer to 10 000 nanometer is micron size. Fibers with an angstrom size diameter are effectively the molecular chain. Nanometer size and sub-micron size fibers can be classed as nanofiber. A micron size fiber is called a micro-fiber. Fibers with a diameter more than a few micrometers are the conventional fibers, including those with millimeter, centimeter or meter order diameters.

Toray has developed a new technology for the fabrication of multi-filament nanofiber composed of nano-order filament with fiber radius 100 times thinner than that of conventional ultra-thin fiber, with each fiber having a diameter of the order of a few micrometers. This technology achieves thinner radius limits than that attained by any extension of conventional technologies. It can be applied for multi-use fibers such as nylon and polyesters, still using existing facilities.

Today Toray has succeeded in developing nylon nanofiber (44 dtex) composed of over 1 400 000 filaments with radius of a few dozens nanometers. Since this nanofiber has a surface much larger than conventional fibers, special functions are associated with the fiber surface enabling the production of revolutionary new materials. Figure 9.4 shows features of nylon nanofiber.

This new technology gives filament-shaped nanofiber. The nanofiber has easy processing ability. Its orientation and shape are easily controlled. Thus application of the nanofiber to various fields can be expected to accelerate quickly.

For example, the nylon nanofiber has a high absorbency twice or three times as much as conventional nylon fibers, and as much as cotton. For the conventional nylon fibers, absorbency at the fiber surface is only one-thousandth that for the inside of the fiber, so the absorbency at the surface is negligible. However, the nylon nanofiber has surface area 1000 times as large as the conventional polymer. Thus the absorbency effect of the surface becomes significant.

Hereafter Toray will undertake basic studies for applying this technology to new materials such as polylactic acid and multi-use polymers, such as polyester and polypropylene. In this way merchandise with new functions can be realized, by coupling the produced nanofiber with higher order processing technology.

Fibrous matter containing more than two molecular chains can be regarded as nanofiber. Fibrous matter containing more than two nanofibers becomes micro-fiber or when even more are added it becomes ordinary fiber. So nanofiber is sometimes termed a fibril or microfibril.

Features

1. Diameter of nanofiber 20-100 nm
2. Large surface area
3. High surface activity
4. Fabricability with multi-use polymer (nylon, polyester, polyolephin)
5. Easy processing because of filament (textile, nonwoven fabrics, cotton)

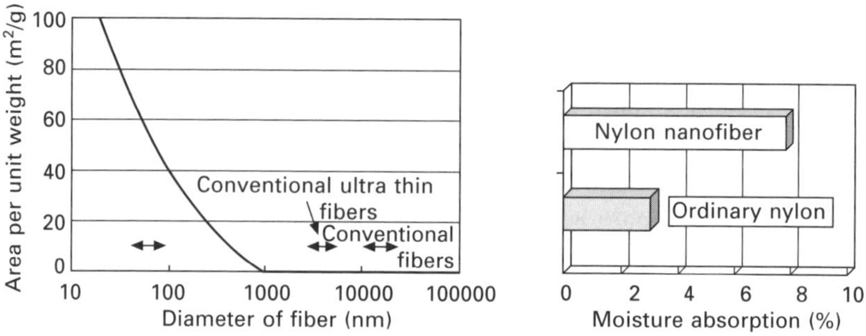


*9.4* Features of nylon nanofiber (Toray).

Nanofabric

Two-dimensional structures formed by molecular chains or nanofibers with diameter less than 10 nm are termed nano-coating. Two-dimensional structures formed by fibers with diameter in range from 10 nm to sub-micron are termed nanofabric. Two-dimensional structures formed by micron-fibers are called micro-fabric.

Nanofiber technology

Nanofiber technology has as its objective the control of structure (such as assembling, hierarchy) of the molecular chains and nanofiber. It includes various technologies for producing micro-fiber and ordinary fibers made up of nanofibers.

Figure 9.5 shows relationships in nanofiber, nanofabric and nanotechnology.

*Nano structured fiber*

'Nano structured fiber' is a fiber given new functions with precise structure designs controlling the inside, outside and surface at nanometer size. These can be:

- radial-nano structure: using nano order refined structure in the fiber radius direction
- axial-nano structure: using nano order refined structure in the fiber axis direction
- nano assembly: nano structure formation using copolymerization and organic-inorganic hybrid technology
- nano interface: using nano order refined structure in the fiber surface and interface
- nano design: using nano order refined structure for the inside of the fiber

Figure 9.6 shows nano structure fiber and its uses, Fig. 9.7 shows radial nano structure and axial nano structure, Fig. 9.8 shows nano assembly, and Fig. 9.9 shows nano interface and nano design.

*Fabrication of nanofiber*

Many researchers in Japanese Institutes now study nanofibers within various fields. These are classified roughly as

(1) gas phase grown nanofiber, such as preparation of carbon nano-tubes,[8]
(2) natural nanofibers such as making self-restorative inductive collagen nanofibers,

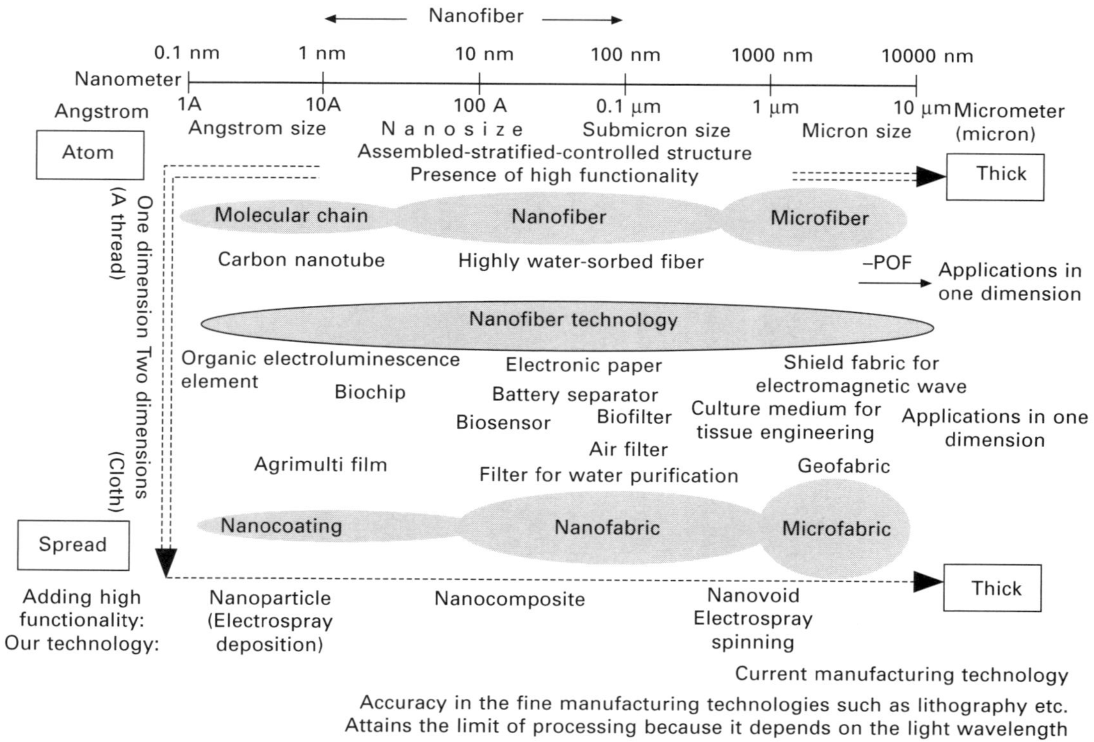


*9.5* Features, properties, development and usage of nylon nanofiber.

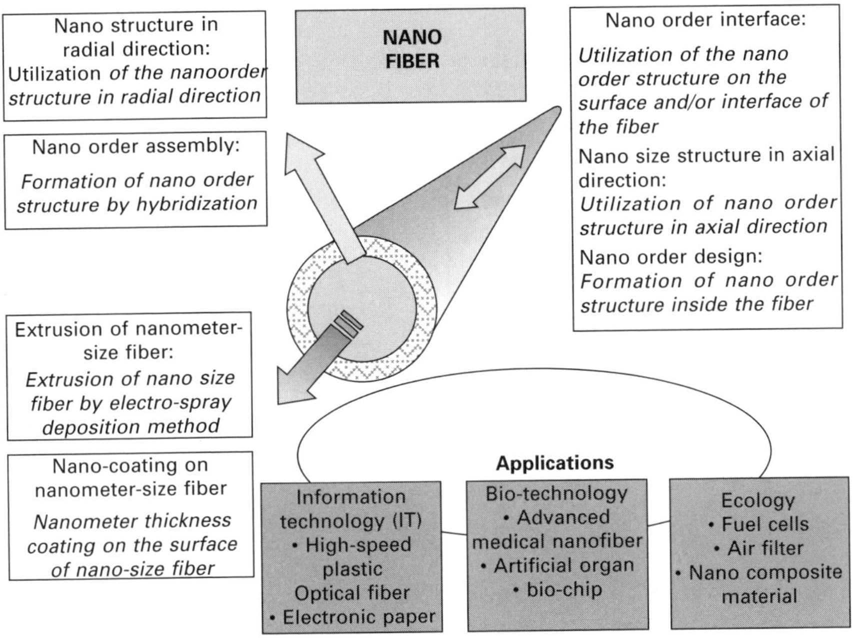


*9.6* 'Nanofiber' technology.

Nanofiber: all the following technologies are designed as 'nono-fiber' technology

(1) Fiber utilizing nano-order structures (highly functional fiber realized by nano order structure design.
(2) Nano size fiber (fiber with the diameter on the order of nanometer).
(3) nanofiber making super molecular nano wire, and
(4) fabricating nano size fiber by spinning[9] (see Fig. 9.10), etc:

- nano spinning: spinning nanometer size fiber by an electro-spray deposition method
- nano coating: Coating nanometer thickness layers on nanometer order fibers.

Fabrication technologies (conventional process technology, nano-processing and nano-measurement technologies) can control the appearance and function of the final product.

Teijin company developed a structurally colored fiber 'MORPHOTEX®' mimicking the brilliant wing color of a tropical butterfly, *morphinae*. The multiple colors are formed by the interference of light passing through the multiple layers of the wing. Teijin has realized such nano order layered thin film structure on a fiber by means of nanotechnology including ultra precision conjugate spinning and polymer technologies.

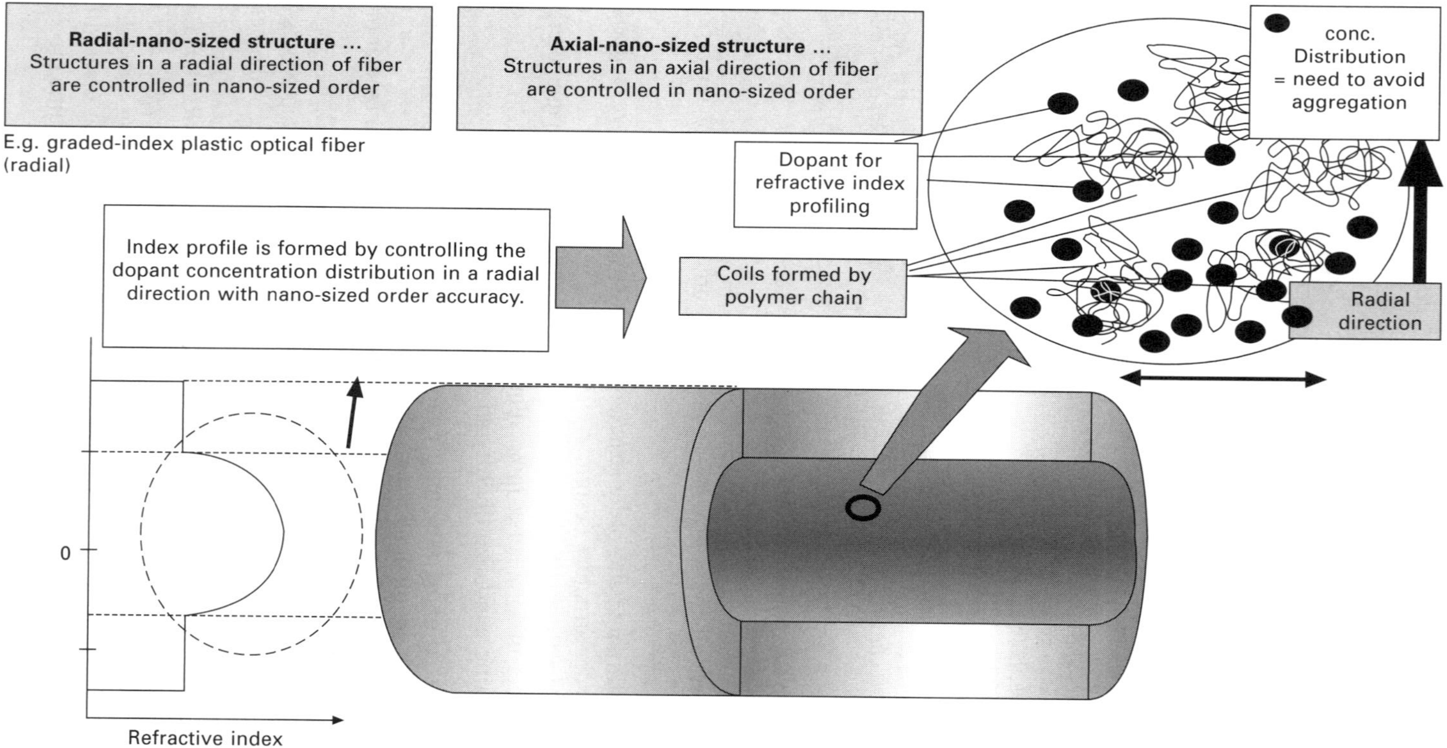


*9.7* Radial and axial nano-sized structures.

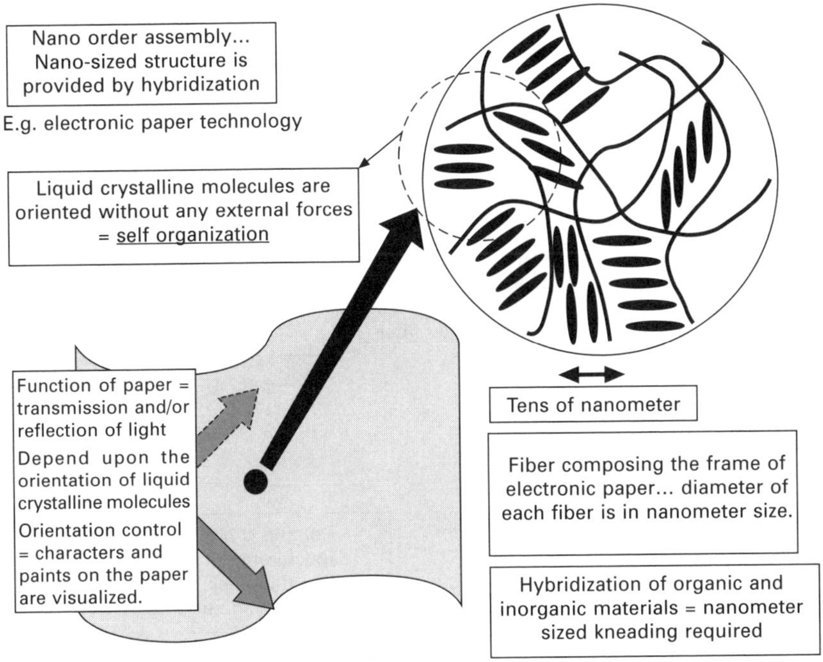


*9.8* Nano order assembly.

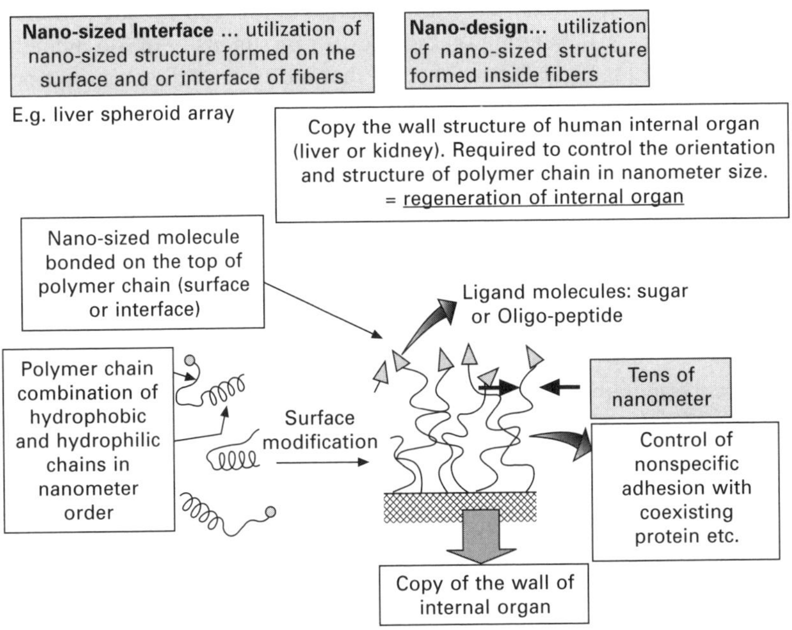


*9.9* Nano-sized interface and nano design.

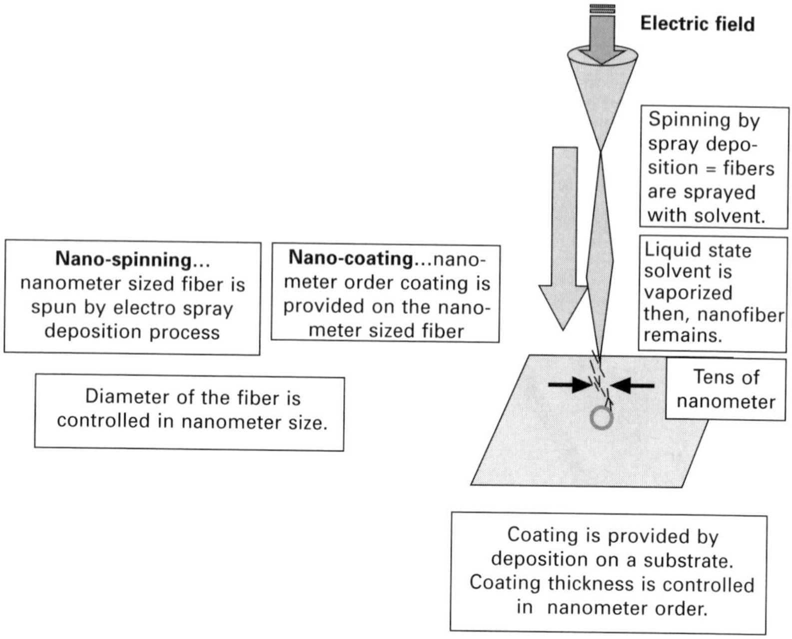


*9.10* Nano spinning and nano-coating.

Morphotex® has a 61-layer structure composed of polyester and nylon in alternating layers 70 nm thick. The various colors of Morphotex® are obtained by light interference, depending on the angle or the intensity of the incident light as shown in Fig. 9.11. Teijin expects that the structurally colored fiber in a filament yarn type will be used mainly for women's outer and inner wears, dresses, sportswear, embroidery threads for logos, car seat materials, interior textile materials, etc. Another interesting field is paint materials (Morphotone®), especially automobile paints, cosmetics, and artificial leather coating.

### *Nanofiber effects which can be produced*

Material properties specific to nano-size materials are:

1. size effects (increment in surface area per volume, increment in reactivity and selectivity attributed to decrease in volume)
2. arrangement of super molecules (molecules arranged regularly, self-organized and with new functions and coherence)
3. recognition of cell and biomaterials (nanofiber with unique structure which cells recognize and combine with) and
4. hierarchal structure effect (effects generated from the nano polymer chain level to organized nano hierarchy structure). Seeds and needs of nanofiber are shown in Table 9.1.

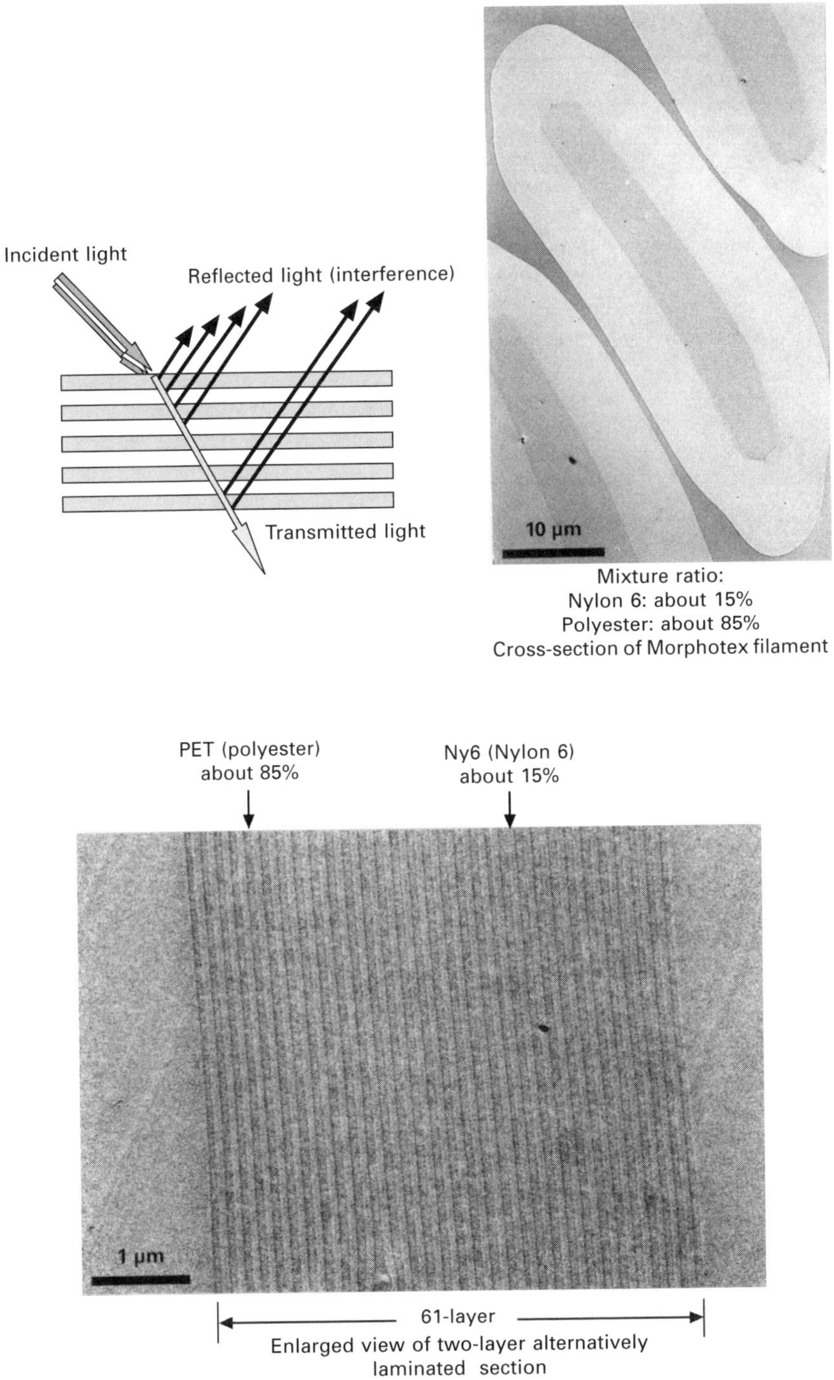


*9.11* Coloring principle of Morphotex® (Teijin).

*Table 9.1* Uses of nanofiber utilizing its properties

| Properties | Various effects | Uses |
|---|---|---|
| ○ Surface area effect | • Fiber surface area per weight is over 1000 times larger than that of usual fibers.<br>• Very large adsorption. | • Adsorption material<br>• Biochemical hazard prevention material<br>• Ion-exchange material |
| ○ Slip effect | • Flow of molecules changes and pressure loss becomes much smaller. | • Air filter<br>• Biochemical hazard prevention material |
| Hole effect | • Sieving by small holes | • Separation material<br>• Sensor |
| ○ Light effect | • Structure color appearance with transparent fiber with diameter less than wavelength of visible light | • Organic EL<br>• Electronic paper<br>• Fashion material<br>• Polarizer |
| ○ Surface tension effect | • Low surface tension force, water repelling of hydrophilic polymer | • Coating material<br>• Paint |
| Amalgamation effect | • Amalgamation in nanometer order | • Electromagnetic wave shielding material<br>• High strength structure material |
| Void effect | • Giving nano-void to fiber | • Humidity keeping material<br>• Fouling prevention material |
| ○ Three-dimensional effect | • Three-dimensional growth of cells on non-weaved textile | • Reclamation medical |
| Sliding effect | • Increment of sliding of materials | • Complex material<br>• Airplane |
| ○ Sub-micron object catching effect | • Catching sub-micron-sized particles | • Biochemical hazard prevention material<br>• Suit<br>• Engine filter<br>• Boiler<br>• Air cleaner<br>• Air conditioner |
| Cell, living body, material recognition effect | • Having unique structure for recognition and bond with cell | • Reclamation medical<br>• Bio-tip<br>• Biosensor<br>• Tailor-made medical |

○: Important effects

### 9.2.3 Establishment of the basic technology – construction of nanofiber analysis technology

Nanofiber technology cannot be achieved by conventional spinning technologies. Self organization of molecules is needed in nano spinning technology and further development in usage. For measurement of the structure and properties of nanofiber, conventional analytical methods do not give sufficient sensitivity and resolution. So developments in analytical methods with high sensitivity and high resolution are required. New structure control technology is necessary for the far larger surface of nanofibers. These basic technologies and fabrication technologies are needed for the birth of a new fiber industry. Figure 9.12 shows development of nanofibers and the ripple effects which will be created.

## 9.3 Creation of new industries

Figure 9.13 shows examples of nanofiber applications in information, bio- and environment technologies.

### 9.3.1 Information technology materials

Increments in density and integration in electric devices require materials which can be assembled at nanometer level. High-speed optical information communication and optical computers can be produced using liquid crystalline polymer nanofibers with nonlinear optical properties and metal-covered nanofiber which pass light in one direction only. Nano-ordered distribution control has been realized in plastic optical fiber.[11,12]

### 9.3.2 Biofiber hybrid material

The orientation of biofibers in living materials requires control at nanometer level. High-ordered structures are formed to perform various deformations such as to stretch, compress, bend, and twist. Using three-dimensional assembly technology with nanofibers the growth of nanofibers from the surface of inorganic crystals, such as apatite can be controlled to give high performance materials with new and revolutionary properties.

### 9.3.3 Biomedical nanofiber

In bio-materials, nanofibers are organized within cells, which transmit communication of substances, energy, and information. For example, cells bound to collagen and fibronectin are recognized as suitable receptors whereby signals are propagated. Nanofibers play a very important role in transmitting

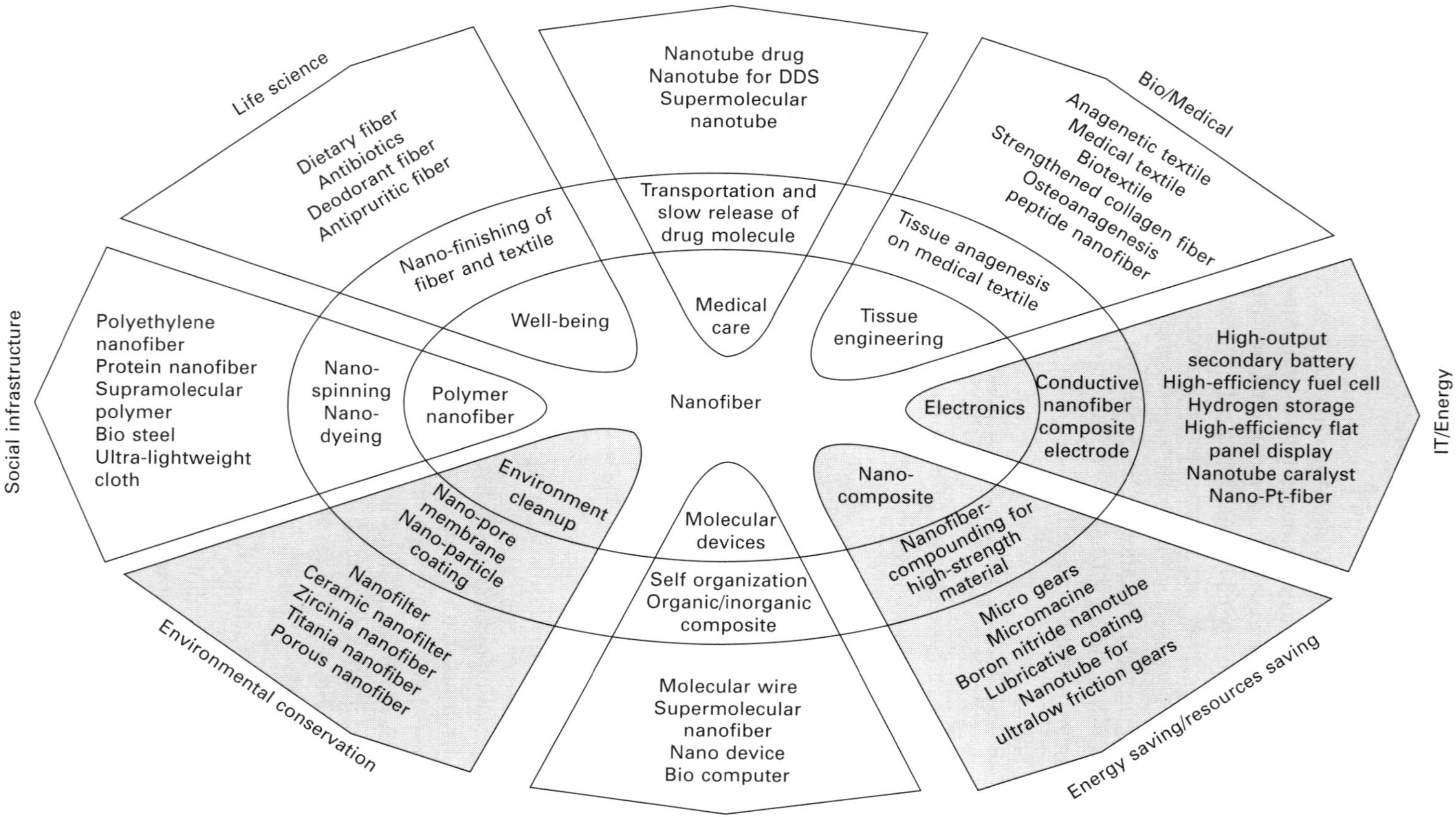


*9.12* Development of nanofibers and their ripple effects.

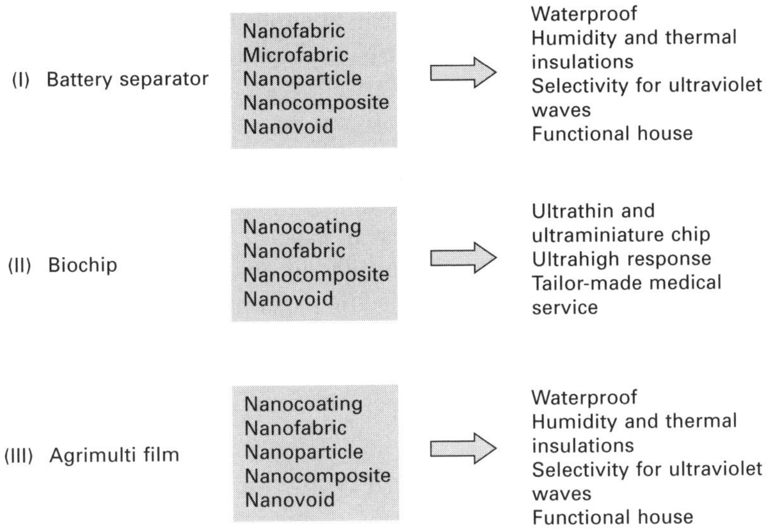


*9.13* Nanofiber application examples for IT (I), bio (II) and environmental technologies (III)

cell information in this way. Understanding the function of bio-nanofibers and constructing models will open up new and advanced medical nanofiber technology.[13,14,15]

### 9.3.4 Bio-nanofiber world – hints for the direction of artificial fibers[5]

The world surrounding humans and the world available to them is made up of nanofibers. Cotton, silk, and polyethylene, skeletal muscles, myofibrils in the human body, and DNA in cells are all composed of nanofibers. Coal, originally derived from plants, is also composed of nanofiber.

Professor Y. Atomi of the University of Tokyo has indicated how clothes from such fibers can have an entirely new function, which can include stimulation of body processes. The effects on the skin can be profound.

### 9.3.5 Ecology materials

Composites of biodegradable polymers and natural nanofibers of cellulose and imogolite (clay) can give green-nano-hybrid which shows high strength within a short time scale, but can be decomposed by enzymes and microbes over a longer time scale. This material can be applied to a structural material in transplantation medicine.

Hollow nanofiber has an extremely large nanospace. Such nanofiber has many functions, such as electron transfer or encapsulation functions, such as removing pollutants from the environment. Controlled release of drugs retained in the nanospace is another application. Non-weaved clothing with nanofibers has extremely new functions.

Electrical paper processing technology using self-organized nanofiber[16] is also useful in ecology. Characteristics which are necessary or achieved using 'nano-size fiber fabric' for electric paper are:

- Quick response time due to thinness of paper.
- Wavelength of visible light is around 600 nm and diameter of micro-fiber is more than 200–500 nm. Thus electric paper made with micro-fiber is less transparent due to light scattering.
- Since nano-size fibers have diameter less than 100 nm, light propagates without scattering. Thus electric paper with nanofiber fabric is transparent.

### 9.3.6 High strength ultra light material

Nanofiber with low defects can be fabricated by controlling structures composed of nanofiber. Nanofiber has a very high aspect ratio, so allows development of high strength materials. Ultra light and high strength materials can be fabricated by dispersing high strength nanofibers into polymer and controlling the interface.

Professor T. Kikutani (Tokyo Institute of Technology) has started the development study of such 'high strength fiber'[17] which was adopted as a National Japan project in 2001. The aim of this study is to achieve a strength of multi-use synthesized fibers which approximate the theoretically calculated strength limit. Fiber strength is increased by 2GPa due to the higher orientation of the molecules.

## 9.4 Researches and global developments of nanofiber

In the United States, the National Science Foundation (NSF), the National Fiber Research Institute (NTC), and the US Army are actively supporting nanofiber research. The US Army alone has provided $50 million ($10 million for five years) to Massachusetts Institute of Technology (MIT) to develop an ultra-lightweight and high functional military. The Institute for Military Nanotechnologies was set up on 1 May 2002, using this money. It is likely that the support to MIT will be provided for a further five years. Additionally non-governmental organizations, such as DuPont, will also participate. In Europe, development of nanofiber is being actively pursued, particularly by Marburg University in Germany. In Asia, China, Taiwan, and South Korea, researchers who have been inspired by the work in the United States are returning to stimulate developments in their own countries. It is an area of fiber technology to keep an eye on in the future.[18]

## 9.5 Further reading

### 9.5.1 General


*C&EN*, 11 August, pp. 29–34, 2003.

Dzenis Y., *Science*, **304**(25), 1917 (2004).

Stegmaier T., Dauner M., Dinkelmann A., Scherrieble A., von Arnim V., Schneider P. and Planck H., *Technical Textiles*, **47** (Nov.), 142–146 (2004).

Kato T., *Science*, **295**, 2414–2418 (2002).

Kikutani T., The 32nd Sen'i Gakkai Summer Seminar Lecture Abstracts, p. 82, The Society of Fiber Science and Technology, Japan (2001).


### 9.5.2 Morpho-structured fibrils


Hongu T. and Phillips G.O., *New Fibers* 2nd edn, p. 81, Woodhead Publishing, Cambridge, (1997).

Iimuro H., Proc. of 1st Inter. Congress on Nanofiber Sci. Tech., June 28, 2004, Tokyo, Japan.

Japanese Patent No. 3036305.

Kumazawa K., Tanaka S., Negita K. and Tabata H., *Jpn. J. Appl. Phys.*, **33**, 2119 (1994).

Kumazawa K., Takahashi H., Tabata H., Yoshimura M., Shimizu S. and Kikutani T., *Sen'i Gakkaishi*, **58**, 195 (2002).

Kumazawa K., Takahashi H., Tabata H., Yoshimura M., Shimizu S. and Kikutani T., *Sen'i Gakkaishi*, **59**, 392 (2003).

Tabata H., The 31th Summer Seminar Proceeding, p. 139, The Society of Fiber Science and Technology, Japan (2000).

Tabata H., Yoshimura M. and Shimizu S., *Bull. Fiber Textile Res. Foundation*, **10**, 8 (2000).

Tabata H., Yoshimura M. and Shimizu S., *Sen'i Gakkaishi*, **57**, 248 (2001).

US Patent No. 6,430, 438.

Yoshimura M., Kagohara K., Tabata H. and Shimizu S., *Sen-i Gakkai-shi*, **56**, 348 (2000).


### 9.5.3 Nanofiber technology


Hongu T., *Strategic excavation of novel advanced industry by nanofiber technology*, p. 26, CMC Publishing Co. Ltd. (2004).

Hongu T. and Tanioka A., *Sen'i Gakkaishi*, **59**(12), 401 (2003).

Tanioka A., *Sen'i Gakkaishi*, **59**(1), 3 (2003).


## 9.6 References


1. Kamiide K., Introduction to Development History in Fiber Industry, *J. Text. Machine. Soc. Jpn.* (1993).
2. Hongu T., Seminar on Trends in Fiber Industry, Chamber of Commerce at Fukui, March, 2002.
3. Toray, 31 October 2002, press conference.
4. Iimuro H., 1st International Congress on Nanofiber Science Technology – Aim for the Practical Application, p. 63, The Society of Fiber Science and Technology, Japan, 2004.

5. Atomi Y., Proceeding of Exhibition Symposia '2004 Expo Fibers for New Era', pp. 42, 175, The Society of Fiber Science and Technology, Japan, 2004.
6. Hongu T., *World of High-Tech Fibers*, Nikkan-Kogyo Shinbun-sha (2001).
7. Hongu T., The 33rd Sen'i Gakkai Summer Seminar Lecture Abstracts, p. 21, The Society of Fiber Science and Technology, Japan (2002).
8. Endo M., *CHEMTECH*, pp. 568–576, ACS, 1988.
9. Tanioka A., *J. Colloid. Interface Sci.*, **250**, 507–509 (2002).
10. Morota K., Matsumoto H., Mizokoshi T., Konosu Y., Tanoka A., Yamagata Y. and Inoue K., *J. Colloid. Interface Sci.*, **279**, 484–492 (2004).
11. Ishigure T., Nihei E. and Koike Y., *Sen'i Gakkaishi*, **53**, 520 (2001).
12. Koike Y., Sato M. and Ishigure T., *Ouyou-Butsuri (Applied Physics Japan)*, **70**, 1287 (2001).
13. Kataoka K., JPN PAT 2001-5240, JPN PAT 2001–226293, etc.
14. Harada A. and Kataoka K., *Science*, **283**, 65 (1999).
15. Harada A. and Kataoka K., *Macromolecules,* **36**(12), 4995 (2003).
16. Kato, T., Science, **295**, 2414–2418 (2002).
17. Kikutani, T., The 32nd Sen'i Gakkai Summer Seminar Lecture Abstracts, p. 82, The Society of Fiber Science and Technology, Japan (2001).
18. Hongu, T., *Survey on the Practical Application of the Nano Fiber Technology*, p. 26, CMC Publishing Co. Ltd (2004).

# Index


A380 aircraft 116
acacia trees 227–8
  *see also* gum arabic
Achilles' tendon 180
acrylic fibers 112–13, 269
adhesion 75
adhesion wetting 189
advanced composite materials (ACMs) 1, 73
  carbon fiber 103–6
Aerocapsule 151
Aeroshelter 55–7
aesthetic sense 156–7
  *see also* high touch fiber
agarose 239
aging population 176, 180–1
air conditioning temperatures 212
air dome 55–7
airbags 79–81, 82
aircraft/aerospace 102
  carbon fiber 107–11
  superfibers 73, 74
  supersonic transport 113–16
Aircube 151
airships 60–1
alginate 239–40
  medical applications 248–50
allografts 260
aloe vera leaf 252–4
Americas Cup 124, 125, 149
ammonia 200, 201
animal hair 8–9
antimicrobial deodorization processing 186–8
antimicrobial fibers 186–8, 195–8
  treatment against MRSA 198–9
anti-weatherability 75
aqueous environments 124–7
aramid fiber
  applications in information and communication technology 59–60, 61
  superfibers 68, 69, 70–1, 72–3
  Technora 78–9
aramid FRP 79, 80
artificial organs
  kidney 52–3, 54, 175, 179–80, 254
  liver 52
ARTIROSA 157–8
Asahi Kasei 44–7
Asia 35–7, 38, 286
aspect ratio 15, 16–17, 20–1
Avicel 241
axial nano structure 275, 277, 278
Azek 213–14

bacterial-produced cellulose 241
balloon catheter 53–5, 56
balloons
  ULDB 61–2, 84–5
  Venus probe 86–7
bamboo 14
bandages 266–7
basic function fiber (first order function) 130, 132, 133, 134, 142, 143
β-glucans 241–3
bicycles 69
biodegradable filter 39–43
biofiber hybrid material 283
biological deodorization 201

biological functions 131, 133, 135
biomedical nanofiber 283–5
biomimetic deodorization 201
biomimetics 12–15, 35, 133, 139, 158–66, 269, 270–1
  challenge to harness nature 159–66
  development 158–9, 160
  high function fibers developed 159, 160
  intelligent materials 14–15, 162–6
  iridescent insects 160–2
  mimicking silk 12–13
  simultaneous polymerization and spinning 13–14
bio-nanofibers 285
bio-spinning 13
black golf shaft 106, 108, 110–11
blood
  clotting 262–3
  virus removal filters for 255
blood poisoning treatment column 139–40, 141
BMM 255
bobbin for a superconductive coil 88, 89
body
  fibers within 179–80
  unpleasant odors 199, 200
Bodyshell 23–4, 147
Boeing 777 aircraft 107–10
building/construction applications 248
  C-FRP space truss 117–20
  *see also* civil engineering
bulkiness 21, 22
bulking, fecal 219, 220
bullet-proof vest 85–6
burns treatment 264–6
business shirts 212–15, 216
butyrate 220
  anti-cancer effects 244

cancer 244
capillarity 189
carbon dioxide 12
carbon fiber 68–9, 99–129
  ACMs 103–6
  aircraft applications 107–11, 113–16, 146
  Americas Cup 124, 125
  applications in various fields 104–5
  aqueous environment applications 124–7
  beds for CT scanning 123–4
  C-FRP space truss 117–20
  changes in global market 99, 100
  classification 100, 101
  clean energy application 120–2
  with CNT core 92
  development of PAN-based carbon fiber around the world 101–2
  future of PAN-based carbon fiber 127–8
  geotextiles 116–17
  in high technology 99–101
  impeller blade of jet engines 102
  Japanese enterprises started from acrylic fiber 112–13
  lightening of rollers 124
  material development 113
  safety-related fields 122–3
  space developments 113
  sports applications 72–3, 106–7, 108, 110–11, 146
  supersonic transport 113–16
  two major characteristics 111–12
carbon nanotube (CNT) 91–2
carboxymethyl cellulose 252, 253
Cargill Dow Polymers (CDP) 39, 41
carrageenan 237–8
cars 69–71, 72
cartilage regeneration 267–8
cellulose 12
  cellullosic membranes 254–5
  dietary fiber cellulosics 240–1, 242
chemical deodorization 201
chemical functions 131, 133, 135
chitin 252
  wound dressings from 267
cholesterol 242–3
chondroitin sulphate 260–1
Chytopoly 199
cigarette deodorant fibers 201–2
Cigernon 202
civil engineering
  carbon fiber 116–17, 118, 119
  composite materials reinforced by superfiber 73–4, 76
  new frontier fibers 58–9
cleanliness, and comfortableness 186–204
clothes
  bullet-proof vest 85–6
  comfortable 212–16

health maintenance 181–3
medical 248
sports *see* sportswear
clothes internal environment 211
CM101 254
collagen 240
medical applications 261–3
colon 244–5
colonic diverticula 244
colo-rectal cancer 244
comfort 17–18
cleanliness and comfortableness 186–204
clothing 212–16
comfort fiber and maintaining health 204, 205, 206, 207, 209
development of comfortable fiber for health 184–6
high comfort function fiber 151–8
commercial aircraft cabin 107
computer, wearable 166
conductive fiber 133
contact angle 189
cool feeling fiber 151–4, 155
corn, fiber from 47
cornice 57
coronary heart disease 242–3
Cortera 44
CORTICO 154–7
character of cloths 155–7
fabrication technologies 155
cotton 10–11
critical overlap concentration (C*) 222
Cross equation 224–5
cross-ply tires 71
cross-section
diameter 16
shape of spinneret and 151
structural control 22–8
CT scanning beds 123–4
cycling wear 149, 150

Daiwabo 215
deodorization processing 186–8
deodorization of unpleasant odors 199–204
DEOFOR 202–4
Dermograft 261
Descente 150
dialyzer, artificial 52–3, 54, 175, 179–80, 254
diameter, classification by 16
dibutyrylchitin (DBC) 267
dietary fiber 218–46
anti-cancer effects of butyrate 244
colonic diverticula 244
definition 218
effective colon performance 244–5
fermentation product effects 219–20
health benefits 220
hydrocolloid fibers 218, 221–7
main hydrocolloids 227–43
methods of measuring 243–4
physical effect 219
prebiotics 221
probiotics and synbiotics 221
Dimplex 150
dispersion wetting 189
diverticulosis 244
Dow Chemical Company 82–3
downhill skiing costumes 150
drawing 13
dressings *see* wound dressings
dry touch 156–7
Du Pont 47
durability 75
dynamic-function-mimetic fiber 163
Dyneema 87–8, 89

ecology materials 285–6
Ecosys 20°C fabric 216
Ecosys 28°C fabric 212, 214
elastic modulus 111–12
electric paper processing technology 286
endotoxin 139–40
energy 16–17
energy storage system, superconductive 88, 89
environment
biodegradable fiber 39–43
environmental conservation fiber 204, 207–8, 209
environmentally-friendly intelligent fiber 28, 29
fiber to guard environment and health 204–12
and health 176–8
enzymic chemical methodology 244

enzymic gravimetric methodology 244
Europe 286
  fiber development 35–7, 38
  superfiber development 68
Eval 151–2, 153
external stimuli 211–12
extreme environments 177
Exu-Dry 265–6

fecal bulking 219, 220
fermentation products 219–20
fiber
  characteristics and shapes 16–18
  classication by cross-sectional diameter 16
  development prospects of the fiber industry 2
  development of usage in 21st century 34
  features characterizing a fiber 8
  flexibility 21–2
  hierarchical structures 18–19
  importance 1–4
  major technical strategy in Japan 32–4
  narrow and broad definitions 15
  structural control in the cross-section 22–8
  utilization in various fields 5
fiber-reinforced concrete (FRC) 74
fiber-reinforced plastic (FRP) 100–1, 103
  C-FRP space truss 117–20
fiber/textile science 19–21
fibrillated cellulose 241
fibrin 180, 181
fibroin 13
fibrous DNA chip 140, 141
fibrous scaffolds 261, 263
fibrous titanium oxide optical catalyst 167–8
Fieldsensor 147
filtration
  medical fibers 247
  membranes for water treatment 135–8
fine filtration membrane 136–8
first order function 130, 132, 133, 134, 142, 143
fishing rods 106, 108, 144–6
five-layered fabric 61–2
flame resistance 75
flexibility 21–2
floating ice 76–7
fluctuations, natural 17–18, 211
fluorocarbon-type POF 25
foam dressings 248–50
forest rhythm effect 18
formaldehyde deodorization 202–4
fossil resource exhaustion 39
4GT (PBT) 44, 46
fourth order function 130, 132, 134, 142, 143
  *see also* intelligent fiber
fractal theory 211
freedom 16–17
Freshcall TMII 202, 203
frontier fiber 47, 48
  *see also* new frontier fibers
fuel cell, mobile 166–7
Fuji Spinning 214–15, 216
function-mimetic fiber 163
FYBOGEL 245

galactomannan seed gums 231–2
gas-barrier property 61–2
gas storage tanks 121–3
gel spinning 65, 67, 68, 87
gelatin 240
gellan gum 236–7
gelling 225–7, 236, 238, 240
geo-membrane 59
geotextiles 74
  carbon fiber 116–17, 118, 119
glass fiber reinforced plastic (G-FRP) 101, 103
global environment 212
golf clubs 106, 108, 110–11, 144–6
Goretex 147, 265
graded index (GI) optical fiber 25, 26
green nano-hybrid 285
guar gum 231–2
guard vest for jockeys 183–4, 185
guluronic acids 239–40
gum arabic 227–30

Healon 258
health 173–217
  benefits of dietary fiber 220
  care products demand rise in aging society 180–1
  cleanliness and comfortableness 186–204

antimicrobial fibers 195–9
moisture absorbency 189–94
new functions of fiber products 186–8
unpleasant odor deodorization 199–204
clothes useful for health maintenance 181–3
comfortable clothing for air-conditioning temperatures 212–16
comfortable fiber for 184–6
definition 173
environment surrounding 176–8
fiber to guard environment and health 204–12
approaches to maintain health 204–5
comfort fiber 204, 205, 206, 207, 209
environmental conservation fiber 204, 207–8, 209
maintaining physical and mental health 208–12
medical applications 206
stimuli relaxation 204, 205, 206, 207, 209
homeostasis 173–5
learning from human body fibers 179–80
*see also* medical fibers
healthcare fiber 204–6, 209
'Healthy Nippon 21 Plan' 175–6
heart rate monitoring 264
heat resistance 68–9, 75
heat retention 183
heat shrinkage 75
heat storage 147, 183
helical structures 160–2
hemodialysis 52–3, 54, 175, 179–80, 254
heparin 52
hierarchical structures 18–19
high function fiber 4, 6, 8–9, 32–5, 130–72, 269
basic function fiber 130, 132, 133, 134, 142, 143
biomimetics and intelligent fibers 158–66
comfort function fiber 151–8
concept of function changing with consumer needs 130–43
higher function fiber 130, 132, 133–8, 142, 143
intelligent function fiber 130, 132, 134, 142, 143
new areas 166–8
present status 130, 131
sportswear using 144–51
super high function fiber 130, 132, 134, 138–41, 142, 143
high Kansei fiber *see* high touch fiber
high performance fiber 4–6, 32–5, 67, 265
extensive applications 67–8
*see also* superfibers
high strength fiber 183
high tenacity polyethylene fiber 87–8, 89
nanofiber 286
high-tech fiber 4–10, 47
enlargement of the frontier 32–5
*see also* high function fiber; high performance fiber
high touch fiber 4, 6, 8–9, 32–5, 269
sportswear 144, 146
*see also* comfort
higher function fiber 130, 132, 133–8, 142, 143
Hinokichi 199
hollow fibers 152, 153
artificial kidney 52, 53, 175, 179–80, 254
microfiltration membrane 136–8
mobile fuel cell 166–7
properties and usage 153
swimming costumes 150–1, 152
hollow nanofiber 285–6
homeostasis 173–5
horse racing 183–4, 185
human-friendly intelligent fiber 28, 29
human hair 13
human interface (HI) textiles 2
hyaluronan (HA) 255–61
hydrocolloids 218, 221–43
agarose 239
alginate 239–40
β-glucans 241–3
carrageenan 237–8
cellulosics 240–1, 242
galactomannan seed gums 231–2
gelatin 240
gellan gum 236–7
gelling characteristics 225–7
gum arabic 227–30
konjac mannan 232–4
source, function and main applications 223–4

thickening characteristics 222–5, 226
xanthan gum 226, 234–6
*see also* dietary fiber
hydrogel dressings 260–1
hydrogen sulfide 200, 201
hydrolysis 40
hydrophilic materials 265
HYGRA 191, 192–4
Hylaform 259
hylans (HY) 257, 258–60
Hylashield 259

ice-boom 76–7
impeller blade of jet engines 102
industrial materials 35
applications and performance required 35, 36
new frontier fibers 55–62
infiltration wetting 189
information 16–17
information and communication technology
nanofiber 283, 285
new frontiers 59–62
optical fiber 24–8, 37–9, 51, 77–8, 86
Institute for Military Nanotechnologies 286
intelligent fiber 14–15, 28, 29, 162–6, 269
intelligent function fiber (fourth order function) 130, 132, 134, 142, 143
Intellitex suit 264
internal environment, stimuli from 208–11
International Ski Federation (FIS) 150
iota carrageenan 237–8
iridescent insects 160–2
ispaghula husk 244–5

Japan 37, 38, 52
carbon fiber enterprises started from acrylic fiber 112–13
change of population composition 176
development of superfiber 68–9
fiber as major technical strategy 32–5
'Healthy Nippon 21 Plan' 175–6
medical care costs 176
research 37–9
Japanese Association for the Functional Evaluation of Textiles (JAFET) 187
jet engines, impeller blade for 102
jockeys, guard vest for 183–4, 185

Kanebo Synthetic Fibers 215, 216
Lactron 40–3
Kansai International Airport 1
kappa carrageenan 237–8
Kevlar 65, 66, 76–8, 82, 83, 147–8
ice boom 76–7
tension member for optical fiber 77–8
kidney, artificial 52–3, 54, 175, 179–80, 254
KINGLIGHT 58
Koletex bandages 266–7
konjac flour 234
konjac mannan (KM) 232–4
Kurabo Industries 215, 216

lactic acid 40
lactic acid bacteria 221
Lactron 40–3
biodegradability 43
characteristics 42–3
development of goods using 43
processes for development of 40–2
lambda carrageenan 237–8
lattice theory 255
LDL cholesterol 242–3
Lencinan 10
light-blocking nets 58
light interference 160–2, 277, 281
Lightron 151
lightweight materials 183, 286
liquid crystalline spinning 65, 67, 69
liver, artificial 52
living environment 212
living spaces
deodorization of 201–4
unpleasant odors in 199–200
locust bean (carob) gum 226, 231–2
low friction materials 147, 183

mannuronic acids 239
Mars Pathfinder 79–81, 82, 83
material (that makes up the fiber) 8–9
matrix engineering 257–61
mattress coverings 265
mechanical functions 131, 133
medical fibers 206, 247–68
alginate fibers 248–50
artificial kidney 52–3, 54, 175, 179–80, 254
balloon catheter 53–5, 56
burns treatment 264–6

collagen 261–3
fibrous scaffolds for tissue engineering 261, 263
hyaluronan 255–61
Koletex 266–7
new frontier fibers 52–5, 56
nonwoven 247–8
spider silk supportive matrix for cartilage regeneration 267–8
superabsorbent fibers 250–1
textile medical sensors 263–4
wound dressings from chitin 267
wound healing and polysaccharide fibers 251–5
cellulosic membranes 254–5
natural systems 252–4
mental fulfillment 17–18
metal replacements 69, 71
methylmercaptan 200, 201
Microart 151
microbial control 186–8
antimicrobial fibers 195–9
microdenier/ultra-fine fibers 22, 35, 39, 135–6
microfabric 275, 276
microfiber 16, 273, 276
microfiltration membrane (MF) 136–8
Microft Lectus 147
microorganisms
anchoring in water purification 124–7
antimicrobial fibers 186–8, 195–9
deodorization 186–8, 199–204
growth and inhibition curves 195–6, 197
microporous materials 265
microstructure 8–9
mobile phones 59–60
modulus 65–7, 68, 75
moisture absorbency 157, 182, 189–94
moisture-permeable material 183
moisture release 191–4
molecular fibers 16
molecular orientation 21, 22
molecular weight 14–15
Morphotex 24, 139, 269, 277–80, 281
MRSA 197, 198–9
multi-component glass optical fiber 25, 26
multi-filament nanofiber 273–4
multi-layer structures in nature 24
multiple functions 139
multi-walled CNT (MWCNT) 91–2

nano assembly 275, 277, 279
nano coating 277, 280
nano composite fiber 92
nano design 275, 277, 279
nano interface 275, 277, 279
nano spinning 277, 280
nano structured fiber 275, 277
nanofabric 275, 276
nanofiber analysis technology 283
nanofiber technology 275, 276
nanofibers 16, 269–88
carbon nanotubes 91–2
creation of new industries 283–6
defining characteristics 273–4
development and ripple effects 283, 284
effects which can be produced 280, 282
fabrication of 275–80
generation of innovative technology using 271–3
promotion of nanofiber development 271, 273
research and global developments 286
nanotechnology 92, 93
nata de coco 241
National Aeronautics and Space Administration (NASA)
Mars Pathfinder 79–81, 82, 83
ULDB 61–2, 84–5
natural fibers 9–10, 39–40
vs synthetic fibers 10–12
natural fluctuations 17–18, 212
natural gas car storage tank 121–2
natural polysaccharides 252–4
natural recycling system of biodegradable polymers 40, 41
net pump action 127
new frontier fibers 32–64
currently growing new frontier field 51–62
industrial materials 55–62
medical developments 52–5, 56
development in response to needs 47–9, 50
enlargement of the frontier 32–5
high-tech fibers, frontier fibers and 47, 48
how to develop new application fields 49–51

international competition 35–43
PTT 43–7
NIF-NaHA 257–8
Nisshinbo 212
non-circular cross-section 22–4
non-homogeneous intelligent materials 14–15
non-transparent effect 23–4
nonwoven materials 247–8
alginate fibers 248–50
Novogel 241
nylon 1, 12, 37, 67–8, 269
moisture absorbing and moisture releasing 191–4
nylon cord 71
nylon nanofiber 274, 276

OASIS 250–1
odors, unpleasant 199–204
deodorization 201–4
1/f fluctuation 17–18, 211
ophthalmic viscosurgery 258
optical catalyst 167–8
optical fiber 24–8
Kevlar as tension member for 77–8
POF 25–8, 37–9, 51
Zylon as tension member for 86
organic electro-luminescence wearable display 166
organism classification chart based on composite structure 163, 164
osteoarthritis 257–8

PAN-based (polyacrylonitrile) carbon fiber 68–9, 99–102
future of 127–8
*see also* carbon fiber
panty hose 191–4
PBO (polyphenylene benzoxazole) fiber 68, 69, 81–7
peace of mind 17–18
performance, required 35, 36
photosynthesis 12
physical deodorization 201
physical functions 131, 133–5
pitch-based carbon fiber 68–9, 99–102
plant fiber 12
*see also* cellulose
*Plantago* spp. 244–5
plasma separation membranes 255
plastic optical fiber (POF) 25–8, 37–9, 51
plasticized polyvinyl chloride (PVC) 265
polyarylate 79
polybutylene terephthalate fiber (PBT) 44, 46
polyester 37, 67–8, 83
polyestersulphone (PSF) 138
polyethylene 65, 66, 68, 69, 147–8
polyethylene glycol (PEG) 261
polyethylene terephthalate fiber (PET) 44, 46, 269
polyglycolic acid 261
polyketone fiber 92–4
polylactic acid (PLA) 39, 40–2, 261
polymerization, simultaneous with spinning 13–14
polymyxin immobilized fiber 140, 141
polysaccharides
hyaluronan 255–61
wound dressings 251–5
polyurethanes 265
porous water absorptive polyester fiber 189–91
practical fibers 16
prebiotics 221
Primacel 241
printed circuit boards 59–60, 79
probiotics 221
Processed Eucheuma Seaweed 238
prolonged action medical bandage 266–7
propane-diol (PDO) 44, 45
PTT (polytrimethylene terephthalate) fiber 39, 43–7
polymer structure and fiber performance 45–7
purification of water 124–7

quartz optical fiber 25, 26
QUUP 191–2, 193

radial nano structure 275, 277, 278
radial tires 70–1
radius of gyration 225
rapid drying 157, 182
rayon 12, 37
reinforced wound dressings 250
reinforcement 94
superfiber as a reinforcing material 69–76
resource exhaustion 39

respiration rate monitoring 264
reverse osmosis (RO) membrane 137, 138
rollers, lightening of 124
rubber 94
rubber matrix composite (RMC) 70

Saroma, Lake 76–7
scaffolds for tissue engineering 261, 263
scaled swimming costume 148–9
Science in Comfort 213, 214
second order function 130, 132, 133–8, 142, 143
SEK mark 186, 187–8, 197
senior citizens 180–1, 182
separation membranes 135–8
serum cholesterol 242–3
shape 8–9
  fiber characteristics and shapes 16–18
Shell Chemicals 39
Shikibo 216
Shingosen 8, 37
shirts 212–15, 216
shish-kebab structure 68
short-chain fatty acids (SCFA) 219, 220
silk 10–11
  biomimetics and 12–13
single mode (SM) optical fiber 25, 26
single-walled CNT (SWCNT) 91–2
ski jump suit 150
ski jumping 58
skin-core structure 192–4
smell
  sense of 200–1
  unpleasant odors 199–204
smooth surface textiles 147, 183
snowbreak nets 58–9
softness 21, 22
Solo 44–7
SOLOTEX 47
soluble fiber 219
Sophista 151–4, 155
space exploration 113
  Mars Pathfinder 79–81, 82, 83
space truss 117–20
spandex 169
specifications for industrial use 74, 75
Speed 148–9
speed competition 149–51
speed skating 149
spider silk 13–14, 159
  supportive matrix for cartilage regeneration 267–8
spinneret 151
spinning, simultaneous with polymerization 13–14
splinting 266
sports goods 72–3, 106–7, 108, 110–11, 144–6
sportswear 182, 189
  functions and performance required 144, 146
  functions from the type of sports 146–8
  guard vest for jockeys 183–4, 185
  using high function fiber 144–51
  high function sportswear 144–9
  speed competition 149–51
  intelligent 28
  purpose of sports and required elements 145
  swimming costumes *see* swimming costumes
ST LINUS 59
standard specifications, need for 74
static-function-mimetic fiber 163
step index (SI) optical fiber 25, 26
stimulus relaxation fiber 204, 205, 206, 207, 209
stratospheric platform 60–1
strength *see* tenacity
structure-mimetic fiber 163
superabsorbent fibers 250–1
substrates 221
Sudan 228–9
super high function fiber 130, 132, 134, 138–41, 142, 143
superconductive materials 88, 89
superfibers 6–8, 32, 34, 35, 65–98, 269
  conventional fibers and 65–7
  definition 65
  development in Europe, US and Japan 68–9
  expected application areas 72
  frontiers of superfiber applications 76–90
  Dyneema 87–8, 89
  Kevlar 65, 66, 76–8, 82, 83, 147–8
  prospects 88–90
  Technora 78–9
  Vectron 79–81

Zylon 65, 66, 81–7
high polyketone fiber 92–4
nanofiber 91–2
needs and development of usage 35, 37
presently available 65, 66
as reinforcing materials 69–76
ultimate strength of conventional synthetic fiber 67–8
supersonic transport (SST) 113–16
sustainability 40
sweat absorbency 157, 182
swimming costumes
hollow fiber 150–1, 152
non-transparent 23–4
scaled 148–9
uneven surface 149
synbiotics 221
synthetic fibers 9, 269
vs natural fibers 10–12, 39–40
ultimate strength of conventional 67–8
Synvisc 259
systematized function 139

tara gum 231–2
tech-textiles 1, 2
Technora 78–9
Teijin 212
Telfa Clear 266
tenacity 65–7, 68, 75
fluctuation of high-tenacity polyethylene fiber 87, 88
PBO fiber 83
strength in lateral direction 39
ultimate strength of conventional synthetic fiber 67–8
ultra high tenacity fibers 147–8
*see also* high strength fiber
tendon adhesion 259
tennis rackets 107, 108, 146
tensile strength 111–12
Terramac 40, 41
Tetoron Powerip 62
textile, defining 15
textile medical sensors 263–4
textiles hygiene processing meeting 186–7
Textrodes 264
thickening 222–5, 226
thickness 16
third order function 130, 132, 134, 138–41, 142, 143
three-dimensional textiles 28
3G 47
3GT polyester fiber 39, 43–7
tires 70–1, 73
tissue engineering
cartilage regeneration 267–8
fibrous scaffolds for 261, 263
hyaluronan 257–61
titanium oxide 23
fibrous titanium oxide optical catalyst 167–8
Tokyo Dome 57
touch *see* high touch fiber
toughness 75
tow 99–100
Toyobo Company Ltd 81, 83, 216
Science in Comfort shirt 213
trimethylamine 200, 201
turbo shell cap 14
2GT (PET) 44, 46

ulcerative colitis 220
ultra high tenacity fibers 147–8
ultra light materials 286
Ultra Long Duration Balloon (ULDB) 61–2, 84–5
ultrafiltration membrane (UF) 137, 138
ultra-fine fibers 22, 35, 39
membrane cloth 135–6
ultra-microfiber 20, 22
Ultressa gas storage tank 122–3
underwear, comfortable 215
uneven surfaces 149–50
United States 52
carbon fiber development 101–2
development of fiber technology 32, 35–7, 38
development of superfiber 68–9
major US companies 39
nanofiber research 286
NASA 61–2, 79–81, 82, 83, 84–5
unmanned balloons 61–2, 84–5

vapor-permeable/waterproof materials 147, 183, 265
Vectran 79–81
Venus balloon probe 86–7

virus removal filters 255
viscosity-shear rate profiles 222–5, 226
viscosupplementation 257–61

water absorbency 189–94
water purification 124–7
water treatment technologies 135–8
waterproof/vapor-permeable textiles 147, 183, 265
wearable computer 166
WELLKEY 189–91
wetting 189
white knitted bandages 266
wind power generation 120–1
windbreak nets 58
Winter Olympic Games 1999 58, 149–50
wool 10–11, 13
wound dressings 248–55
  alginate 248–50
  burns 265–6
  from chitin 267
  hydrogel 260–1
wound healing
  cartilage regeneration 267–8
  collagen 263
  hyaluronan 255–61
  polysaccharide fibers 251–5
  scaffolds for tissue engineering 261, 263

X-rays 123–4
xanthan gum 226, 234–6

yachts 124, 125, 149
YG28°C 215

zirconium carbide 147
Zylon 65, 66, 81–7
  balloon for Venus probe 86–7
  tension member for optical fiber 86
  unique spinning 84–5